A Handbook of Determinative Methods in Clay Mineralogy

A HANDBOOK OF DETERMINATIVE METHODS IN CLAY MINERALOGY

Edited by
M.J. WILSON
Head, Department of Mineral Soils
Macaulay Institute for Soil Research
Aberdeen

Blackie
Glasgow and London

Published in the USA by
Chapman and Hall
New York

Blackie & Son Limited,
Bishopbriggs, Glasgow G64 2NZ
7 Leicester Place
London WC2H 7BP

Published in the USA by
Chapman and Hall
in association with Methuen, Inc.
29 West 35th Street, New York, NY 10001-2291

First published 1987

British Library Cataloguing in Publication Data

A Handbook of determinative methods in clay mineralogy.
1. Clay minerals
I. Wilson, M.J.
549'.67 QE389.625

ISBN 0–216–91801–4

Library of Congress Cataloging-in-Publication Data

A Handbook of determinative methods in clay mineralogy.

Includes bibliographies and index.
1. Clay—Analysis. 2. Mineralogy, Determinative.
I. Wilson, M. J. (Michael John), 1939–
QE471.3.H36 1987 549'.6 87-11714
ISBN 0–412–00901–3 (Chapman and Hall)

Phototypesetting by Thomson Press (I) Ltd., New Delhi
Printed in Great Britain by Bell & Bain (Glasgow) Ltd.

Preface

This book emerged gradually from discussions on clay mineralogy with visiting scientists at the Macaulay Institute. It was realized that to many such scientists clay mineralogy was a vital, but peripheral, component of their work, yet they were often in the position of having to study clay mineralogy from scratch. It became apparent that there was a need for an introductory text covering practical aspects of the main investigative methods in clay research. At the same time, such a text could also provide a useful compilation for scientists already familiar with the difficulties of investigating clay minerals.

All but one of the contributors to this book are, or have been, associated with the Macaulay Institute. The Institute has a long and distinguished history of research into clay mineralogy and has numbered on its staff such eminent figures as D.M.C. MacEwan, G.F. Walker, V.C. Farmer and R.C. Mackenzie. Special mention must be made of Dr Mackenzie, however, because he fully appreciated the complex nature of clays, and was instrumental in initiating and encouraging a multi-technique approach in the characterization of these materials at the Institute. In a very real sense this book is a tribute to his work.

Special thanks are due to the technical staff of the Institute who, in uncertain circumstances, did a magnificent job in providing much of the original material on which this book is based. In particular the authors are indebted to D.M.L. Duthie, D.R. Clark and Mrs L. Forsyth (X-ray diffraction), Miss J.L. Bunch and Miss C.J. Bruce (thermal analysis), A.R. Fraser and I. Black (infrared spectroscopy), Miss L. Mielewczyk (electron microscopy), Miss S. Buchan (chemical analysis) and Mrs S. Ritchie and Miss M. Thom (selective dissolution methods). We are also grateful to the Director, Professor T.S. West, for allowing us to undertake and complete this book with the minimum of formalities, and to Miss Y. Bisset for the not inconsiderable labour of typing the manuscript.

MJW

Contributors

D. C. Bain, Macaulay Land Use Research Institute, Craigiebuckler, Aberdeen AB9 2QJ, UK.

A. C. Birnie, Macaulay Land Use Research Institute, Craigiebuckler, Aberdeen AB9 2QJ, UK.

P. L. Hall, Schlumberger, PO Box 153, Cambridge CB2 3BE, UK.

W. J. McHardy, Macaulay Land Use Research Institute, Craigiebuckler, Aberdeen AB9 2QJ, UK.

B. D. Mitchell, Macaulay Land Use Research Institute, Craigiebuckler, Aberdeen AB9 2QJ.

P. H. Nadeau, Statoil, Forus, Postboks 300, Stavanger, N-4001 Norway.

E. Paterson, Macaulay Land Use Research Institute, Craigiebuckler, Aberdeen AB9 2QJ, UK.

J. D. Russell, Macaulay Land Use Research Institute, Craigiebuckler, Aberdeen AB9 2QJ, UK.

B. F. L. Smith, Macaulay Land Use Research Institute, Craigiebuckler, Aberdeen AB9 2QJ, UK.

R. Swaffield, Macaulay Land Use Research Institute, Craigiebuckler, Aberdeen AB9 2QJ, UK.

J. M. Tait, 9 Papdale Road, Kirkwall, Orkney KW15 1JT, UK.

M. J. Wilson, Macaulay Land Use Research Institute, Craigiebuckler, Aberdeen AB9 2QJ, UK.

Contents

1 Clays: their significance, properties, origins and uses

P.L. HALL

1.1 Introduction

To most of us the word 'clay' usually means a fine-grained material which when mixed with water becomes extremely plastic and mouldable. Yet clays have a much wider impact on our lives than many of us commonly realize. They form an important constituent of soils, where they exert a dominant influence on soil structure and plant nutrition. Clays are used for brick-making, as fillers and coating agents in paper manufacture, and of course in ceramics. Their properties are complex and of great scientific interest, as indicated by the enormous amount of research on clays which has been undertaken by geologists, mineralogists, chemists, engineers and others over a considerable period of time. In this chapter, an introduction is given to the nature of clays and their fundamental properties, how they occur and are formed in nature, and some ideas of their use and importance.

The term 'clay' is often (e.g. in soil mechanics) used in the restricted sense of a particle-size classification term for the natural constituents of soils and sediments. In this usage, clays describe a chemically very heterogeneous and structurally complex assemblage of colloidal particles, having mean diameters ranging from a few microns (μm) down to a few hundredths of a micron (a micron is one thousandth of a millimetre). Chemically, this complex assemblage consists of a considerable quantity of hydrous aluminium silicates, with lesser amounts of finely divided quartz, feldspars, carbonates, oxides and hydroxides (mainly of aluminium and iron) and organic matter. 'Clay minerals' refers collectively to the group of fine-grained hydrous silicates of aluminium (and also to some extent of magnesium or iron). When examined under the scanning electron microscope, these minerals are seen to consist of readily identifiable particles, which can have a variety of geometric shapes (flat or crinkly plates, ribbons, tubes or pods). Despite this variety of morphology, clays are closely interrelated in terms of their basic crystal structures, and also in the characteristic physical and chemical properties resulting from their crystal chemistry.

1.2 Structure and chemistry of clays

The crystal structures and chemical compositions of the clay minerals, and their associated minerals, are not discussed extensively in this book. Many excellent reviews and books concerning the structural and chemical aspects of clay minerals are available, including Brindley and Brown (1980), Brown *et al.* (1978), Dixon and Weed (1977) and Nemecz (1981), and the interested reader should consult these texts for details. By way of introduction, it is merely noted that there are two basic building blocks common to soil clay mineral structures:

(i) A sheet of silicate tetrahedra, consisting of two layers of oxygen atoms, and containing silicon in four-fold (tetrahedral) coordination.

(ii) An octahedral sheet, consisting of two layers of oxygen atoms or hydroxyl groups, between which aluminium, magnesium or iron are bonded in six-fold coordination. If the layer contains mainly trivalent aluminium ions which occupy two-thirds of the available octahedral interstices, this is known as a gibbsite-type or dioctahedral layer. Conversely, if the layer contains mainly divalent magnesium ions, these will occupy all of the octahedral sites, and the layer is called a brucite-like or trioctahedral layer.

The structural units of clays therefore consist of either alternating tetrahedral and octahedral sheets (OT or 1:1 structure), as in the kaolinite group of clay minerals; a sandwich of one octahedral sheet between two tetrahedral sheets (TOT or 2:1 structure), as in illite and the smectite clay minerals, of which the most common member is montmorillonite; or an arrangement in

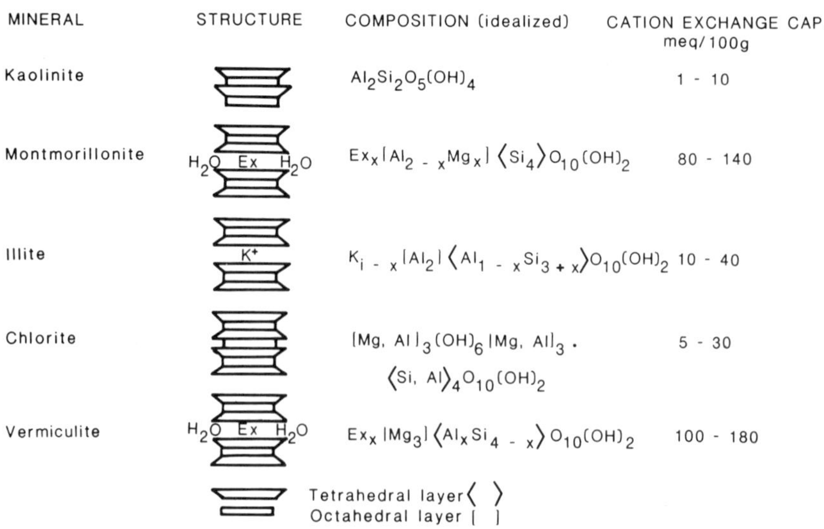

Figure 1.1 Schematic view of structural units of the principal clay-mineral groups, together with typical chemical compositions (half formula unit) and cation exchange capacities. The symbol Ex denotes an exchangeable cation (from Berner, 1971)

which the three-layer TOT units alternate with a brucite layer (2:1:1 structure) as in chlorite. The basic structural arrangements of the more common clay minerals are illustrated schematically in Figure 1.1.

The various clay minerals that will be considered in this book are as follows.

1.2.1 *Kaolinite group minerals*

These consist of uncharged, dioctahedral layer units with a 1:1 structure of ~ 7 Å in thickness and of formula:

$$Al_4Si_4O_{10}(OH)_8$$

The individual mineral species considered are:

(i) Kaolinite itself, which has varying degrees of crystallinity and a distinct, platy form

(ii) Dickite, which is similar to kaolinite, but with a different stacking arrangement of the unit layers

(iii) Halloysite, which consists of a poorly ordered arrangement of kaolinite-like units, with variable amounts of water between the layers, generally between 0.6 to 4 H_2O per formula unit, and often with a tubular form.

1.2.2 *Serpentine group minerals*

These may be regarded as the trioctahedral analogues of the kaolinite group minerals, and again have a unit layer thickness of ~ 7 Å, but with a formula approximating to:

$$Mg_6Si_4O_{10}(OH)_8$$

Small amounts of other ions may also be present. The main species considered are:

(i) Chrysotile, an asbestos mineral with a tubular, fibrous structure

(ii) Lizardite, a fine-grained mineral with a platy form

(iii) Antigorite, a mineral with an unusual corrugated layer structure

(iv) Berthierine, a more iron-rich mineral with a platy form and a formula close to:

$$(Fe^{2+}_4Al_2)(Si_2Al_2)O_{10}(OH)_8$$

1.2.3 *Illite/mica group minerals*

These have a di- or trioctahedral 2:1 layer structure of ~ 10 Å in thickness. The structure is charged, due to ionic substitutions in both the tetrahedral and the octahedral sheets, and contains interlayer non-exchangeable cations (usually K) that compensate for the charge deficit on the unit layers. The chemistry of the various members of the group is variable, the general formula being:

$$(X_2)(Z_8)(Y_4 \text{ or } Y_6)O_{20}(OH)_4$$

where cations are located in the interlayer (X), octahedral (Y) and tetrahedral (Z) sites, respectively. The individual minerals considered are:

(i) Illite: a group name covering fine-grained, dioctahedral mica-like clays that usually contain less K, more water, less tetrahedral Al and have a larger degree of octahedral substitution than the coarse-grained muscovite micas.

(ii) Muscovite/phengite are coarse-grained dioctahedral micas with compositions approximating to:

$$K_2(Si_6Al_2)(Al_4)O_{20}(OH)_4$$

and

$$K_2(Si_7Al_1)(Al, Fe^{3+})_3(Mg, Fe^{2+})O_{20}(OH)_4$$

respectively. These minerals usually have a two-layer polytype structure.

(iii) Biotite is a coarse-grained, trioctahedral mica, of variable composition, but with an ideal formula of:

$$K_2(Si_6Al_2)(Mg, Fe^{2+})_6O_{20}(OH)_4$$

(iv) Glauconite and celadonite are green, fine-grained dioctahedral, often hydrous, micaceous minerals that contain more iron and less aluminium than do illites. Their composition approximates to:

$$K_{2-x}(Na, Ca)_x(Si_7Al_1)(Fe^{3+}Al)_2(MgFe^{2+})_2O_{20}(OH)_4$$

1.2.4 *Chlorite group minerals*

These are sometimes dioctahedral but usually trioctahedral, with a 2:1:1 layer structure of $\sim$ 14 Å in thickness. The 2:1 layer is negatively charged because of ionic substitutions, and is balanced by a positive charge on the interlayer hydroxide sheet. Chemically, the group is extremely variable, but it should suffice to concentrate on only three types:

(i) Mg-rich chlorite (e.g. clinochlore) which is trioctahedral with a general formula close to:

$$(Si_{8-x}Al_x)(Mg_{12-y}Al_y)O_{20}(OH)_{16}$$

(ii) Fe-rich chlorite (e.g. chamosite) which is also trioctahedral with a general formula close to:

$$(Si_{8-x}Al_x)(Fe_{12-y}Al_y)O_{20}(OH)_{16}$$

(iii) Al-rich chlorite (e.g. sudoite) which is mainly dioctahedral with a general formula close to:

$$(Si_{8-x}Al_x)(Al_{10-y}Mg_y)O_{20}(OH)_8$$

1.2.5 *Vermiculite group minerals*

These sometimes have a dioctahedral, but usually a trioctahedral 2:1 layer structure with a negative charge > 1.20 per formula unit. The charge is compensated for by the occurrence of interlayer, exchangeable cations, usually in a hydrated form. The layer spacing varies between ~10 and 15 Å, depending generally on the nature of the exchangeable cation and, to a lesser extent, on the relative humidity. The general formula for a trioctahedral vermiculite is:

$$M_xH_2O(Si_{8-x}Al_x)(Mg,Fe)_6O_{20}(OH)_4$$

where M represents the exchangeable cation. Vermiculite often occurs in large crystals that are derived by alteration of biotite or phlogopite.

1.2.6 *Smectite group minerals*

These have a dioctahedral or a trioctahedral 2:1 layer structure, with isomorphous substitution leading to a negative layer charge of less than 1.2 per formula unit. As with vermiculite, this charge is compensated for by interlayer exchangeable cations. Layer spacings vary between ~10 and 15 Å, and are generally dependent on the nature of the exchangeable cation and the relative humidity. Smectites nearly always occur in fine-grained particles of clay size. The main dioctahedral minerals are:

(i) Montmorillonite, which is predominantly octahedrally charged, with an ideal general formula of:

$$M_xSi_8(Al_4{-}Mg_x)O_{20}(OH)_4$$

It should be stressed that this formula *is* ideal and that in practice montmorillonites usually exhibit some tetrahedral substitution and also contain some octahedral ferric iron. Tetrahedral substitution should always amount to < 50% of the total charge deficit.

(ii) Beidellite, which is predominantly tetrahedrally charged and contains more aluminium and less magnesium than montmorillonite. The ideal formula is:

$$M_x(Si_{8-x}Al_x)Al_4O_{20}(OH)_4$$

Again, the real situation is that the composition of beidellite is somewhat variable, but nevertheless tetrahedral substitution should always account for > 50% of the total charge deficit.

(iii) Nontronite, which is an iron-rich smectite with an ideal formula of:

$$M_x(Si_{8-x}Al_x)Fe_4O_{20}(OH)_4$$

Few nontronites actually have this composition. Octahedral substitution is common, but does not account for > 50% of the total charge deficit. Some nontronites are known where Si is substituted for by Fe^{3+} in the tetrahedral sheet.

The main trioctahedral smectites are as follows:

(iv) Hectorite, the trioctahedral 'equivalent' of montmorillonite in that, ideally, the layer charge arises only from octahedral substitution, as in the following formula:

$$M_xSi_8(Mg_{6-x}Li_x)O_{20}(OH, F)_4$$

As seen above, lithium and fluorine are essential constituents of hectorite.

(v) Saponite is again a magnesium-rich clay, but because it is tetrahedrally substituted it is more closely related to beidellite. The ideal formula for saponite is:

$$M_x(Si_{8-x}Al_x)Mg_6O_{20}(OH)_4$$

but in practice there is usually some octahedral substitution of Mg, usually by Fe^{3+}.

1.2.7 *Interstratified or mixed-layer minerals*

These are usually constructed of two different layer types arranged on a fine scale in a regular or an irregular manner. In principle, the layer structures of all these minerals are so similar that interstratification of any combination of minerals could be envisaged. In practice, interstratified minerals nearly always contain an expansible component, either smectite or vermiculite. Regularly interstratified minerals generally consist of two components in equal proportions and have been given definite mineral names. The ones discussed in this book are:

(i) Rectorite: a dioctahedral mica-smectite
(ii) Hydrobiotite: a trioctahedral biotite-vermiculite
(iii) Corrensite: a trioctahedral chlorite-smectite
(iv) Tosudite: a dioctahedral chlorite-smectite
(v) Aliettite: a trioctahedral talc-saponite.

The most common types of irregularly interstratified minerals are illite-smectite and chlorite-smectite. Kaolinite-smectite also occurs and is becoming more widely recognized.

1.2.8 *Sepiolite and palygorskite*

These both have a fibrous morphology and a chain structure, and in this respect differ fundamentally from the layer-silicate minerals that have been discussed up to now. The structure consists of ribbon-like bands aligned parallel to the *c*-axis, the ribbons themselves having a 2:1-type structure. The alternating arrangement of the bands is such that, running parallel with them, are open channels containing cations and water molecules. The sepiolite/palygorskite minerals are both magnesium rich, although palygorskite is more aluminous. For sepiolite, the ideal formula is:

$$Si_{12}Mg_8O_{30}(OH)_4(OH_2)_4 \cdot 8H_2O$$

and for palygorskite:

$$Si_8(Mg_2Al_2)O_{20}(OH)_2(OH_2)_4 \cdot H_2O$$

A deficiency of octahedral cations for sepiolite and isomorphous substitution in palygorskite leads to cation-exchange capacities ranging from 20–45 meq/100 g for the former and 5–20 meq/100 g for the latter.

1.2.9 *Allophane and imogolite*

These minerals are non-crystalline, or strictly speaking, paracrystalline (Wada, 1982) hydrous aluminosilicate minerals. The term paracrystalline in this context suggests long-range order in one or two directions. Allophane consists of 30–50 Å diameter hollow spheres whose walls are made up of one Al-octahedral and one Si(Al)-tetrahedral sheet. Imogolite consists of a tubular structure of 20 Å diameter, which is composed of a curved Al octahedral sheet with orthosilicate groups attached to the inside of the tube. The $SiO_2:Al_2O_3$ ratio of allophane varies from ~ 1.3 to 2, whereas for imogolite it remains constant at ~ 1.

1.2.10 *Accessory minerals*

Accessory minerals, particularly oxides, hydroxides, oxyhydroxides, etc. of iron and aluminium are common in clay materials. Some of those most likely to occur are shown in Table 1.1. Many other minerals commonly occur in

Table 1.1 Some oxide and hydrous oxide minerals likely to occur as accessory minerals in clay materials.

Mineral	Formula
Hematite	α*-Fe_2O_3
Maghemite	γ-Fe_2O_3
Magnetite	Fe_3O_4
Goethite	α-FeOOH
Lepidocrocite	γ-FeOOH
Feroxyhyte	δ-FeOOH
Gibbsite	$Al(OH)_3$
Diaspore	α-AlOOH
Boehmite	γ-AlOOH
Birnessite	$(Na_{0.7}Ca_{0.3})Mn_7O_{14} \cdot 28H_2O$
Opal CT	SiO_2

*Greek letter prefixes denote particular types of packing of the structural anions.

minor quantities in clays, particularly quartz, feldspars, zeolites, phosphates, carbonates, etc. Details of their chemistry and structure may be found in standard mineralogical texts.

1.3 Properties of clays

1.3.1 *Ion exchange*

One of the fundamental properties of clays is the electrical charge on their unit particles which means that clays will sorb cations and/or anions that neutralize the layer charge, but which are exchangeable. This means that they can readily be replaced by other anions or cations when brought into contact with these ions in aqueous solution (Grim, 1968; van Olphen, 1977). Except under extremely acidic conditions, the layer charge is predominantly negative, a fact which may be verified by investigating the electrochemical properties of dilute, aqueous dispersions of clay (Swartzen-Allen and Matijevic, 1974). This means that clays tend to sorb cations, and are said to have a certain *cation exchange capacity*, or CEC, which can be expressed as a certain number of chemical equivalents per unit weight of dry clay (normally quoted in milli-equivalents per hundred grams). As shown in the previous section, the principal origin of negative electrical charge in clay particles is an isomorphous substitution of cations in the lattice by cations of lower charge, e.g. the substitution of aluminium for silicon in the tetrahedral sheet, and magnesium and/or ferrous iron for aluminium, or sometimes lithium for magnesium in the octahedral layers.

Smectites contain extensive isomorphous substitution of this kind, and frequently have CEC values > 100 milli-equivalents/100 g. Their exchange cations are mainly between the adjacent tetrahedral layers. Depending on the relative humidity (or, in contact with a liquid, the ionic concentration) smectites may adsorb water and hydrate, the crystal lattice swelling in a direction perpendicular to the plane of the surface oxygens. However, neither the CEC of a clay nor the degree of swelling it can accommodate is simply related to the amount of isomorphous substitution. For example, at fairly high levels of substitution of silicon by aluminium, potassium ions can be strongly ionically bonded in the interlayer region in a non-swelling and non-exchangeable state—as in illite and glauconite, which are essentially fine-grained micas. Thus illites, while having a higher negative layer charge than smectites, have lower exchange capacities.

Other clays have much less isomorphous substitution and correspondingly lower CEC values. It has been suggested that broken bonds at particle edges may also contribute to exchange capacity, particularly for clays of the kaolinite group. However, there is considerable evidence to suggest that, except under fairly alkaline conditions, the charges due to edge defects are positive, leading to anion rather than cation adsorption at these sites. There are two possible mechanisms for creation of positive edge charges:

(i) Protonation of exposed Al-OH groups or

(ii) The dissolution of some of the clay at $pH < 4$ followed by edge-adsorption of aluminium ions. Whichever mechanism is dominant, the existence of positive edge charges explains many of the colloidal properties of clay dispersions, as will be discussed in Section 1.3.2.2.

The negatively charged clay surfaces do not have an equal affinity for all cations, but in fact exhibit a selectivity which depends upon such factors as the ionic charge, size and state of hydration of the cations. In general, cation-exchange selectivity increases with cation charge and (for a given charge) increases with ionic radius. Thus, the order of increasing affinity for the alkali-metal ions is Li, Na, K, Rb, and Cs, and for the alkaline-earth ions Mg, Ca, Sr, Ba (Swartzen-Allen and Matijevic, 1974). In both series, increasing adsorption selectivity corresponds with decreasing polarizability and energy of hydration in aqueous solution, i.e. the cations which hold on to their hydration shells most firmly are the least strongly adsorbed by clays. The relative adsorption selectivity of monovalent and bivalent ions varies somewhat with clay type and layer charge. Bivalent ions and polyvalent ions are in general quite strongly adsorbed. Also some monovalent ions, e.g. ammonium or the organic alkylammonium ions, are very strongly adsorbed and permit only limited swelling on account of their large sizes and relatively low hydration energies.

For various exchange forms of smectites, there is a clear correlation between the degree of selectivity of the ion, the extent of swelling and the state of aggregation of the clay in suspension. Cations which are least preferred by the surface are associated with limited particle aggregation and extensive swelling, while strongly bound ions promote extensive aggregation and limited swelling. The behaviour of clay–water systems is discussed in more detail later (Section 1.3.2).

Ion-exchange reactions in clays are important for a number of reasons:

(i) As a source of nutrients in soils for plant growth. This particularly applies to calcium, magnesium and potassium ions, although there is some evidence that plants can utilize potassium in soils even when it is present in a non-exchangeable form (Newman, 1984).

(ii) The mechanical properties of a clay soil (strength, yield point, plasticity) strongly depend upon whether Na^+ or Ca^{2+} is the principal counter-ion present (Grim, 1968).

(iii) Ion-exchange processes may play a role in the immobilization of undesirable cations such as organic pesticides or components of radioactive wastes.

(iv) In recent years, there has been considerable interest in the fact that certain large cationic oxyaluminium polymers can produce thermally stable 'pillars' which can prop apart the unit layers of a smectite clay permanently. This makes it easier to 'tailor' the catalytic and molecular sieve properties of clays to specific applications, as in the case of zeolites (Pinnavaia, 1983).

1.3.2 *Clay–water interactions*

1.3.2.1 *Hydration behaviour at relatively low water content.* The behaviour of clays in the presence of water is complex and extremely important. This section considers the adsorption of water by clays from a relatively dry state, i.e. interactions taking place when water molecules become attached to the particle surfaces and/or their associated exchangeable cations.

There has been controversy for many years regarding the structure of water immediately adsorbed at clay surfaces. In particular, it has not been clear to what extent its density and structure resemble or differ from those of bulk water or ice, and to what distance the effect of the surface modifies its properties (see, for example Low (1961) and Martin (1962) for earlier reviews of the subject). In recent years a wide range of techniques have been employed to investigate these questions, and a brief outline of the current state of knowledge is given in this section. Further details may be found in reviews by Farmer (1978), MacEwan and Wilson (1980), and Sposito and Prost (1982).

Clay hydration from a dry state is an exothermic process, as can be readily tested by feeling the heat on the side of a beaker that is generated when a quantity of powdered clay is wetted. The adsorption of water by clays can occur either by hydration of the crystal surfaces or hydration of the exchange cations. Hydration of the siloxane surfaces of the tetrahedral sheets essentially corresponds with hydrogen bonding between a water proton and a surface oxygen, and that of the gibbsitic surfaces of the octahedral sheets to hydrogen bonding between a surface hydroxyl proton and a water oxygen atom. Hydration of exchange cations essentially amounts to formation of a coordination complex. It is not always possible to determine experimentally which are the adsorbing sites, since a water molecule may be bonded to both a surface oxygen atom and an exchange cation, and also a molecule adsorbed at the surface may be influenced in its orientation and bonding strength by the nature and presence or absence of neighbouring cations. However, two cases may be distinguished:

(i) In kaolinite, water can be adsorbed only on external surfaces, of which there are two kinds (siloxane and gibbsitic), and on particle edges. The enthalpy of adsorption of this water is quite small, and it can largely be removed by heating at a small increment above ambient temperature.

(ii) In smectites, water may also be held in the interlayers between the individual structural units, or in the case of minerals such as sepiolite, in tubular openings. Removal of this water requires rather more energy input, in some cases heating to temperatures considerably in excess of 100°C for complete removal.

The important case of water adsorbed by smectites such as montmorillonite, in which both water–cation and water–siloxane surface interactions are significant, will be discussed. The different stages of hydration which occur can be inferred by combining X-ray diffraction data at controlled humidity

(Glaeser and Méring, 1968) and gravimetric adsorption measurements (e.g. van Olphen 1965; Keren and Shainberg, 1980), and depend on both the nature of the exchange cation and the magnitude and location of layer charge. Essentially, however, it is clear that at low relative humidity the interlayer cations are unhydrated and embedded in cavities between adjacent siloxane surfaces, so that the initial adsorption occurs at external surfaces only. At slightly higher relative humidities the interlayer cations become hydrated, causing a stepwise expansion of the interlayer spacing. Further adsorption of water finally results, at high relative humidities, in multilayer adsorption and/or the filling of external micropores. However, monovalent and bivalent exchange forms behave rather differently. X-ray diffraction studies indicate that the Na^+-exchanged clay exhibits a series of hydration states, corresponding to one, two, three or more molecular sheets of interlayer water, the number of sheets increasing with increasing relative humidity. In contrast, in the case of the Ca^{2+}-exchange-form clay, the higher hydration energy of the calcium ion and the stability of the octahedral $Ca(H_2O)_6^{2+}$ solvation complex causes initial expansion to a bilayer complex, and later a trilayer complex. Also, on complete immersion in water calcium montmorillonite does not expand beyond three or possibly four interlayer water sheets, whereas sodium montmorillonite undergoes essentially unlimited osmotic swelling to large interlayer spacings.

The osmotic swelling of Na^+-montmorillonite results in interlayer spacings which vary inversely as the square root of the ionic concentration, as predicted by the Gouy–Chapman theory of electrical double layers (Norrish, 1954). This is illustrated in Figure 1.2. Osmotic swelling is exhibited only by lithium- and sodium-exchange forms of montmorillonite. Exchange cations of higher

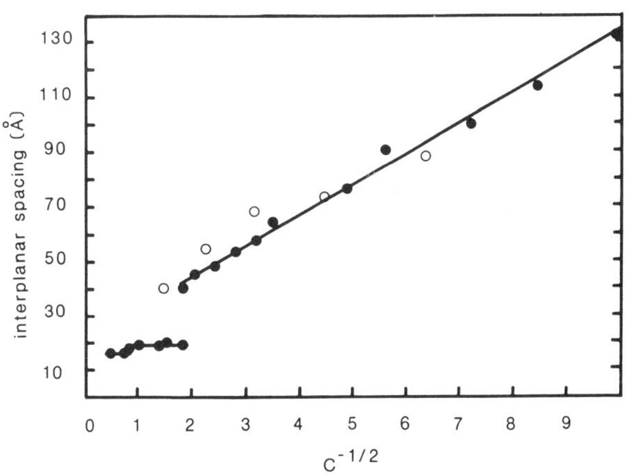

Figure 1.2 Dependence of the interplanar spacing in Na^+-montmorillonite on electrolyte concentration, C, showing the regions of stepwise crystalline hydration and continuous osmotic swelling (Norrish, 1954)

charge, as well as larger monovalent cations, e.g. potassium, permit only limited swelling of smectites up to a few water layers. It is clear that there exist energy barriers to extensive layer separation, which cannot be overcome either if the ion–surface Coulomb attractions are strong, or if the hydration energies of the cations are weak.

Spectroscopic studies such as infrared (Farmer and Russell, 1971, Sposito and Prost, 1982), NMR (Fripiat *et al.*, 1984), electron-spin resonance (Hall, 1980) and quasi-elastic neutron scattering (Hall, 1982) have probed the vibrational and diffusive behaviour of the interlayer water and give a reasonably consistent picture of the nature of interlayer water in smectites. The structure is essentially liquid-like, but generally there are two distinct populations of water molecules:

(i) Strongly hydrogen-bonded water, associated with cations and sites of surface negative charge, which have much less translational and rotational mobility than bulk water.

(ii) More weakly bonded water, between the cation solvation shells, which is essentially liquid-like, although having a rather smaller translational diffusion coefficient than bulk water, and an overall preferential orientation due to surface hydrogen bonding.

Both NMR (Fripiat *et al.*, 1984) and neutron scattering studies (Olejnik and White, 1972) indicate that beyond ~ 10 Å from the siloxane surfaces, the adsorbed water is dynamically similar to bulk water. Nevertheless, the existence of montmorillonite gels with stable, large interplatelet separations up to 1200 Å (Cebula and Ottewill, 1981) suggests that in the dispersed state there is some influence, possibly weak electrostatic fields, exerted over larger distances.

1.3.2.2 *High water contents: properties of clay colloids.* The osmotic swelling to relatively large interplatelet spacings, exhibited by Na^+ and Li^+ exchange-forms of montmorillonite, can be explained on the basis of electrical double-layer theory, which is discussed in detail in the book by van Olphen (1977). Basically, the negatively charged clay platelets cause a net attraction of cations and a net repulsion of anions, causing the region near each platelet to have a positive electrostatic charge, which tends to repel adjacent platelets. There are also counteracting attractive van der Waals' forces operating at short distances. The net result of this is an energy minimum at an interplatelet separation which is a function of the electrolyte concentration (see Figure 1.3). The characteristic double-layer thickness can be shown to vary as the square root of the ionic concentration, and measurements of the forces between dispersed lithium montmorillonite platelets are in reasonable agreement with the theory (Lubetkin *et al.*, 1984).

Nevertheless, the structures of most dilute clay dispersions do not consist of single, well-separated and aligned platelets, but of aggregates of platelets arranged in a combination of configurations, such as face-to-face (FF), edge-

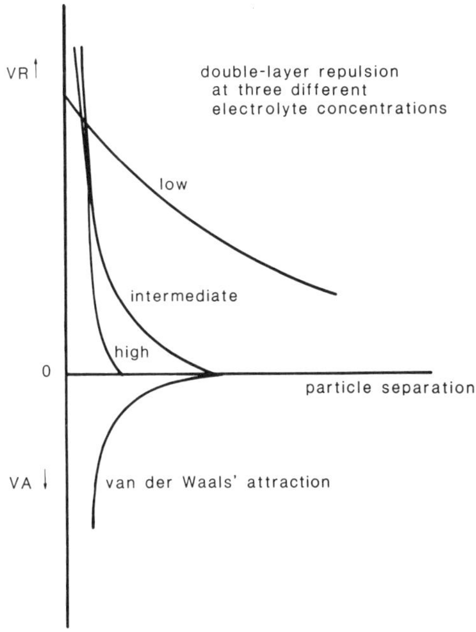

Figure 1.3 Repulsive and attractive energy as a function of particle separation at three different electrolyte concentrations (van Olphen, 1977)

to-face (EF) and even edge-to-edge (EE). The rheological properties of clay colloids depend crucially on the arrangement of their constituent particles. Where FF associations predominate, the colloid is said to be *aggregated*; where EE and EF associations are important, a special case of aggregation occurs, known as *flocculation.* Flocculation is associated with voluminous aggregates or 'flocs' in dilute suspensions, and a fairly rigid, viscous 'card-house' structure in concentrated suspensions. Where the particles are well dispersed, the suspension is said to be *peptized* or *deflocculated.* The arrangement of clay particles which will occur depends crucially on the electrolyte concentration and pH. These factors control the double-layer thickness and the charge at particle edges, and consequently the relative magnitude and attractive/repulsive character of FF, EF and EE configurations (van Olphen, 1977). Two cases will be discussed: kaolinite and montmorillonite suspensions.

The simpler case is that of kaolinite suspensions, which are flocculated under acidic conditions and dispersed under alkaline conditions, arising from the attractions between negative surface and positive edge charges at low pH, and the mutual repulsion of negative face and edge charges at high pH (Jepson, 1984). Deflocculation can be promoted by adding a polyanion (oxalate or phosphate) which increases the negative edge charge. Sodium montmorillonite particles undergo a similar charge reversal under alkaline conditions. However, in the presence of neutral or acidic electrolyte solutions they exhibit a more complex behaviour, being flocculated at low and high salt concentrations but dispersed at intermediate salt concentrations. This phenomenon is due to the extensive thickness of the double layer at low salt concentrations which promotes strong EF attraction, while at intermediate salt concentrations compression of the double layers reduces these forces, while not significantly influencing the repulsive EE and FF interactions. At high salt concentrations, the double layers are still further compressed, and the repulsive forces again become weaker than the attractive forces (van Olphen, 1977).

One major practical consequence of double-layer phenomena is the occurrence of landslides when rainfall leaches the salt out of clay-rich soils, causing a weakening of the interparticle association and consequently the mechanical strength of the soil. Other factors which may cause a breakdown of the aggregate structures of clays are shearing forces, responsible for the behaviour of quicksands.

1.3.3 *Interaction of organic matter with clays*

Many organic molecules, like water, can be readily adsorbed by clay minerals. In some cases, particularly for non-polar organic molecules, the interaction forces are relatively weak, corresponding with physical adsorption only. However, polar or ionic organic species can enter into a wide variety of chemical reactions to form complexes with clay minerals. Smectites, vermiculite and kaolinite group minerals can be expanded by penetration of such molecules between the unit layers to form an intercalation complex. Depending on the nature of the organic species, the bonding in organo-clay complexes occur via:

(i) Hydrogen bonding

(ii) Ion-dipole forces, including formation of coordination complexes to exchangeable cations

(iii) Bonding via 'water bridges' between organic molecules and hydrated cations

(iv) Cation exchange

(v) Anion exchange

(vi) d-π bonding between adsorbed arenes and transition-metal exchange ions.

There is a very extensive literature on the formation, characterization and properties of organic derivatives of clay minerals, which is reviewed by Grim (1968), Weiss (1970), Theng (1971), MacEwan and Wilson (1980) and Lagaly (1984) among others. In this section, a few examples are given of the type of behaviour which is observed.

1.3.3.1 *Intercalation complexes of kaolinite with polar organic molecules.* In kaolinite, there are a number of polar organic molecules, notably formamide, acetamide, hydrazine, urea, dimethylsulphoxide [$(CH_3)_2SO$] and some organic salts (such as potassium acetate) which are able to disrupt the interlayer bonding between the adjacent siloxane and hydroxyaluminium surfaces, and to penetrate the interlayer space to form a complex by hydrogen bonding to both surfaces. The interlayer spacing of kaolinite, normally ~7 Å, can be expanded in this manner to 10 Å or more, depending on the identity of the guest molecule. Recently, Thompson and Cuff (1985) have determined a three-dimensional crystal structure for the kaolinite–$(CH_3)_2SO$ complex, in which the oxygen atom of the $(CH_3)_2SO$ is triply hydrogen bonded to the gibbsitic sheet above an octahedral vacancy, while one of the two methyl groups is keyed into the siloxane surface. The formation of a regular structure requires significant lateral displacement of adjacent kaolinite units to bring the octahedral vacancies and ditrigonal cavities opposite one another, which explains the disaggregation and loss of crystallinity which is observed on desorption of $(CH_3)_2SO$ from kaolinite. One striking practical use of kaolinite intercalation complexes was discovered by the makers of thin-walled porcelain in Ancient China. Kaolinite was stored in the presence of urine (causing slow formation of the urea intercalation complex) and subsequently ground to produce much finer individual particles (Weiss, 1963).

1.3.3.2 *Intercalation of montmorillonite by polar organic molecules.* Neutral, polar organic molecules which are miscible with water, such as alcohols or carbohydrates, can replace water in the interlamellar space of montmorillonite, being either coordinated to the exchange cations or held more loosely between cation solvation sheaths. These intercalation complexes can be prepared by reacting montmorillonite with the organic species in either liquid or vapour form. In many cases, the complexes form a well-ordered structure which can be determined by X-ray or neutron diffraction. One such well-known complex is formed when ethylene glycol (ethane-diol) replaces water in the interlamellar space of montmorillonite to form a bilayer complex with an (001) spacing of ~17 Å. The position of the cations in interlamellar complexes of aliphatic chain compounds may be either central (as in the ethylene-glycol complex) or adjacent to the surface, depending on the strength of the ion/adsorbate interactions. Similar complexes are formed by glycerol, and have been used both as a means of distinguishing smectites from other clay types in X-ray diffraction identification (see Chapter 2), and in surface-area determination.

1.3.3.3 *Ionic organic species.* Organocations, such as the alkylammonium ions, can replace the inorganic exchange cations of smectites, and are strongly bound, producing an interlayer space which is organophilic in character, and swells little in water, but may swell extensively in polar organic solvents. Neutral amines may similarly be adsorbed, and may protonate in the interlamellar space to produce alkylammonium ions, on account of the strongly acidic (and consequently, partly dissociated) nature of the interlayer water of smectites (Fripiat *et al.*, 1965). Weiss and coworkers (Weiss, 1970; MacEwan and Wilson, 1980) have extensively studied the structural conformation of long-chain aliphatic amines and alkylammonium ions in smectite interlayers. Above a certain chain length, the adsorbed molecules tend to change from a flat-lying to a tilted configuration, the critical chain length depending upon the layer charge of the clay, and therefore being a sensitive method for layer-charge determination.

In contrast to organocations, organic anions such as tannates are adsorbed at particle edges, and can be efficient peptizing agents when they promote charge reversal. They are widely applied in oilfield drilling fluids (van Olphen, 1977).

1.3.3.4 *Clay–polymer interactions.* This is an extremely important subject, which is extensively treated in the book by Theng (1979). High-molecular-weight and/or non-ionic polymers tend to be adsorbed at external particle surfaces, and can act as 'bridges' between more than one clay particle. Polycations appear to exhibit Langmuir-type adsorption isotherms, and are often strongly adsorbed beyond the CEC of the clay, resulting in charge reversal. Polyanions are adsorbed at particle edges, and have the same influence on edge-to-face association as does raising the pH. The conformation of neutral polymers adsorbed by clays is still not well understood, although it is clear that uncharged polymers may either inhibit interparticle association, or cause flocculation via particle bridging. In soils, the interaction between clays and natural, anionic, polymeric organic residues, such as the humic substances, have a major controlling influence on structure and permeability. It is thought that the interaction of humic and fulvic acids with clays occurs via bridges involving polyvalent exchange cations, or polymerized aluminium or iron oxyhydroxides coating clay-particle surfaces, anion exchange at particle edges, or strong ligand bonding at specific surface sites (Greenland, 1971).

1.4 Formation and occurrence of clays

This section discusses the environments at or near the Earth's surface in which clays are formed, and the mechanisms of formation which occur in each environment. It also addresses the more complex questions of which type of clay will form under a given set of circumstances, why it will form, whether or not a particular clay will be stable, dissolve, or be transformed into another phase. Three major environments are distinguished, in which clays exist or can be formed:

(i) The weathering environment, at the interface between the Earth's surface and atmosphere, including exposed rocks, gravels and soils

(ii) The sedimentary environment, at or near the bottom of river deltas, estuaries, lakes or oceans

(iii) The diagenetic/hydrothermal environments, characterized by somewhat higher temperatures and/or pressures than the sedimentary or weathering environments.

The possible mechanisms of clay mineral formation may be summarized as follows:

(i) Neoformation by precipitation from solution

(ii) Neoformation by crystallization from amorphous colloidal material

(iii) Transformation from a non-clay precursor, which involves considerable structural alteration

(iv) Layer transformation from a parent clay mineral (or related phyllosilicate, e.g. mica) in which substantial structural order is preserved.

The interrelationship between these environments may be described in terms of the rock cycle or sedimentary cycle, shown in simplified form in Figure 1.4. Particles formed in the weathering environment may be transported by erosion and the action of rivers to the sedimentary environment. Upon burial and consequent increases in pressure and temperature, sediments undergo a complex series of chemical and biological changes, referred to as *diagenesis*. To some extent, similar reactions can occur at shallower depths in hydrothermal settings. At a later stage, very deep burial, or the intrusion of igneous rocks, causes metamorphic alteration of sedimentary rocks, or eventual melting to give igneous rocks. After subsequent uplift to the surface, the entire process begins again. It is obvious that clays which formed in one environment can be 'inherited' by another environment, so that their present location does not imply local formation.

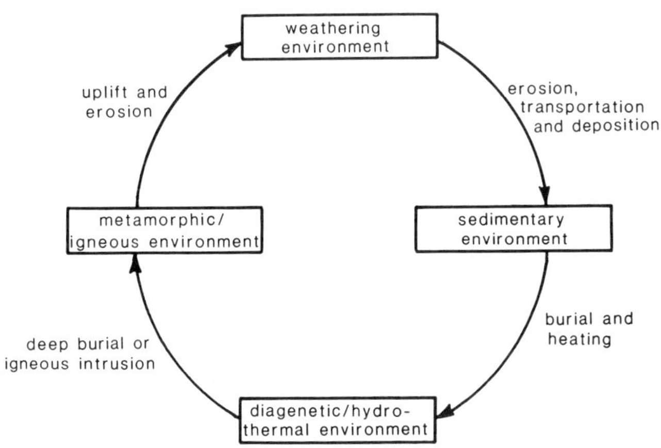

Figure 1.4 Simplified illustration of the rock cycle (Eberl, 1984)

In any of the three environments, the identity of neoformed clays will be controlled by the composition of the local aqueous solution at the time of crystallization. The solution composition is in turn controlled by the degree of availability of various ions, oxides and hydroxides and their respective solubilities. The latter can be influenced by temperature, pressure, and the proton and electron activities (pH and Eh) of the solution. In principle, if the components (both solid minerals and solution species) and the possible reactions between them are known, the phases that will form at equilibrium can be predicted by chemical thermodynamics (Berner, 1971). This approach predicts the fields of stability for various possible phases in terms of the composition of the solution (see Figure 1.5). For clay-forming reactions, one of the key controlling factors is the content of dissolved silica (strictly, silicic acid, H_4SiO_4). Except under strongly acidic or alkaline conditions, silicic acid is generally much more soluble than any aluminium-containing species.

Two of the assumptions of the thermodynamic approach are that the system is closed, i.e. it does not exchange matter and energy with its environment on the time-scale of the reaction, and also that an equilibrium state is reached. In some circumstances, one or both of these assumptions may not be valid. Where the rates of reaction are too slow for equilibrium to be reached, kinetic

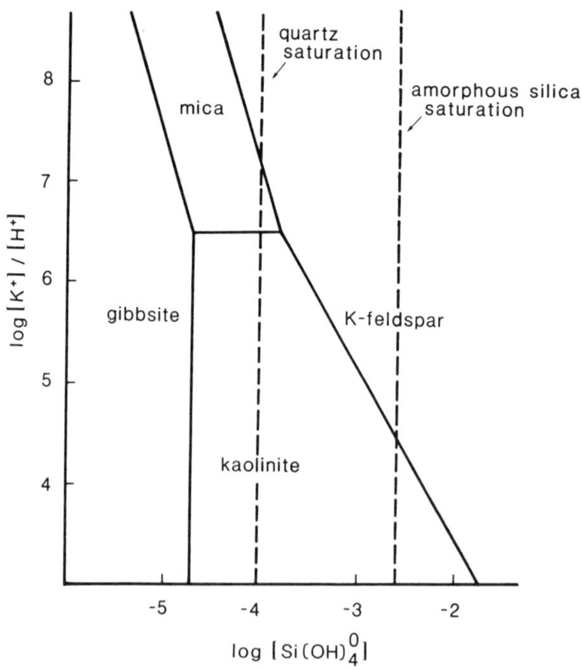

Figure 1.5 A thermodynamic stability diagram, showing fields of stability of minerals as a function of solution composition (from Curtis, 1983)

rather than thermodynamic considerations will prevail, and phases which are metastable may either be formed or persist outside of their field of stability. A useful guide to the likely behaviour in these circumstances is Ostwald's Step Rule, which states that the first phase to appear in a system at constant temperature, pressure and composition may not be the thermodynamically stable phase, but a metastable phase whose formation is much more rapid than the equilibrium phase. The initial phase is then converted with time, possibly via more intermediate phases, as equilibrium is gradually approached by a number of steps. This behaviour is observed naturally in the diagenetic alteration of silica, in which amorphous silica produced by micro-organisms or alteration of volcanic glass is converted to one or more phases of opal before crystallizing as quartz (Singer and Muller, 1983). The solubility of amorphous silica is more than an order of magnitude greater than quartz, so that the stage reached in silica diagenesis is of fundamental importance in determining the phases which are stable in contact with siliceous solutions (see Figure 1.5). Ostwald's Step Rule was observed to operate in laboratory hydrothermal simulations of clay diagenesis, in which several stepwise series of reaction products were observed, depending to some extent on both the starting material and the temperature (Eberl, 1984).

1.4.1 *Clays in the weathering environment*

Weathering describes the processes which break down rock (chiefly igneous rocks) at the Earth's surface to form discrete particles (Selley, 1982). Weathering processes may be divided into physical (freeze/thaw effects and other sources of mechanical disintegration); chemical (oxidation, dissolution and reprecipitation) and biological (particularly bacterial processes). Biological processes are an extremely important contributing factor to weathering processes and soil formation, although they are, at present, little understood.

The major constituents of igneous rocks are quartz, feldspars, micas, and the ferromagnesian minerals such as olivines, pyroxenes and amphiboles. The ferromagnesian minerals often weather to produce chlorite clays, whereas feldspars are converted to kaolinite or smectite, depending on the conditions. Micas can weather to produce vermiculite or mixed-layer mica-vermiculites such as hydrobiotite, either as macroscopic crystals or as clay-grade minerals. The insoluble residue of the weathering process consists of quartz with various quantities of feldspars and micas (Selley, 1982).

The nature of chemical weathering products often depends on the severity of the leaching conditions. For example, in Hawaii, smectite is formed on the dry, leeward side of the island, and gibbsite on the windward side where the annual rainfall is very much higher. Under intermediate conditions, the formation of kaolinite can be favoured. This may be understood in terms of the higher solubility of silica in comparison with aluminium hydroxide: phases which are progressively more aluminium rich being created as more and more silica is leached away (Berner, 1971; Eberl, 1984).

In some cases, clay minerals are derived from the alteration of volcanic ashes and glasses. At higher concentrations of alkali and alkaline-earth metals and at lower Al/Si ratios, montmorillonite will be formed preferentially, while at higher Al/Si ratios and lower concentrations of other elements, phases such as allophane, imogolite and halloysite can occur (Grim, 1968).

1.4.2 *Clays in sediments*

As stated previously, the clays in this environment are principally inherited from the weathering zone, from where they are transported by erosion and river action to the sedimentary environment of marine deltas or ocean floors. Since this environment differs considerably from the weathering environment in its ionic composition and concentration, equilibrium is disturbed, resulting in some degree of slow alteration of the detrital clays. These changes consist principally of ion-exchange reactions, mainly replacement of Ca^{2+} by Na^{+}, K^{+} and Mg^{2+} (Curtis, 1977). Uptake of potassium may cause some degree of irreversible collapse and dehydration of expanded interlayers. This appears to be true of vermiculitic soil clays, with a relatively high tetrahedral layer charge, but not volcanic smectites. Montmorillonite may continue to form from submarine volcanic glasses, but it is unclear whether the mineral is stable in a sea-water environment, or undergoes slow alteration to produce a chloritic phase (Grim, 1968; Berner, 1971).

1.4.3 *Clays in sedimentary rocks*

Sedimentary rocks such as sandstones and mudrocks are formed from ocean-floor sediments by burial to depths of up to several kilometres, accompanied by the development of high pressures and moderately high temperatures. The nature of the diagenetic alteration processes which take place in this environment has been recently reviewed by Singer and Muller (1983).

Mudrocks contain upwards of 40% of clays, and may be classified as shales or mudstones, depending on whether the clay-mineral microfabric has a high preferential orientation, leading to bedding-plane fissility, or a random orientation. These differences may be explained by the presence or absence of organic peptizing agents during initial sediment deposition (Moon and Hurst, 1984). Clay minerals in both sandstones and mudrocks may be detrital (allogenic) or neoformed (authigenic). Detrital clays are merely inherited from the sedimentary environment, and are predominantly dioctahedral with some replacement of octahedral Al^{3+} by Fe^{3+}. Conversely, authigenic clays are dioctahedral but iron-poor (kaolinite and illite), or trioctahedral and Fe^{2+}-rich, with phases such as berthierine and trioctahedral chlorite (Curtis, 1985).

One exception is the formation of glauconite, a dioctahedral, iron-rich illitic phase, under relatively oxidizing conditions (Ireland *et al.*, 1983).

In sandstones, authigenic clays are predominantly formed by crystallization from solution, and the thermodynamic stability diagrams previously discussed remain valid if drawn for the appropriate pressure and temperature. Since sandstones are fairly permeable, their pore-fluid compositions can be considerably altered by the influx of large volumes of water expelled from adjacent compacting mudrocks. This water may be relatively acid, particularly in the temperature range 80–120°C, where kerogen, the solid precursor of liquid and gaseous hydrocarbons, begins to break down, evolving CO_2 and carboxylic acids (Hunt, 1979). Under these conditions, aluminium will have a significant mobility, whether as a dissolved ion or as an organic complex (Surdam and Crossley, 1985), and kaolinite precipitation will be favoured (Curtis, 1983). Under other conditions, a wide range of clay types and morphologies have been observed, including fine ribbons of illite (McHardy *et al.*, 1982). Authigenic clays in the pores of sandstones may adopt various conformations, described as pore-lining, pore-bridging and pore-filling, and lead to a variety of problems in hydrocarbon exploration and production, including adverse effects on permeability (Selley, 1985).

In mudrocks, the major change in clay mineralogy with increasing depth is the progressive conversion of smectite to illite. This is of great practical importance, since the release of interlayer water from the smectite provides an extra fluid source at depth, which can cause elevated pore-fluid pressures and mechanical weakening of the shales and mudstones. These processes are not only a source of difficulties or hazards during drilling for oil and gas, but may also influence the driving forces which cause oil and gas to migrate from their sources (impermeable organic-rich mudrocks) to extensive reservoirs in more porous and permeable rocks such as sandstones and chalks (Burst, 1969). The change from smectite to illite has long been considered, though not proven, to occur via a layer transformation mechanism (see for example Eberl (1984) and the subsequent discussion). However, more recent high-resolution transmission electron microscopy (TEM) studies tend to favour a neoformation mechanism in which authigenic illite precipitated from solution replaces dissolving smectite (Nadeau *et al.*, 1984; Ahn and Peacor, 1986).

1.5 Uses of clays

A brief summary will be given of the main industrial uses of clays. For more details, see Jones (1983), Odom (1984) and Jepson (1984).

1.5.1 *Uses of kaolin*

Kaolin refers to the clay which consists principally of the clay mineral kaolinite. Of the deposits in Cornwall, UK, about 3 million tons are produced annually, of which the bulk is used in the paper industry as a coating and filling agent. About 15% is sold for pottery and porcelain manufacture, while the remainder is used as extenders in paints, plastics and rubbers. In the paper industry, the particle size and brightness of kaolinites are important, the latter being adversely affected by small quantities of iron impurities.

1.5.2 *Uses of smectites*

Smectites have numerous industrial uses, of which the most important are:

(i) In foundries, as a binding agent in the preparation of moulding sands for metallurgical casting

(ii) In oilwell drilling, as a constituent of drilling fluids, either in the form of Na-montmorillonite (in aqueous fluids) or as an organophilic clay (in water-in-oil emulsion-based fluids)

(iii) In civil engineering, to hinder the movement of water or chemical wastes; for excavation support; for lubrication of cables or pipes, and as a waterproofing agent

(iv) As a backfill in radioactive waste repositories

(v) As an adsorbent: in industry for the removal of oil and grease from floors: domestically, in pet litter. It is also used for decolourizing vegetable oils, clarifying wine, and in water filtration and purification

(vi) As a catalyst in various hydrogenation reactions

(vii) In the manufacture of bricks, tiles and other building materials

(viii) Organophilic forms of smectites are used in the manufacture of inks cosmetics and polishes; as suspending and thickening agents in paints; to control the gelling properties of lubricating greases, and to add thioxotropic (shear thinning) properties to fibreglass resins.

1.6 Introduction to analytical techniques

Of the analytical techniques described in this book, X-ray powder diffraction, infrared spectroscopy, thermal methods and chemical analysis represent the traditional, well-established tools of the clay mineralogist. X-ray diffraction provides knowledge of the type or types of clay present, and the degree of crystalline order; infrared spectroscopy provides details of the vibrational modes of functional groups—which give additional structural information, and is particularly useful in studying the nature and manner of bonding of adsorbed species. Chemical analyses can then allow a distinction between different members of a group of structurally similar clays, e.g. the smectites. Although X-ray and infrared methods provide details at the interatomic level, what they observe is nevertheless an average over an enormous number of individual particles, and cannot resolve questions relating to detailed microfabrics of soils or sediments, i.e. the *in situ* interrelationships between clays on a fine scale. Also bulk chemical methods cannot determine the extent of local heterogeneities of chemical composition. In more recent years, electron microscopy and microanalytical techniques have provided very powerful insights into these areas, notably into the details of clay formation and transformation in various environments. Scanning electron microscopy provides direct imaging of particle morphologies at scales down to only a few nanometres, and when used in back-scatter mode, can be a very sensitive means of identifying different minerals, since the brightness of the resulting image depends strongly on mean atomic number. Transmission electron microscopy can resolve down to a few tenths of a nanometre, and so it can directly image the spacings and spatial variation between atoms in layer structures such as clays. In addition, the X-rays which are emitted when electrons strike the specimen have energies which depend upon the elements present, and can provide a very localized means of quantitative chemical analysis. Nevertheless, electron microscopy and microanalysis are most powerful when used in conjunction with more traditional techniques such as X-ray diffraction. Apparent discrepancies between direct imaging data and data provided by other techniques such as powder diffraction can help to re-examine or challenge long-held assumptions about the nature of the very fine-grained constituents of soils, sediments and rocks, their transformations and their role in natural processes (e.g. Nadeau *et al.*, 1984). The details of all of these techniques and their applications are dealt with in the chapters which follow.

References

Ahn, J.H. and Peacor, D.R. (1986) Transmission and analytical microscopy of the smectite-to-illite transition. *Clays & Clay Miner.*, **34**, 165–179.

Berner, R.A. (1971) *Principles of Chemical Sedimentology*. McGraw-Hill, New York, 240 pp.

Brindley, G.W. and Brown, G. (1980) *Crystal Structures of Clay Minerals and their X-ray Identification*. Mineralogical Society, London.

Brown, G., Newman, A.C.D. Rayner, J.H. and Weir, A.H. (1978) 'The structure and chemistry of soil clay minerals,' in *The Chemistry of Soil Constituents*, (D.J. Greenland and M.H.B. Hayes eds.) Wiley, Chichester, 29–178.

Burst, J.F. (1969) Diagenesis of Gulf Coast clayey sediments and its possible relation to petroleum migration. *A.A.P.G. Bull.*, **53**, 73–93.

Cebula, D.J. and Ottewill, R.H. (1981) Neutron diffraction studies on lithium montmorillonite–water dispersions. *Clays & Clay Miner.*, **29**, 73–75.

Curtis, C.D. (1977) Sedimentary geochemistry: environments and processes dominated by involvement of an aqueous phase. *Philos. Trans. Roy. Soc. Lond.*, **A286**, 353–372.

Curtis, C.D. (1983) Link between aluminium mobility and destruction of secondary porosity. *A.A.P.G. Bull.*, **67**, 380–393.

Curtis, C.D. (1985) Clay mineral precipitation and transformation during burial diagenesis. *Philos. Trans. Roy. Soc. Lond.*, **A315**, 91–105.

Dixon, J.B. and Weed, S.B. (1977) *Minerals in Soil Environments*, Soil Sci. Soc. Am., Madison, Wisconsin, USA.

Eberl, D. (1984) Clay mineral formation and transformation in rocks and soils. *Philos. Trans. Roy. Soc. Lond.*, **A311**, 241–257.

Farmer, V.C. (1978) 'Water on particle surfaces,' in *The Chemistry of Soil Constituents*, (D.J. Greenland and M.H.B. Hayes eds.) Wiley, Chichester, 405–448.

Farmer, V.C. and Russell, J.D. (1971) Interlayer complexes in layer silicates: the structure of water in lamellar ionic solutions. *Trans. Faraday Soc.*, **67**, 2737–2749.

Fripiat, J.J., Jelli, A., Poncelet, G. and Andre, J. (1965) Thermodynamic properties of adsorbed water molecules and electrical conduction in montmorillonites and silica. *J. Phys. Chem.*, **69**, 2185–2197.

Fripiat, J.J., Letellier, M. and Levitz, P. (1984) Interaction of water with clay surfaces. *Philos. Trans. Roy. Soc. Lond.*, **A311**, 287–299.

Glaeser, R. and Méring, J. (1968) Homogeneous hydration domains of the smectites. *C. R. Acad. Sci., Paris, Série II*, **267**, 436–466.

Greenland, D.J. (1971) Interactions between humic and fulvic acids and clays. *Soil Science*, **111**, 34–41.

Grim, R.E. (1968) *Clay Mineralogy*, McGraw-Hill, New York, 596 pp.

Hall, P.L. (1980) The application of electron spin resonance spectroscopy to studies of clay minerals. II. Interlamellar complex—structure, dynamics and reactions. *Clay Miner.*, **15**, 337–349.

Hall, P.L. (1982) 'Neutron scattering techniques for the study of clay minerals,' in *Advanced Techniques for Clay Mineral Analysis*, (J.J. Fripiat, ed.) Elsevier, Amsterdam, 51–75.

Hunt, J.M. (1979) *Petroleum Geochemistry and Geology*, W.H. Freeman, San Francisco, 617 pp.

Ireland, B.J., Curtis, C.D. and Whiteman, J.A. (1983) Compositional variation within some glauconites and illites and implications for their stability and origins. *Sedimentology*, **30**, 769–786.

Jepson, B. (1984) Kaolins: their properties and uses. *Philos. Trans. Roy. Soc. Lond.*, **A311**, 411–432.

Jones, T.R. (1983) Properties and industrial uses of clays which swell in organic solvents. *Clay Miner.*, **18**, 399–410.

Keren, R. and Shainberg. I. (1980) Water vapor isotherms and heat of immersion of Na/Ca montmorillonite systems. III. Thermodynamics. *Clays & Clay Miner.*, **28**, 204–210.

Lagaly, G. (1984) Clay–organic interactions. *Philos. Trans. Royal Soc. Lond.*, **A311**, 315–332.

Low, P.F. (1961) Physical chemistry of clay water interaction. *Adv. in Agronomy*, **13**, 269–327.

Lubetkin, S.D., Middleton, S.R. and Ottewill, R.H. (1984) Some properties of clay–water dispersions. *Philos. Trans. Roy. Soc. Lond.*, **A311**, 353–368.

MacEwan, D.M.C. and Wilson, M.J. (1980) 'Interlayer and intercalation complexes of clay minerals', in *Crystal Structures of Clay Minerals and their X-ray Identification*, (G.W. Brindley and G. Brown eds.), Mineralogical Society, London, 197–248.

McHardy, W.J., Wilson, M.J. and Tait, J.M. (1982) Electron microscope and X-ray diffraction studies of filamentous illitic clay from the Magnus Field. *Clay Miner.*, **17**, 23–39.

Martin, R.T. (1962) 'Adsorbed water on clay: a review', in *Clays & Clay Miner., Proc. 9th Nat. Conf.*, Pergamon, New York, 28–70.

Moon, C.F. and Hurst, C.W. (1984) 'Fabric of mud and shales: an overview', in *Fine-grained Sediments: Deep Water Processes and Facies*, (D.A.V. Stow and D.J.W. Piper eds.), Geological Society/Blackwell, Oxford, 579–593.

Nadeau, P.H., Wilson, M.J., McHardy, W.J. and Tait, J.M. (1984) Interparticle diffraction: a new concept for interstratified clays. *Clay Miner.*, **19**, 757–769.

Nemecz, E. (1981) *Clay Minerals*. Akadémiai Kiadó, Budapest, Hungary.

Newman, A.C.D. (1984) The significance of clays in agriculture and soils. *Philos. Trans. Roy. Soc. Lond.*, **A311**, 375–389.

Norrish, K. (1954) The swelling of montmorillonite. *Trans. Faraday Soc.*, **18**, 120–134.

Odom, I.E. (1984) Smectite clay minerals: properties and uses. *Philos. Trans. Roy. Soc. Lond.*, **A311**, 391–409.

Olejnik, S. and White, J.W. (1972) Thin layer of water in vermiculites and montmorillonites: modification of water diffusion. *Nature (Phys. Sci.)*, **236**, 15–16.

Pinnavaia, T.J. (1983) Intercalated clay catalysts. *Science*, **220**, 365–371.

Selley, R.C. (1982) *An Introduction to Sedimentology*. Academic Press, London, 417 pp.

Selley, R.C. (1985) *Elements of Petroleum Geology*. W.H. Freeman, New York, 449 pp.

Singer, A. and Muller, G. (1983) 'Diagenesis in argillaceous sediments', in *Diagenesis in Sediments and Sedimentary Rocks, 2*. (G. Larsen and G.V. Chilingar eds.), Elsevier, Amsterdam, 115–212.

Sposito, G. and Prost, R. (1982) Structure of water adsorbed on smectites. *Chem. Rev.*, **82**, 573–593.

Surdam, R.C. and Crossley, L.J. (1985) Organic–inorganic reactions during progressive burial: key to porosity and permeability enhancement and preservation. *Philos. Trans. Roy. Soc. Lond.*, **A315**, 135–156.

Swartzen-Allen, S.L. and Matijevic, E. (1974) Surface and colloid chemistry of clays. *Chem. Rev.*, **74**, 385–400.
Theng, B.K.G. (1971) *The Chemistry of Clay–Organic Reactions*, Hilger, London.
Theng, B.K.G. (1979) *Formation and Properties of Clay–Polymer Complexes*, Elsevier, Amsterdam.
Thompson, J.G. and Cuff, C. (1985) Crystal structure of kaolinite: dimethylsulfoxide intercalate. *Clays & Clay Miner.*, **33**, 490–500.
van Olphen, H. (1965) Thermodynamics of interlayer adsorption of water in clays. I. Na vermiculite. *J. Colloid & Interface Sci.*, **20**, 822–837.
van Olphen, H. (1977) *An Introduction to Clay Colloid Chemistry*, Wiley, New York, 318 pp.
Wada, K. (1982) 'Amorphous clay minerals—chemical composition, crystalline state, synthesis and surface properties', in *Intern. Clay Conf.*, 1981 (H. van Olphen and F. Veniale eds.). Elsevier, Amsterdam, 385–398.
Weiss, A. (1963) A secret of Chinese porcelain manufacture. *Angewandte Chem., Intern. Ed.*, **2**, 697–703.
Weiss, A. (1970) 'Organic derivatives of clay minerals, zeolites and related minerals', in *Organic Geochemistry*, (G. Eglinton and M.T.J. Murphy eds.) Longman, Harlow, Essex, 737–781.

2 X-ray powder diffraction methods

M.J. WILSON

2.1 Introduction

X-ray powder diffraction is undoubtedly the most widely used technique for identification and characterization of clay minerals. The principles of the method and the instrumentation involved have been discussed in depth in many excellent books (for example Klug and Alexander, 1974; Cullity, 1956; Nuffield, 1966) and there would seen to be little point in repeating these discussions. However, for the sake of completeness, an introduction to the method is required in this chapter, and in the following sections a very much simplified description of principles and instrumentation, with specific reference to the examination of clay minerals, will be presented. As a reference work on the X-ray diffraction of clay minerals, the monograph published by the Mineralogical Society (Brindley and Brown, 1980), is totally indispensable. This monumental work is the most complete available source of X-ray diffraction information on clay minerals, although for the newcomer to the field it might appear to be rather formidable. The present chapter aims to provide a more gentle introduction to X-ray identification of clay minerals, without being too superficial and in a way that will give encouragement to analysts faced for the first time with this 'difficult' material.

2.2 X-ray diffraction by clay minerals

2.2.1 *Principles*

As was discussed in Chapter 1, clay minerals consist of tiny crystals which are themselves made up of ordered arrays of atoms, arranged in a periodic or repetitive way. When an incident X-ray beam impinges upon such an array, general scattering occurs, such that the scattered waves interfere with and destroy one another. In certain specific directions, however, the scattered waves are in phase with one another and combine to form new wave-fronts. This constructive interference is known as diffraction. The directions in which diffraction occurs depend upon the size and shape of the unit cell of the crystal, whereas the intensity of diffraction is determined by the actual atomic array or the nature of the crystal structure.

X-ray diffraction can be conveniently visualized as a reflection of the

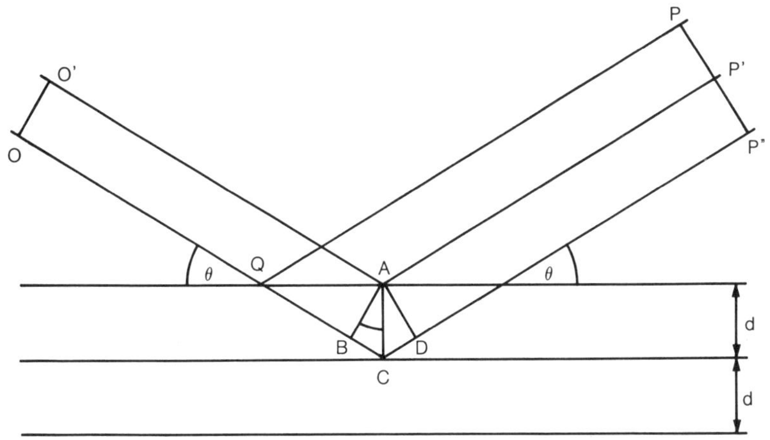

Figure 2.1 X-ray diffraction from a layer structure

incident beam by parallel, closely spaced planes of atoms within a crystal. The condition for reflection is the well-known Bragg equation:

$$n\lambda = 2d \sin \Theta$$

where n is an integer, λ is the wavelength of the X-rays, Θ is the glancing angle of incidence and d is the interplanar spacing of the crystal. Minerals with layer structures are particularly suitable for illustrating the basic aspects of X-ray diffraction.

Figure 2.1 shows a beam of parallel X-rays of wavelength λ falling at angle Θ onto the face of a crystal possessing a layer structure of parallel planes of atoms with spacing d. The rays of the incident beam are in phase with one another so that they have a common wave-front OO′. Rays OQ and O′A, which are scattered by the top plane of the crystal in the directions QP and AP′, also at angle Θ, again form a common wave-front, because their path length is the same, and so they arrive at PP′ in phase with each other. When a ray penetrates to the next plane of atoms in the crystal, the path length of the incoming and scattered rays, O″CP″, will differ from the path length of the rays scattered from the surface layer of atoms by an amount, BCD, which is equal to 2BC. As BC $= d \sin \Theta$ in the right-angled triangle ABC, the path difference can be written as $2d \sin \Theta$, which is the Bragg condition for reflection. However, the Bragg equation also indicates that reflection will occur when the path difference equals an integral number of wavelengths, n, so that there are a number of solutions for Θ for particular values of d and λ. Recasting the equation to solve for diffraction angle

$$\Theta = \sin^{-1}\left(\frac{1\lambda}{2d}\right)$$

$$\Theta = \sin^{-1}\left(\frac{2\lambda}{2d}\right)$$

$$\Theta = \sin^{-1}\left(\frac{3\lambda}{2d}\right) \ldots \text{etc.}$$

This means that, with respect to diffraction from the layer-structured crystal illustrated in Figure 2.1, as the crystal is rotated through an arc, it will pass through positions at which the phase differences are 1, 2, 3 ... wavelengths, and where there will be a series of reflections deriving from the set of planes of spacing d. Taking a specific clay mineral as an example, *chlorite* has a layer repeat of about 14.1 Å, and thus yields an integral series of reflections at 14.1, 7.05, 4.70, 3.52 Å etc. corresponding with 7.28, 14.58, 21.92 and 29.46° 2Θ using CoKα radiation ($\lambda = 1.790$ Å). In clay mineralogy such reflections are usually termed basal reflections, and in the above example one would refer to the 001, 002, 003 and 004 basal reflections of chlorite. As the basal reflections derive from the fundamental layer-repeat of clay minerals, they are of great diagnostic significance and, consequently, the aim of many preparative methods is to enhance their intensity as much as possible, as will be shown later.

So far, only X-ray diffraction from one set of planes deriving from a layer-repeat has been considered but, of course, diffraction will also occur from many other lattice planes within a layer-structured crystal. For clay minerals these planes may be referred to in a general way as non-basal (or hkl) planes, as they occur at right angles or are inclined to the basal planes. The complete X-ray powder pattern of a clay mineral, therefore, consists of basal and non-basal reflections, and when the mineral is highly crystalline this pattern should be completely diagnostic for identification purposes, as for most other minerals. Unfortunately, however, many clay minerals are anything but highly crystalline, and are subject to various kinds of structural disorder, which can particularly influence the diffraction from the non-basal planes.

These disorders most often arise from variable hydration of expanding minerals, interstratification of different layer types, and stacking disorder of a single layer type. Stacking disorders may be related to the extremely small size of clay-mineral crystallites, which in extreme cases can be regarded as two-dimensional crystals. This means that the a and b plane lattice of one layer bears no fixed relationship to the a and b plane lattices of neighbouring layers, so that a randomly stacked (or turbostratically disordered) layer structure results. The perpendicular distance between the layers should, however, still be fixed. The X-ray diffraction pattern of a material disordered in this way would be characterized by relatively strong *basal* reflections, but by broad and diffuse *non-basal* reflections, which strictly speaking would now be given hk indices. The form of the hk reflections is often characteristic and consists of an asymmetrical band of scattering terminated sharply on the low-angle side, the reasons for this being fully discussed in Brindley and Brown (1980). This type of disorder is highly characteristic of *smectites*, as will be shown later. Other types of disorder are found in *kaolinites*, which by and large consist of much thicker crystals.

One hkl reflection is of particular diagnostic significance and should always be measured whenever possible. This is the 060 or 06 reflection, which occurs at ~ 1.5 Å, and is often stronger and sharper than the other hkl reflections. Measurement of this reflection will yield the *b* parameter of the clay mineral and will indicate whether it belongs to the dioctahedral or trioctahedral subgroups. In general, the 060 reflection of dioctahedral minerals occurs between 1.49 and 1.51 Å, whereas for trioctahedral minerals a spacing of between 1.53 and 1.55 Å is usually recorded.

2.2.2 *Instrumentation*

Nowadays the X-ray study of clay minerals almost inevitably requires an X-ray powder diffractometer. There are still some advantages in using powder camera methods, but the relative ease and speed of diffractometry in the acquisition and measurement of good-quality diffraction patterns has led to this method being almost universally adopted as standard in powder diffraction laboratories. The principles of X-ray diffractometry have been fully described in the text books mentioned previously, and only a brief explanation of the main features of the method will be attempted here, followed by a discussion of important factors in the study of clays.

2.2.2.1 *The X-ray diffractometer.* A diffractometer consists essentially of an X-ray generator, a goniometer for rotating the sample and measuring diffraction angles, and an X-ray counter tube and counting circuits to detect, amplify and measure the diffracted radiation. The diffractogram represents a permanent record of intensity against diffraction angle, and is obtained directly by a strip-chart recorder or by some other output device.

(a) The X-ray generator and tube: the X-ray generator is designed specifically for high output power, with full rectification and high current stability. The maximum power is usually about 3 kW, and is obtained through a variable transformer controlled from a multiposition switch on the control unit. This supplies continuously variable tube voltages from about 20 to 60 kV, the set values being displayed on a meter. Full-wave rectification is achieved by silicon rectifiers, and a smoothing capacitor ensures a constant potential output, usually stabilized to 0.03 or 0.01%, against mains fluctuations of + 10 to − 15% and temperature variations of 15°C. The tube current in similarly stabilized, and is varied by means of an adjustment switch on the control unit, usually between 5 and 80 mA.

A most convenient arrangement is for the generator to be built around a four-window X-ray tube, mounted in such a way that the table top can be used to support powder cameras or another goniometer. The most commonly used tubes are of the hot-cathode type, which are permanently sealed and evacuated. In this type of tube, a heated filament acts as a cathode, and emits electrons which are drawn by a potential-gradient across the tube to strike a perpendicularly situated target, which then emits characteristic X-rays. The intensity of the emitted X-ray beam is improved by focusing the electron stream to a line, typically 1 mm wide by 10 mm long in a normal focus tube. The focal spot can be viewed along its length or perpendicular to its length, giving X-ray beams with distinctly different characteristics. For a normal focus tube with a 6° take-off angle, the projected focal spot dimensions are typically 1.0 by 1.0 mm when viewed end-on (the point focus), and 0.1 by 10 mm when viewed perpendicular to its length (the line focus). The point-focus beam is more intense than the line-focus beam, but nevertheless the latter is preferred for use with the diffractometer. The advantage of the line focus is that the narrow effective source-width greatly improves the resolution of the diffraction lines, the loss of intensity being avoided by using samples with a large surface area. Greater beam intensity is also achieved by making the windows of the X-ray tube especially transparent to X-rays with materials of low atomic number, such as thin-sheet beryllium metal.

(b) The goniometer: for standard powder diffractometry the goniometer rotates the specimen and the X-ray counter in a fixed 1:2 ratio. The rotation of the X-ray counter at twice the angular speed of the specimen is crucial for the so-called parafocusing arrangement of the goniometer, whereby the X-ray beam diverges from the focus in the X-ray tube, strikes the flat surface of the specimen and is brought back to a focus at the receiving slit in front of the X-ray counter. The sample surface lies tangentially to the circumference of the focusing circle, which includes both the focus of the X-ray tube and the receiving slit, and always bisects the diffraction angle 2Θ (Figure 2.2).

The optical arrangement for the diffractometer is designed specifically for use of the line focus described above, which yields a very narrow X-ray beam.

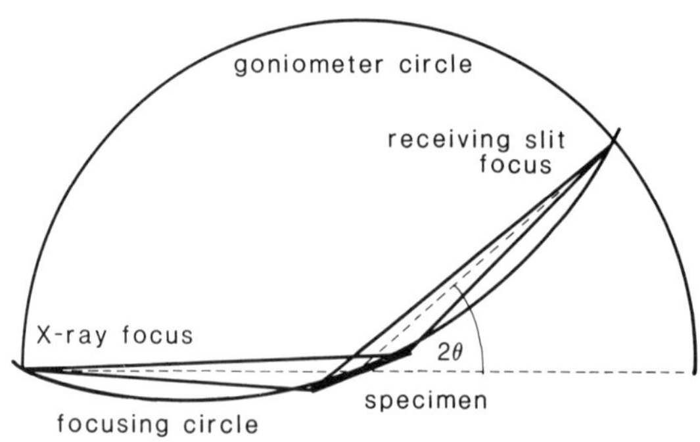

Figure 2.2 The parafocusing arrangement of the goniometer on an X-ray diffractometer

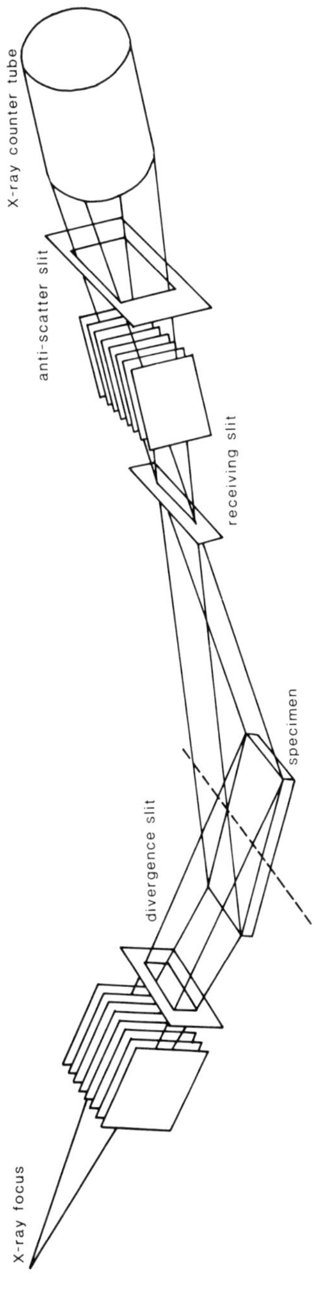

Figure 2.3 The X-ray optical arrangement for a Philips diffractometer

The actual angular aperture of the beam is defined by a divergence slit, which limits the beam to the specimen area (Figure 2.3). Various slit widths are available for particular angular ranges. If an inappropriately wide divergence slit is used, there is a risk that radiation scattered from the specimen holder will be detected and give rise to spurious peaks. The width of the diffracted beam is defined by the receiving slit, and the antiscatter slit near the window of the detector excludes non-diffracted radiation and reduces the background radiation of the diffraction pattern. The axial divergence of the incident and diffracted X-ray beams is also collimated by two Soller slit arrangements, consisting of closely spaced, highly absorbing metal foils.

(c) X-ray detectors, control electronics and output devices: until recently, gas-filled proportional counters were standard in X-ray powder diffractometry, but solid-state detectors are now being introduced. Gas-filled detectors make use of the initial ionization of a gas (usually xenon) following absorption of X-rays which, in a suitable electrical field, results in the emission of an electrical pulse to the external circuit. The amplitude or height of the pulse is directly proportional to the energy of the initiating X-ray quantum—hence the term ‘proportional counter’. The quantum counting efficiency of the proportional counter is actually inferior to that of the scintillation counter, but the latter has much poorer energy resolution. This is important because most diffractometers use electron energy-selective methods to discriminate and count only a selected range of pulses whose amplitudes lie within certain limits, a process known as pulse-height analysis. In this way, unwanted radiation can be effectively screened, and its contribution to the background of the powder pattern can be minimized. After discrimination, the pulse rate is analysed on a scaler and is reduced to measurable proportions. The scaler output is fed into a ratemeter, which converts the digital information into a continuous electric current which activates a single pen recorder. The ratemeter can be used to vary the time constant, that is, the rate at which the pulse-integrating circuit is discharged, resulting in a smoothed diffractometer trace where a large time constant is used, or in a noisy trace where a small time constant is used. Until recently the most commonly used output device was a strip-chart recorder, but with the strong trend towards automatic and computer-controlled operations, the option of a printer in conjunction with a graphics video terminal has become widely available.

2.2.2.2 *Factors of importance in the study of clays. (a) Choice of radiation*: the most widely used radiations for clay studies are undoubtedly copper and cobalt. Previously, the former was used almost universally, because the loading characteristics of copper tubes enabled higher radiation intensities to be obtained. Recent design and manufacture improvements for cobalt tubes enable comparable, although still slightly lower intensities to be produced. Cobalt radiation does however, have a definite advantage over

Table 2.1 Some characteristics of X-radiation used in clay studies.

Element	Kα (Å)	Kβ (Å)	Excitation[1] potential (kV)	Transmission (%)[2]	Tube[3] rating (IV)
Cu	1.5418	1.3922	8.98	95	2000
Co	1.7903	1.6208	7.71	92	1800
Fe	1.9399	1.7566	7.10	90	1500

[1] The minimum voltage applied to an X-ray tube required to excite the K spectrum.
[2] Percent transmission through a 300-μm thick beryllium window.
[3] Ratings of Philips normal focus tubes at a loading of 40 kV, 40 mA.

copper, in that its longer wavelength can lead to the easier detection of the high-spacing reflections that are a feature of some clay minerals. In addition, cobalt is preferable for examination of iron-rich materials, because copper radiation is heavily absorbed by such materials, and generates secondary fluorescent radiation—leading to diffraction patterns with high background. Nevertheless, it should be emphasized that copper radiation is quite adequate for X-ray analysis of most clay materials. The only other radiation worthwhile considering for clay studies is that of iron. Its wavelength is longer than that of cobalt, and it is particularly useful for investigating specimens rich in manganese, which fluoresce strongly under cobalt. The target loadings for iron tubes are, however, somewhat lower than for cobalt tubes, so that one would have to accept some loss in intensity. Information relevant to the discussion in this section is shown in Table 2.1.

(b) Monochromatic radiation: for most routine X-ray powder diffraction studies, it is highly desirable that monochromatic radiation is used. This is best achieved by the use of a crystal monochromator, which in many diffractometers is sited in the diffracted beam. This arrangement is now widely used and undoubtedly leads to low-background, high-quality powder patterns, although at the expense of some loss in overall intensity. For most clay studies, however, it is not essential to use a crystal monochromator, and the traditional arrangement of inserting an appropriate metal foil in the primary beam—between the X-ray source and the sample—is perfectly adequate. The metal foil is selected for its characteristic absorption edge, and effectively absorbs or filters out the shorter wavelength Kβ radiation of the target metal, but is relatively more transparent to the longer wavelength Kα radiation (Figure 2.4). The intensity of the continuous spectrum is also reduced, and it is reduced even further by the pulse-height discrimination process mentioned previously. A full explanation of the absorption edge phenomenon may be found in the standard texts referred to at the beginning of the chapter, and here it is sufficient to point out the appropriate β filters for the three most commonly used radiations (Table 2.2). Parrish and Taylor (1956) suggest that

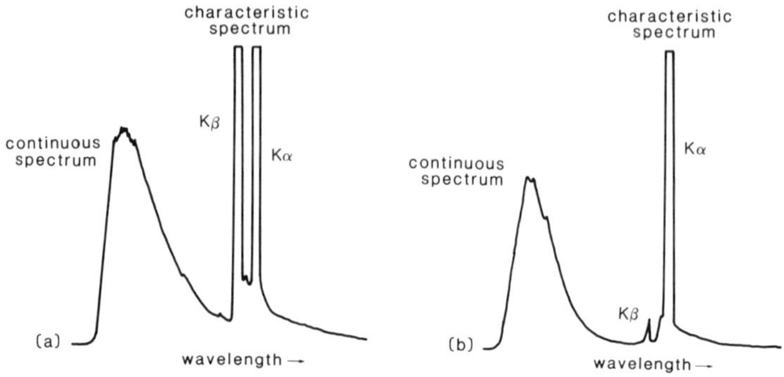

Figure 2.4 The continuous and characteristic spectrum of a metal target (a) without and (b) with an appropriate B filter

Table 2.2 Filter characteristics for X-radiation used in clay studies.

Target	Filter	Thickness (mm)[1]	% Kβ absorbed
Cu	Ni	0.0170	98.4
Co	Fe	0.0166	98.9
Fe	Mn	0.0168	99.2

[1] Thickness required for 50% transmission of Kα.

a filter thickness which increases the Kα:Kβ intensity ratio to 100:1 is quite satisfactory for most purposes.

(c) Selection of diffractometer slit system: to obtain sharp diffraction patterns with narrow line width, it is essential to collimate the X-ray beam through a system of parallel slits of selected width. However, the inherent poorly crystalline nature of many clay minerals means that a *choice* of slit widths which is correct for well-crystallized materials, is unnecessarily restrictive in terms of the overall intensity of the diffraction patterns. Nevertheless, one should be aware of the important factors to be taken into account when choosing slit widths. With the divergence slit, if the investigation aims to evaluate intensities over the diffraction pattern as a whole, it is essential that the correct slit width is used, so that the incident beam is intercepted by the specimen and not by the specimen holder. For the 2-cm wide aluminium holder used with the Philips diffractometer, the minimum angles at which the various divergence slits are used are shown in Table 2.3. Note that the commonly used 1° divergence slit should not be employed below ~ 18°2Θ when the sample is packed in the specimen holder, or below 13°2Θ when 2.6 cm-wide glass slides are used. The approximate sample length (L),

Table 2.3 Divergence slit data for a Philips diffractometer.

Divergence slit (°)	Minimum[1] angle (°2Θ)	Maximum d spacing(Å) measured with: CuKα	CoKα	FeKα
1/4	4.6	19.24	22.34	24.16
1/2	9.1	9.66	11.22	12.14
1	18.4	4.83	5.61	6.07
2	32.2	2.42	2.81	3.04
4	78.0	1.22	1.42	1.54

[1]Smallest angle where X-ray beam will not overlap a 20-mm wide flat specimen.

irradiated by a collimated X-ray beam, may be calculated from:

$$L = \left(\frac{\gamma}{57.3}\right)\left(\frac{R}{\sin \Theta}\right)$$

where γ is the angular aperture of the divergence slit in degrees, R is the goniometer radius in centimetres, and Θ is the angle between the X-ray beam and the sample. The use of an inappropriately wide divergence slit may result in the generation of a high background towards the low angles, making it impossible to detect genuine high-spacing peaks. The problem of choosing the correct divergence slit seems to have been averted by the recent introduction of a theta-compensating slit, so constructed that the irradiated specimen area remains constant over the entire scan range. So far, however, this device has not been evaluated in the investigation of very poorly ordered or interstratified clay minerals, which may yield weak, broad diffraction effects at high spacings. It is often observed that such reflections are detectable using an 'incorrect' 1° divergence slit, but cannot be observed when the correct 1/4° slit is used.

Although the common use of a 1° divergence slit over the whole scan range may be deplored from a purist point of view, for routine clay studies employing glass slides it does offer the advantages of convenience and higher peak intensities in the important 14–20 Å range, although often against a strongly sloping background. However, it is essential to realize that when a 1° divergence slit is used at low angles, it then becomes impossible to compare the intensities of reflections recorded in the low-angle range with those recorded at higher angles.

So far as the appropriate widths of the receiving and antiscatter slits are concerned, the general rule for clay studies is that these widths may be wider than those normally recommended for well-crystallized materials, as long as this does not lead to greatly increased line widths. Brown and Brindley (1980) point out that, although a receiving slit of width 0.2°(2Θ) is generally suitable in clay studies, broader slit widths are permissible for poorly ordered materials that in any case yield very broad reflections. The use of broader slit widths for such materials greatly increases intensity, with little loss in resolution or increase in background. However, one should be aware of possible spurious effects at low angles.

2.3 Sample preparation

2.3.1 *Oriented mounts*

To identify clay minerals it is essential to record their diagnostic basal reflections, the intensities of which are enhanced if the platy clay minerals of the sample lie generally parallel to the plane surface of the diffractometer. Both the nature of the mount substrate and the technique employed for making the mount need to be considered. A thorough compilation of these aspects, together with a detailed comparison of the advantages and disadvantages of the various methods, has been made by Hauff *et al.* (1982), and only a brief review of some aspects will be attempted here. Probably the most widely used method is to pipette the clay–water suspension onto a microscope slide cut to fit the diffractometer sample-holder. The method is easy, quick, and produces a good and fairly reproducible orientation. There are, however, a number of disadvantages and problems, the most serious of which is differential settling of clay minerals during sedimentation onto the glass slide, resulting in layering according to grain size and density (Gibbs, 1965). This is, of course, particularly serious for quantitative work, and can lead to overestimation of the fine-grained smectite minerals which tend to concentrate nearer the mount surface. Nevertheless, the author's experience is that, for most clay materials, differential settling is not such a serious problem as one might think. In any case, it can be minimized by rapidly drying-down the clay–water suspension onto a hot-plate heated to 60°C. Hauff *et al.* (1982) list ten other disadvantages of the glass-slide method, of which the most substantive are:

(i) That ordinary glass slides may crack and warp if auxiliary heating treatments are performed

(ii) Smectitic clays may crack, peel or curl, and

(iii) Chemical treatments are not as easy to perform as with porous tiles.

Although these points are generally valid, glass slides may be heated up to ~ 500°C without undue deformation, and problems with smectite clays may be avoided if more dilute suspensions are used.

Oriented mounts can also be made on glass slides by smearing on a clay–water paste, although problems arising from uneven surfaces, thickness variations and cracking may be encountered. Tien (1974) described a simple frame designed to hold slides, and to achieve a uniform thickness and smoothness of clay smears.

Another widely used method for preparing oriented mounts is to centrifuge or suck down a clay suspension onto a porous ceramic tile, as described, for example, by Kinter and Diamond (1956), Rich (1969) and Shaw (1972). The outstanding advantages of this method are that it is easy to study the effects of different cation saturations, solvation and washing treatments, and that the tiles can withstand temperatures greater than 550°C without deformation. On the other hand, it takes a longer time to prepare large numbers of samples, and if the clay film is thin the crystalline material in the tiles—which usually includes quartz and feldspars—may contribute to the diffraction pattern of the sample.

Suction of clay suspensions onto membrane filters is another possible technique that may offer certain advantages. For example, cellulose filters yield excellent orientation and even surfaces, but cannot be used in heating experiments. The latter objection does not hold for silver membrane filters but, again, these tend to be expensive. Evidently, there is no single recommended method for preparing oriented clay mounts and the chosen method largely depends on the purpose of the investigation and the material to be examined. If the prime purpose is identification and characterization, then it is probable that reasonable results can be obtained from all the above methods.

2.3.2 *Randomly oriented preparations*

Completely randomly oriented clay materials are very difficult to prepare. Almost inevitably, the platy clay minerals tend to adopt some preferred orientation close to the sample surface, which must be smooth and even. Certainly, the common procedure of back-filling the sample-holder against a glass slide, followed by gentle downward pressure, still results in some preferred orientation. Filling from above, by passing the sample through a sieve and allowing the particles to free-fall into the holder, again reduces preferred orientation but does not eliminate it. An arrangement for enabling side packing of the sample-holder is probably the easiest way of ensuring near-random orientation, although again, perfect randomness is not achieved (Brindley, 1980). Other methods that have been suggested include embedding the sample in a thermoplastic cement (Brindley and Kurtossy, 1961), sorption onto flat polyester foam (Thomson *et al.*, 1972) and spray-drying (Jonas and Kuykendall, 1966; Hughes and Bohor, 1970). The author has found that mixing clay with powdered cork yields diffraction patterns with peak intensities similar to those observed in powder photographs. If only a small amount of clay is available (< 10 mg), the method of Paterson *et al.* (1986), which involves drying an acetone–clay suspension onto a glass slide, yields patterns that are at least comparable with those obtained by back-filling an aluminium holder.

2.3.3 *Ethylene glycol/glycerol solvation*

As will be described later, such treatment is used extensively to provide essential supplementary information for clay-mineral identification. Various methods of applying these organic liquids to clay films have been suggested; the four main ones being as follows:

(i) Addition of the solvating agent directly to the clay suspension, followed by thorough mixing, and allowing the resultant slurry to dry onto a glass slide

(ii) Placing dried, sedimented aggregates over the organic liquid in a sealed desiccator, followed by heating for 16–24 hours at 60°C for ethylene glycol and at 100°C for glycerol

(iii) Spraying a fine mist of the organic liquid (15% in water or 6% in ethanol) directly onto a dried, sedimented aggregate. Alternatively, drops of the solution may be added directly to the clay film, care being taken to ensure that it is thoroughly wet

(iv) By direct contact between the clay film and ethylene glycol or glycerol. Placing the oriented aggregate face down onto an ethylene glycol or glycerol-saturated filter paper is one method of ensuring adequate contact.

According to Novich and Martin (1983), the most reproducible of these methods is (i), because the organic complex is produced when the clay is in its maximum possible hydration state. The other methods involve a partial dehydration stage of the clay, because the clay has been air-dried, which may lead to non-reproducible spacings. In practice, for most investigations any of the above methods will yield adequate results. However, where the objective is to record precise spacings—and different smectites do yield slightly different spacing with these treatments (Srodon, 1980)—it is important to use the method that produces the optimum solvation and the greatest regularity in the sequence of basal reflections. The capacity of clays to swell with organic liquids is sometimes increased by chemical pretreatments (Na-dithionite citrate and HCl) and by mechanical grinding (Cowking *et al.*, 1983). This reinforces the point that pretreatments of clay materials should be kept to a minimum if possible.

2.3.4 *Heating and control of relative humidities*

Ideally, observation of the responses of clays to heating treatments and variations in relative humidity should be made using a controlled heating-stage (Brown *et al.*, 1972) or specimen chamber. Such facilities may not be available, however, and in these circumstances simple techniques may be adopted, provided that one is aware of possible pitfalls, such as the rapid rehydration of some clays. Heating at 300°C is often useful, and this can be done by heating a clay-coated slide for 2 hours on a hot-plate, preferably sited near the diffractometer. The specimen is taken from the hot-plate and inserted directly into the diffractometer sample holder, and then rapidly scanned through the angular range where the major basal spacings occur (3–15°2Θ, CoKα). This method is usually (although not always) sufficiently rapid to avoid problems with rehydration. If such problems are encountered, then saturation with a less hydrateable cation, such as potassium, is recommended.

Glass slides can also be heated to 500–600°C, but at these temperatures the slide may warp, which sometimes leads to spurious high-spacing peaks on the diffraction pattern.

For humidity control, reproducible results can be obtained by equilibrating sedimented aggregates at specific relative humidities over selected saturated salt solutions, and then sealing the glass slide from the atmosphere with self-adhesive plastic film. At very low relative humidities it is more difficult to achieve satisfactory results with this technique.

2.4 X-ray identification of clay minerals

2.4.1 *Kaolinite minerals*

Little difficulty should be experienced in identifying these minerals when they occur in a fairly pure form. All (apart from hydrated halloysite) are characterized by intense first- and second-order basal reflections at ~7.1 Å and ~3.57 Å, and by prominent groups of non-basal reflections. Both types of reflections, but especially the latter, become broader and more diffuse with increasing structural disorder. In practice, kaolinite minerals may be associated with other types of clays. Often, in these circumstances, auxiliary treatments are needed to positively identify a particular kaolinite mineral, or even to establish the presence of a kaolinite group mineral.

2.4.1.1 *Kaolinite.* An X-ray pattern typical of well-crystallized kaolinite, and with the most intense diagnostic spacings listed, is shown in Figure 2.5(a). Both basal and non-basal reflections are sharp and well-defined. A particularly important point is that the closely spaced $11\bar{1}$ and $02\bar{1}$ reflections between 4.18 and 3.84 Å and the group of 131 and 201 reflections on either side of the 003 peak at 2.38 Å are clearly resolved. In highly disordered kaolinites these reflections are much more poorly defined, and may be replaced by a continuous diffractions band (Figure 2.5(c)); in addition, the basal reflections become somewhat broader and less intense. Many, perhaps most, kaolinites show diffraction characteristics falling somewhere between these extremes (Figure 2.5(b)). The nature of kaolinite disorder has been systematically investigated and explained from a theoretical point of view by Plançon and Tchoubar (1977) but, unfortunately, because of the inherently complex nature of this disorder, it is difficult to use the findings in practical investigations. Here, many workers still make use of a so-called kaolinite 'crystallinity' index. The index devised by Hinckley (1963) is still widely used, and, for randomly oriented clays, consists of measuring the heights of the $1\bar{1}0$ and $11\bar{1}$ peaks above a line drawn from the trough between the 020 and $1\bar{1}0$ peaks to the background just beyond the $11\bar{1}$ peak. These measurements are combined to form the numerator and are divided by the height of the $1\bar{1}0$ peak above the

No.	d (Å)	hkl
1	7.14	001
2	4.46	020
3	4.35	$1\bar{1}0$
4	4.18	$11\bar{1}$
5	3.84	$02\bar{1}$
6	3.57	002
7	2.56	$1\bar{3}0$, $20\bar{1}$
8	2.53	$13\bar{1}$
9	2.49	$13\bar{1}$, 200
10	2.34	$20\bar{2}$, $1\bar{3}1$
11	2.29	131

No.	d (Å)
1	7.14
2	4.46
3	4.35
4	4.18
5	3.57
6	2.56
7	2.49
8	2.34
9	2.29

No.	d(Å)
1	7.16
2	4.48
3	3.57
4	2.56
5	2.49
6	2.34

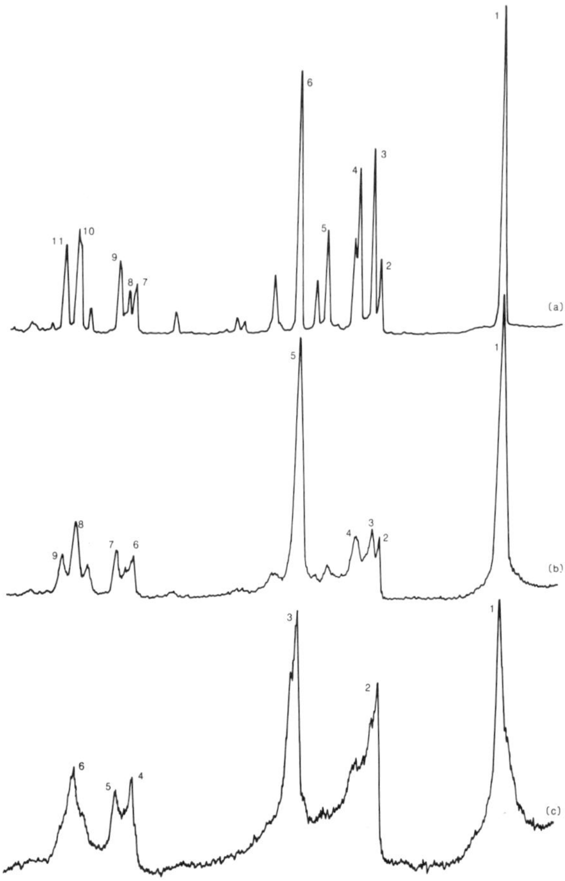

Figure 2.5 X-ray diffractometer patterns of randomly oriented kaolinites from (a) Keokuk, USA—highly crystalline, (b) Sedlec, Czechoslovakia—moderately disordered, and (c) Sotland—highly disordered. Partial diffraction data showing some diagnostic reflections are indicated

general background (Figure 2.6). This index is useful for the characterization of commercial clay deposits, but may be difficult or impossible to apply to very poorly ordered soil kaolinites. For this reason, Hughes and Brown (1979) devised another empirical crystallinity index for kaolinites, which they applied to tropical soil clays. This index is the ratio of the intensity above background, of the reflection at 4.6 Å, to the intensity above background of the reflection at ~ 2.43 Å.

2.4.1.2 *Dickite.* The X-ray diffraction pattern of dickite is superficially similar to that of well-crystallized kaolinite, and no doubt the two are often

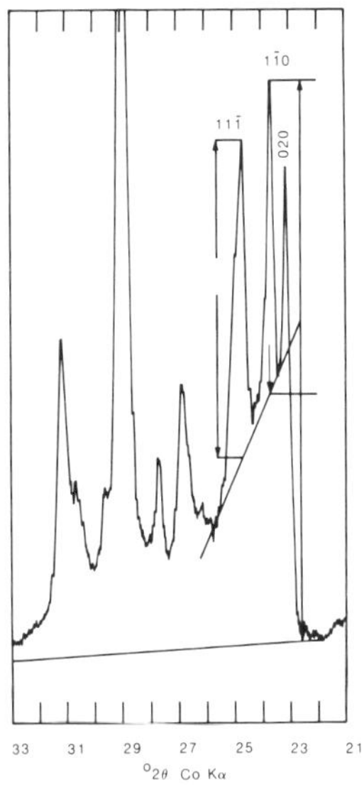

Figure 2.6 Measurements used for the calculation of the Hinckley crystallinity index for kaolinites, h $1\bar{1}0$ + h $11\bar{1}$ above sloping background and h $1\bar{1}0$ above general background

confused. However, a careful comparison of the patterns brings out many points of difference, particularly with regard to the non-basal reflections. It must be noted, however, that many of these reflections are weak, and could well be missed if the sample is not prepared in a completely randomly oriented form. The strongest diagnostic reflections for dickite, which are not found in kaolinite, would appear to be at 3.79 and 3.43 Å, and for kaolinite (not found in dickite) at 4.18, 3.84, 3.74, 3.15 and 2.75 Å. These comments apply to highly ordered dickite, and may not be appropriate for the more unusual disordered dickites described by Brindley and Porter (1978).

2.4.1.3 *Halloysite.* The X-ray identification of pure halloysite is straightforward, even when the mineral occurs in its dehydrated state. The diffraction pattern typically shows a rather broad 001 reflection at 7.2–7.4 Å, often with distinct asymmetry towards the low-angle side, and a similarly broad 002 peak at 3.6 Å (Figure 2.7). In addition, the pattern is characterized by the almost complete absence of well-defined, non-basal reflections, which are generally replaced by continuous bands of scattering. This is because—to a first approximation—the halloysite structure is highly disordered, there being little or no crystallographic relationship from one silicate layer to the next. A diagnostic feature of the X-ray pattern of halloysite with the common tubular morphology—that has been prepared as a sedimented aggregate—is that the 11, 02 reflection at ~4.42 Å may be more intense than the basal reflections.

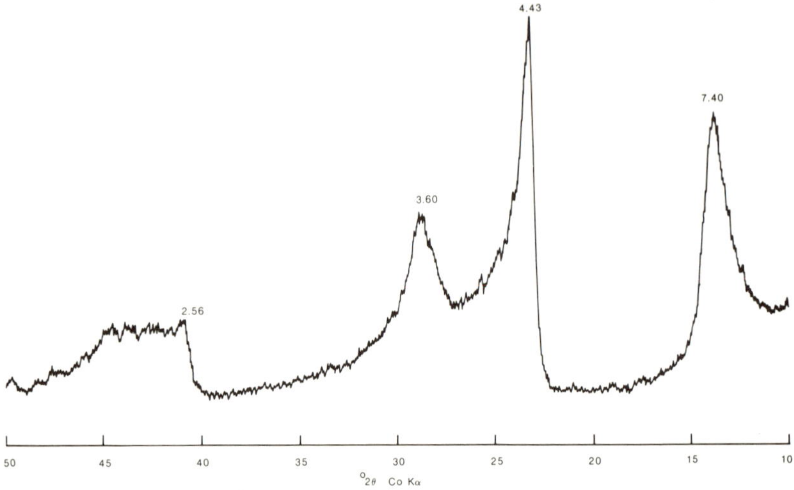

Figure 2.7 X-ray diffractometer pattern of dehydrated halloysite

This results from the non-planar morphology of the mineral, which means that it is impossible to prepare the usual clay-mineral-type aggregate with well-oriented basal planes. A halloysite-type pattern without an intense 4.42 Å reflection suggests that the mineral has a non-tubular morphology. In its fully hydrated state, halloysite yields a broad reflection at ~ 10 Å, which will contract to ~ 7.2 Å with gentle heating to about 100°C. Halloysites vary somewhat in their tendency to dehydrate; some collapse to the 7 Å form after drying at temperatures only slightly above ambient, and in these circumstances, it is essential to preserve samples in a wet or field-moist condition if one needs to investigate the *in situ* hydration state of the mineral. Other halloysites, particularly spheroidal halloysites from volcanic soils, appear to be more resistant to dehydration. It must be stressed, however, that dehydration of halloysite cannot be reversed, except through the use of interlameller complexes.

2.4.1.4 *Auxiliary treatments.* These are useful for confirming the presence of kaolinite minerals in a mixture of other clay minerals—perhaps the most common situation encountered in practice—and for distinguishing different kaolinite minerals within the same sample. Simple methods such as heating and acid treatments have been used routinely for many years, and in many instances yield diagnostic information. For example, heating at 550–600°C for 1 hour will destroy the crystallinity of nearly all kaolinites, thus causing the basal reflection at 7 Å to disappear from the diffraction pattern. This is a useful confirmatory test, except when chlorite is present in the mixture. In such a case, the 001 and 002 reflections of kaolinite and chlorite, respectively, may virtually coincide, and heat treatment will not give a positive identification of kaolinite. However, nearly all the commonly encountered chlorites (and serpentine minerals) can be easily decomposed by hydrochloric acid (e.g. 6 N HCl at 90°C for 30 minutes), whereas kaolinite minerals will resist such treatment. But it is often unnecessary to go to these lengths to prove the simultaneous presence of kaolinite and chlorite. Thus, while it is usually (but not always) difficult to separate the 001 kaolinite/002 chlorite reflections on the X-ray diffractogram, it is frequently found that the 002 kaolinite (3.57 Å)/004 chlorite (3.54 Å) reflections form a distinct, easily separable, double peak system. Unfortunately, this observation does not hold for all kaolinite/chlorite mixtures.

Over the past twenty years, other auxiliary treatments, based upon the formation of intercalation complexes, have been developed, and have proved useful in distinguishing different kaolinite minerals within the same sample. Before this, it was extremely difficult to characterize a clay containing, for example, a combination of well-ordered kaolinite, disordered kaolinite and halloysite; the diffraction pattern of such a mixture may be indistinguishable from that of moderately ordered kaolinite. Now, however, such a mixture can be differentiated by using intercalation procedures. The first step is to make the desired complex and the second is to displace the complex by water washing. Various substances have been used to make intercalation complexes, including, for example, potassium acetate, hydrazine hydrate, formamide and dimethyl sulphoxide.

Wada (1965) showed that dehydrated halloysite will form a complex with potassium acetate and can then be converted to a stable water complex with a 10.1 Å spacing, whereas kaolinite will only form a partial complex. The procedure of Range *et al.* (1969), using the intercalation properties of hydrazine, is particularly useful, and several types of kaolinite can be

Table 2.4 Characterization of various reaction types of kaolinite (after Range *et al.*, 1969).

Reaction with hydrazine	Type I[1]: fast	Type II: fast	Type III[2]: slow	Type IV: no reaction	Halloysite: fast
Basal spacing (Å)					
after hydrazine treatment	10.42	10.42	9.5–9.6	7.2–7.3	10.0–10.2
After leaching hydrazine compound with water	7.2	10.04	10.2	7.2–7.3	9.95
After replacement of water by ethylene glycol	7.2	10.85	10.85	7.2–7.3	10.85
After replacement of water by glycerol	7.2	7.2	7.2	7.2–7.3	11.10

[1] Well-crystallized kaolinite.
[2] Highly disordered kaolinite.

distinguished on the basis of observed spacings, after treatment with hydrazine, water, ethylene glycol, and glycerol (Table 2.4). At present, the significance of these various reaction types of kaolinite is not clear, but the author has found that hydrazine complexes are a useful and practical way of differentiating between well-crystallized kaolinite, highly disordered kaolinite and halloysite. The initial hydrazine complex is formed by sealing the hydrazine–kaolinite mixture in a glass ampoule for 7 days at 65°C. Recently, Churchman *et al.* (1984) have shown that formamide intercalation can be used as a simple and rapid test to distinguish halloysite from kaolinite in clays. With halloysite, complex formation is usually complete in less than 1 h and yields a 10.4 Å peak. With kaolinite, there is no significant complex formation until at least 4 h of contact with the formamide. The complex is made by spraying the X-ray diffraction mount with an aerosol of formamide.

2.4.2 *Serpentine minerals*

The serpentine minerals are structurally similar to kaolinite, and for this reason there is at least a superficial resemblance in their X-ray powder patterns. There are, however, so many differences in detail that there is little likelihood of misidentifying kaolinite when serpentine minerals form the dominant component of a clay. However, confusion could arise where serpentine minerals occur in subordinate amounts, and it is therefore important to be aware of their diagnostic X-ray diffraction characteristics, as well as their responses to auxiliary treatments. Most serpentine-type minerals (unfortunately not all) yield sharp basal reflections at significantly higher spacings than the kaolinite minerals. Thus, the strongest and most easily observed basal reflection of chrysotile, antigorite and lizardite—the three most common serpentinite minerals likely to be encountered in practice—occurs at ~ 7.3 Å, and can be readily distinguished from the kaolinite 001 basal reflection at 7.14 Å. Furthermore, the 002 reflection of serpentine occurs at ~ 3.65 Å, which is even more significantly displaced from the kaolinite 002 peak at 3.57 Å. Other diagnostic differences arise from the trioctahedral structure of the serpentine minerals, and their 020 and 060 reflections are found at ~ 4.6 and 1.55 Å respectively, compared with reflections at 4.47 and 1.49 Å for dioctahedral kaolinite. Serpentine minerals also respond differently to heating and acid treatments. Thus, they are generally (although not always) more thermally stable than kaolinites and do not decompose at 550–600°C. They are more acid-labile than kaolinites and will dissolve in warm HCl and, finally, they do not form intercalation complexes. It is therefore relatively easy to identify the presence of serpentine minerals by X-ray diffraction, even in complex mixtures, although identification of the specific type of serpentine mineral is very difficult.

2.4.2.1 *Chrysotile.* This is a fibrous mineral, the various forms being clinochrysotile (the most common monoclinic form), orthochrysotile (ortho-

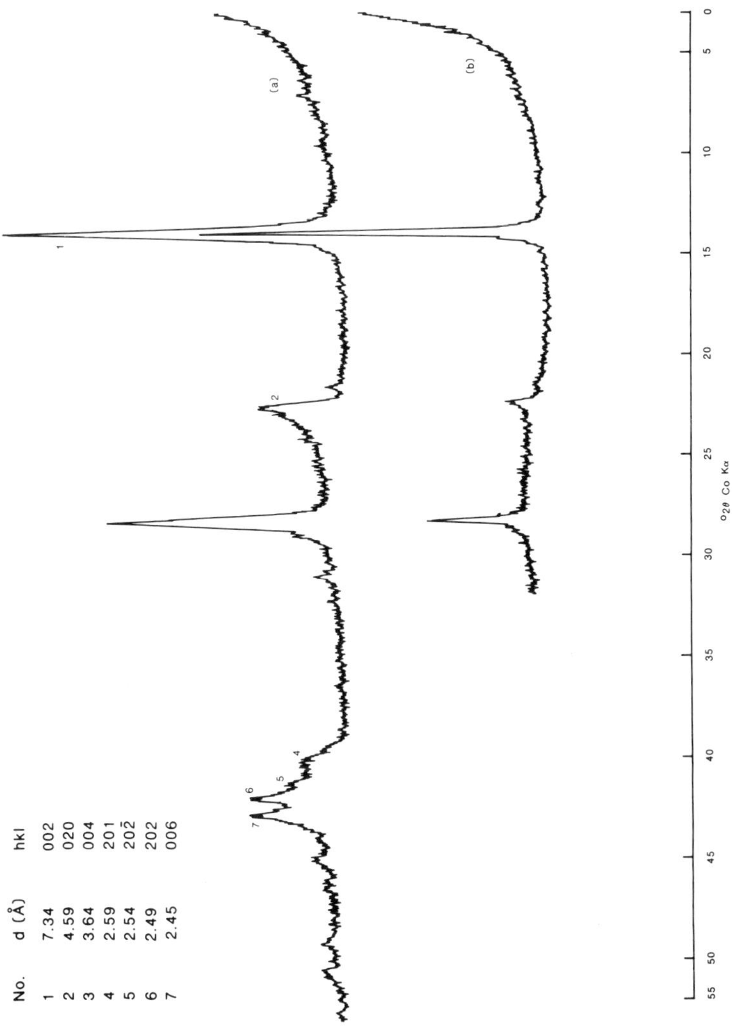

Figure 2.8 X-ray diffractometer pattern of chrysotile (a) in random orientation, and (b) after sedimentation onto a glass slide. Partial diffraction data are shown

rhombic form) and parachrysotile (with elongation of fibres along the *y*-axis rather than the *x*-axis). These varieties cannot be readily distinguished by X-ray power diffraction alone, and certainly not when present in a mixture with other minerals. As in halloysite, the tubular nature of chrysotile means that it should be impossible to prepare an oriented aggregate where only the basal reflections can be observed, but the effect is not so distinct (Figure 2.8).

2.4.2.2 *Antigorite.* The structure of antigorite is extremely complex, and is based upon a repeating wave structure yielding a large repeat distance along the *a*-axis (Bailey, 1980). The mineral occurs in plates or lath-like particles when dispersed to clay size. Its X-ray diffraction pattern is characterized by an abundance of lines but, because the *a*-axis repeat may vary from sample to sample, it is found that not all antigorites yield exactly the same pattern. Nemecz (1981) suggests that there are some features characteristic of all antigorites which enable their differentiation from lizardite. Thus, the 2.53 Å line of antigorite can be readily separated from a strong line at ~2.50 Å of lizardite. Similarly, in antigorite there is a line at 1.56 Å, forming a conspicuous pair with the 060 reflection at 1.54 Å. In chrysotile and lizardite, this pair of reflections occurs at 1.53 and 1.50 Å.

2.4.2.3 *Lizardite.* This is the most abundant serpentine mineral in massive fine-grained specimens. It has a platy morphology when viewed under the electron microscope. There are a number of different lizardite polytypes with widely varying degrees of structural disorder, and for this reason, X-ray powder patterns are somewhat variable. In general, the pattern of lizardite shows fewer lines than that of antigorite and can be distinguished from the latter by the criteria suggested by Nemecz (1981), as outlined above. There is more of a similarity with the chrysotile pattern, although because of the different morphologies of the two minerals there could be differences in the intensity relationships of basal and non-basal reflections in sedimented aggregates. Also, lizardite shows some reflections with no counterpart in the chrysotile pattern, for example the reflection at 3.90 Å.

2.4.2.4 *Berthierine.* This is an iron-rich serpentine mineral which occurs in a variety of sedimentary deposits. Berthierine is difficult to identify by X-ray diffraction methods alone, because the patterns vary considerably due to different polytypes, types of disorder and oxidation states. The major point of difference, as compared with the other serpentine minerals, is that the 001 and 002 reflections occur at ~7.05 and 3.52 Å respectively. This difference is probably great enough to distinguish berthierine where it occurs in a mixture with other serpentine minerals or with kaolinite minerals. Where berthierine occurs in a mixture with iron-rich chlorites, however, it might be impossible to make an unambiguous identification by X-ray methods alone. If berthierine occurs in a reasonably pure form, it is possible to identify the monoclinic polytype on the basis of a medium-strong reflection near 2.40 Å, the orthogonal form giving no such reflection near this position (Brindley, 1961).

2.4.3 *Illite and mica minerals*

The illite–mica minerals all yield a strong reflection at ~ 10 Å, and are readily identifiable by X-ray diffraction, at least in a general sense. To go beyond the general, however, it is necessary to be clear about the meaning of 'illite' because there is, at present, no agreed definition of terms. The name illite was originally suggested by Grim *et al.* (1937) as a non-specific name for 'the clay mineral constituents of argillaceous sediments belonging to the mica group', but differing from ideal muscovite in containing more Si, Mg and H_2O and less interlayer K. As research proceeded, however, it became clear that many of the clays described as illite did, in fact, contain interstratified smectite layers. These layers lead to distinct changes in X-ray diffraction characteristics after treatment with, for example, ethylene glycol. Idcally, illite should be a non-expansible mineral. This question has recently been treated in detail by Srodon (1984), who developed techniques to enable detection of very small amounts of interstratified expansible layers (< 1%) in illitic clays. Using these techniques, and defining illite as a non-expanding mineral, Srodon and Eberl (1984) concluded that, in the series of standard illitic samples that they examined, there was only one truly non-expanding mineral. All other illites that they investigated were found to contain expansible layers ranging from 9 to 20%. Therefore, to insist that illites by definition contain no expansible layers, leads to the undesirable situation where most of the material traditionally regarded as illite (for example, illite from Fithian, Illinois, USA) falls outwith the definition. This situation becomes even more unsatisfactory, because at the present time it is uncertain whether or not the detected expansible layers represent another phase that can be regarded as an impurity. In fact, there is evidence that many illites consist of aggregates of extremely thin particles some tens of Ångstroms in thickness, and that the interfaces between these particles are capable of adsorbing ethylene glycol. The composition of these interfacial regions may not necessarily be chemically different to the composition of the bulk material, but nevertheless the bulk material is perceived by X-ray diffraction as interstratified illite–smectite (Nadeau *et al.*, 1985, Nadeau and Bain, 1986). Until some consensus emerges regarding a usable definition of illite and the nature of the interstratified expansible layers, it is proposed that a working definition be adopted, whereby illite is considered to contain < 10% expansible layers. This proposal has the advantage that the diffraction pattern of an oriented preparation is still dominated, in most cases, by a peak at ~ 10 Å with its higher orders.

Illitic clays containing 10% or more of expansible layers are considered under interstratified minerals, and are discussed later. It must be emphasized that this dividing line is purely arbitrary, and that it is possible, indeed likely in the author's opinion, that there is a continuum, between illites with only a few percent expansible layers, and illitic clays approaching 50% expansible layers. Illitic clays also vary considerably in their composition, structural polytype and environment of formation, and these factors are taken into account in the following discussion of diagnostic X-ray diffraction criteria.

2.4.3.1 *Illite (1M and 1Md polytypes).* Illites of 1M and 1Md polytypes are abundant in the argillaceous rocks characteristic of a large part of the sedimentary column. In general, these illites typically show a rather broad 10 Å reflection, often asymmetrical towards the low-angle side, with prominent higher orders at ~ 5, 3.33 and 1.99 Å. The 002 reflection is usually about one third of the intensity of the 10 Å reflection, indicating an aluminous dioctahedral nature; this can be easily confirmed by measuring the 060 peak which occurs at ~ 1.50 Å. 1M illites yield a number of non-basal reflections of medium intensity that are not found in 2M illites. These include reflections at 3.66 Å (11$\bar{2}$), 3.07 Å (112) and 2.68 Å (023). With the 1Md polytype, these reflections, together with others, either disappear from the pattern or become much weaker. In addition, many of the reflections below ~ 2.50 Å become broad and diffuse (Figure 2.9). At present, it is not clear what the 1Md polytype represents and how it can be interpreted. Brindley (1980) pointed out that there is no obvious general rule that the missing reflections correspond to the hkl reflections, as in the case of kaolinite, and that there was, therefore, no

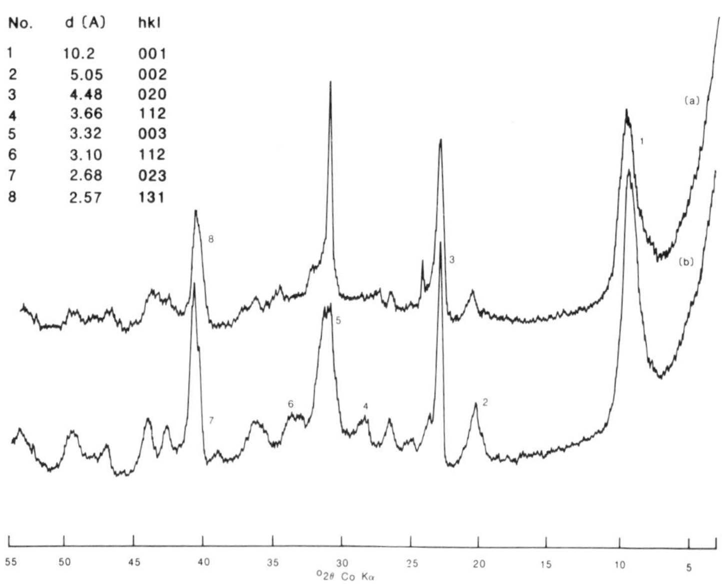

Figure 2.9 X-ray diffractometer traces of randomly oriented (a) 1Md illite, and (b) 1M illite. Partial X-ray diffraction data showing some diagnostic reflections are indicated

particular pattern of disorder. The simplest explanation could be that 1Md illite corresponds with aggregates of very thin 1M crystallites. Such crystallites are typical of many illitic materials, and have been well-characterized by the transmission electron microscope studies of Nadeau *et al.* (1984). This work could also have a bearing on the so-called 'Kubler index' (Kubler, 1964) which has been extensively used to describe illite 'crystallinity' during the passage from diagenesis to anchimetamorphism. This index is simply the 001 peak measured in mm or °2Θ at half peak height, a measurement that may be standardized by using quartz as an internal reference. However, as pointed out by Srodon and Eberl (1984), the breadth of the 10 Å peak is mostly a function of interstratified smectite. According to the view just presented, this in turn could represent interparticle diffraction. Thus, the observed decrease in the width of the 10 Å reflection with increasing depth merely reflects the growth of thicker illite crystals, which proceeds to the point where interparticle diffraction is no longer detectable.

2.4.3.2 *$2M_1$ illite/mica.* The 10 Å reflection of $2M_1$ illite/mica is usually more narrow and sharper than that of the 1M form (Figure 2.10), although, of

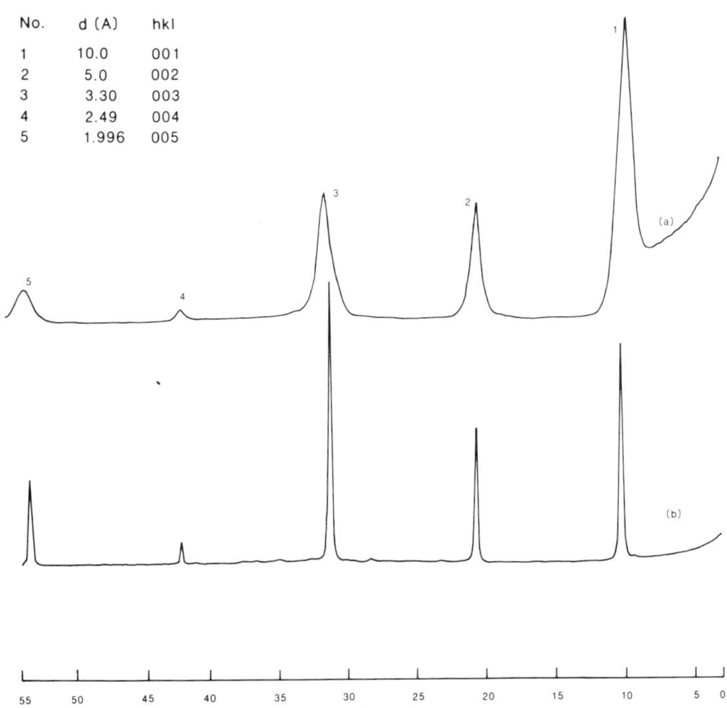

Figure 2.10 X-ray diffractometer traces of well-oriented (a) 1M illite and (b) $2M_1$ mica

No.	d (Å)	hkl	No.	d (Å)	hkl
1	10.01	002	7	3.35	006
2	5.02	004	8	3.21	114
3	4.46	110	9	2.99	025
4	3.89	$11\bar{3}$	10	2.87	115
5	3.73	023	11	2.80	$11\bar{6}$
6	3.50	$11\bar{4}$	12	2.57	116
			13	2.01	0.0.10

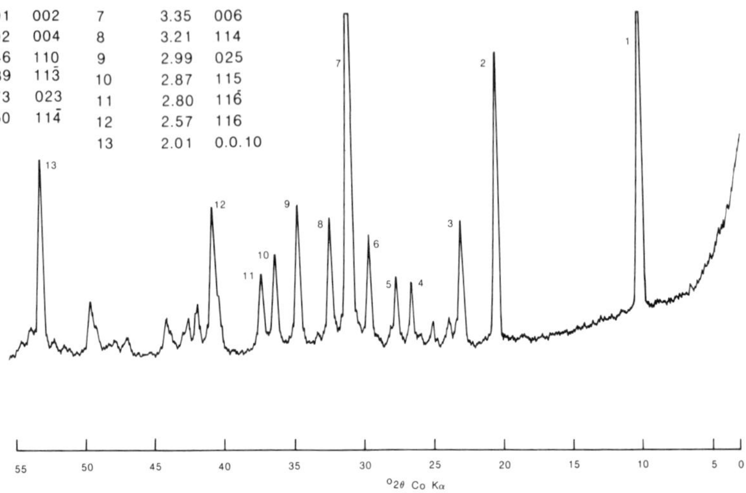

Figure 2.11 X-ray diffractometer trace of randomly oriented $2M_1$ muscovite. Partial diffraction data showing some diagnostic reflections are indicated

course, no absolute distinction can be made on this basis. There are a number of non-basal reflections, however, that are characteristic of $2M_1$ mica, and which should be readily observable in a randomly oriented sample. These include reflections at 3.89 Å ($11\bar{3}$), 3.73 Å (023), 3.50 ($11\bar{4}$), 3.21 Å (114), 2.87 Å (115) and 2.80 Å ($11\bar{6}$) (Figure 2.11). Where it is desired to determine the $1M/2M_1$ ratio, an effective procedure would seem to be that described by Maxwell and Hower (1967), which is based on the intensity ratio of a unique $2M_1$ peak to a peak representing all dioctahedral mica polytypes. Some of the most intense $2M_1$ reflections listed above may coincide with feldspar peaks, and so the weaker $11\bar{6}$ reflection at 2.80 Å was chosen as the numerator and the 2.58 Å peak as the denominator. The latter receives nearly equal intensity contributions from all dioctahedral polytypes. The ratio obtained is then compared with the 2.80 Å/2.58 Å ratio for muscovite—assumed to consist entirely of the $2M_1$ polytype—and plotted on an empirical working curve determined from known $2M_1/1M$ mixtures.

2.4.3.3 *Glauconite and celadonite.* These minerals are considered jointly because they yield almost identical X-ray diffraction patterns. Their distinctive green colour means that there is little difficulty in making a general identification, once the essential mica-like nature of these minerals—as manifested by their strong 10 Å reflections—has been established. However, glauconite is subject to the same complications as illite in that it often contains interstratified smectite layers, ranging from 5 to 65% (Thompson and Hower, 1975). For discussion purposes, the same expedient is adopted here for

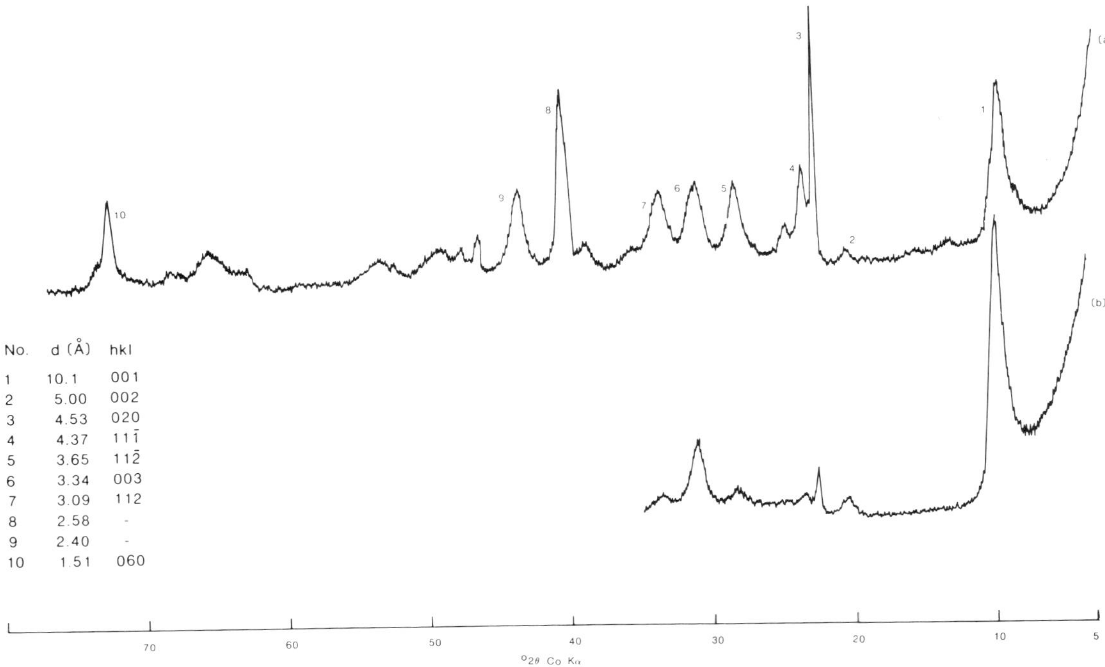

No.	d (Å)	hkl
1	10.1	001
2	5.00	002
3	4.53	020
4	4.37	$11\bar{1}$
5	3.65	$11\bar{2}$
6	3.34	003
7	3.09	112
8	2.58	-
9	2.40	-
10	1.51	060

Figure 2.12 X-ray diffractometer traces of glauconite in (a) randomly oriented, and (b) oriented forms. Partial diffraction data showing some diagnostic reflections are indicated

glauconite as for illite, the mineral being arbitrarily defined as containing $< 10\%$ expansible layers. Both glauconite and celadonite occur in the 1M form, and yield well-defined 112 and 112 reflections at 3.63 and 3.09 Å respectively (Figure 2.12). Another typical feature of the diffraction pattern is the very low intensity of the 002 reflection at ~ 5 Å, due to the high octahedral iron content (Odom, 1984). According to Buckley *et al.* (1978), glauconite and celadonite can be distinguished on the basis of their 060 spacings, when these are plotted against ferric iron content. The 060 spacing of celadonite occurs mainly in the 1.507–1.509 Å range, whereas glauconite yields an 060 reflection between 1.510 and 1.517 Å. Subsequent work seems to indicate that there may, in fact, be a slight overlap in the 060 spacings of the two minerals at 1.510 Å (Buckley *et al.*, 1984).

2.4.3.4 *Other illite–mica minerals.* Although clay-size trioctahedral illite is uncommon, it does occur (Walker, 1950), especially in young soils developed on glacial deposits. For all practical purposes, trioctahedral illite may be regarded as comminuted biotite and may be suspected when the 002 reflection at 5 Å is very weak relative to the 001 reflection at 10 Å. Confirmatory evidence may be obtained by carefully measuring the 060 spacing, which should occur at ~ 1.535 Å, and by observing the response to acid treatment which generally destroys trioctahedral minerals.

The possibility that carefully measured basal spacings may also indicate the existence of relatively unusual mica species, should also be taken into account. For example, the sodium mica, paragonite, is characterized by a basal spacing of ~ 9.6 Å, which may be quite distinct from the normal K-mica spacing on a diffractometer trace. Again, ammonium illite characteristically yields a slightly enlarged 001 spacing at 10.16 Å (Sterne *et al.*, 1982), a small difference from the 001 of K-mica, but one which becomes more obvious with the higher 001 orders.

Detailed studies have been made relating the chemical composition of mica minerals to *b*-parameter measurements, to intensities of basal reflections and to basal spacings (Brown and Brindley, 1981). Apart from the general diagnostic characteristics outlined, however, it is unlikely that this work will find extensive application to clay samples, which are often impure. Nevertheless, it is always advisable to take careful note of such diffraction features, especially when samples come from the same geological environment, because systematic or unexpected variation may yield information relevant to the influence and mechanism of important geochemical processes.

2.4.4 *Chlorites*

The preliminary identification of clay chlorites by X-ray diffraction is relatively straightforward, as all species yield a basal reflection at about 14.2–14.4 Å, with a readily observable series of higher orders at least up to the 005 reflection. Moreover, the fundamental 14 Å spacing is unaffected by treatment with ethylene glycol or glycerol, or by heating to $\sim 500°C$. Such simple observations are sufficient for the identification of chlorite, even in many complex clay mixtures, although the presence of subordinate amounts of chlorite in a clay that contains large amounts of kaolinite or interstratified minerals can sometimes lead to difficulties. In such a case, where the 14 Å spacing may be obscured by low-angle scattering, and where the 002 chlorite reflection at 7 Å coincides with the 001 reflection of kaolinite, it is often helpful to look for higher-order basal reflections. For example, the 003 reflection at ~ 4.73 Å is diagnostic for chlorite and it is often possible to separate the 004 reflection of chlorite at ~ 3.54 Å from the 002 reflection of kaolinite at 3.57 Å on the diffractometer trace.

Where chlorite is relatively abundant in a clay material, the X-ray pattern, in conjunction with auxiliary treatments, may yield additional chemical and structural information. The chemical composition of chlorites is exceedingly variable: many different species having been proposed, but for XRD purposes it is sufficient to consider only a small number of types that are commonly associated with clay materials. These are the magnesium-rich and iron-rich trioctahedral chlorites (clinochlore and chamosite respectively) and the aluminous dioctahedral chlorites (sudoite and donbassite). Diagnostic features of chlorite X-ray diffraction patterns involve determination of the 060 spacings, and of the relative intensities and spacing of the basal reflections.

2.4.4.1 *060 spacings.* A general distinction between tri- and dioctahedral chlorites can of course be made on the basis of the 060 spacing, which yields the b-parameter. For trioctahedral chlorites this spacing usually occurs between 1.535 and 1.548 Å, while for dioctahedral chlorites it occurs between 1.489 and 1.509 Å. In addition, the dioctahedral chlorites resist acid treatments much more than their trioctahedral counterparts. These trioctahedral chlorites are, for example, easily decomposable by treatment with 1 N HCl at 95°C for half an hour. Further information on the substitution of Mg by Fe^{2+} and equivalent ions in trioctahedral chlorites may be obtained by measurement of the 060 spacing, and use of the equation:

$$b = 9.210\,\text{Å} + 0.037\,(Fe^{2+}Mn)$$

after Brindley (1961). Similar equations taken from the work of other authors are given in Brown and Brindley (1980).

2.4.4.2 *Intensities of 001 reflections.* It has long been known that the odd order basal reflections of high-iron chlorites are much less intense than those of low-iron ones, so that a distinction between clinochlore and chamosite, for example, can be made merely by a visual inspection of their diffraction patterns (Figure 2.13). In principle, careful measurement of the intensities of the 001 reflections of chlorite can yield information on the content of iron and

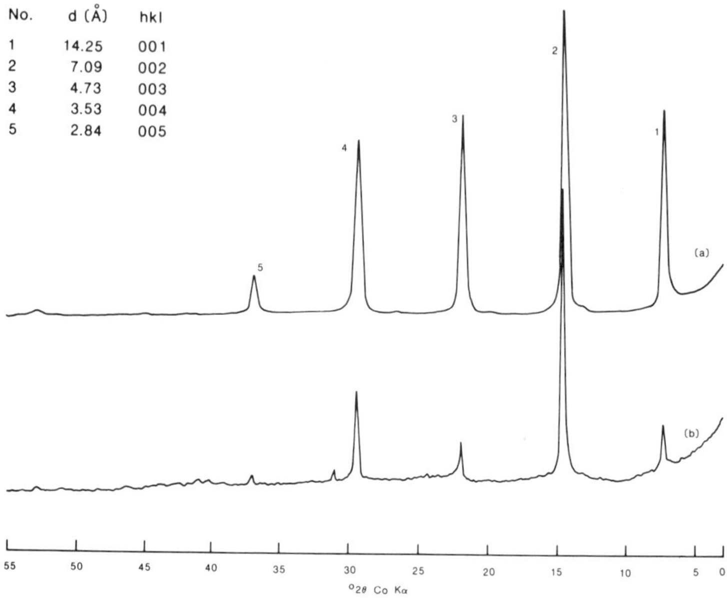

Figure 2.13 X-ray diffractometer traces of well-oriented (a) Mg-rich clinochlore, and (b) Fe-rich chamosite

equivalent heavy atoms. It can also shed light on the distribution of these atoms in the octahedral planes of the talc and brucite sheets that make up the chlorite structure. The procedure is too complicated to describe in brief, and the reader is referred to Brindley and Brown (1980) for a full discussion. The authors caution that only approximate results are to be expected with these methods, in view of the difficulties of measuring intensities accurately.

2.4.4.3 *Basal spacings.* Changes in the chemical composition of chlorites can also be determined by carefully measuring the 001 spacing. In particular, this spacing varies with the abundance of tetrahedral Al, approximately according to the equation.

$$d(001) = 14.5\,\text{Å} - 0.31\,x$$

(Nemecz, 1981), similar equations having been given by other authors. It is important to determine the 001 spacing with as much accuracy as possible, and for this purpose the use of higher order 001 reflections is recommended, in conjunction with slow speed scans. With well-oriented preparations it should be possible to measure the 005 reflection, although it may be more convenient to use the 004 peak, which is more intense.

2.4.4.4 *Polytypes.* Theoretically, up to 24 different stacking arrangements are possible for the constituent talc and brucite sheets of the chlorite structure, but in practice only four need to be considered. In a study of 303 different chlorites, Bailey and Brown (1962) found that 80% belonged to the IIb polytype, 12% to the Ib (90°) and the remainder about equally to the Ib and Ia polytypes. A full discussion of polytype nomenclature can be found in Bailey (1980), but essentially I and II refer to the positions of cation sites in the brucite sheet, with respect to a fixed set of hexagonal axes above the lower hexagonal OH rings. The letters a and b refer to the different superpositions of the interlayer cations on the upper talc sheet, and a further distinction can be made on the basis of the unit cell. Thus, one of the above polytypes has an orthorhombic-shaped unit cell (90°), but with monoclinic symmetry, while the remainder are based on a monoclinic unit cell with $\beta \sim 97°$. The importance of determining chlorite polytypes is that the Ia and Ib forms crystallize at relatively low temperatures, typically during diagenesis, while the IIb form is associated with the higher temperatures of metamorphism. Fortunately, polytype identification by X-ray powder diffraction is easy if the chlorite sample is reasonably pure. The basic types can be distinguished by observing the positions and intensities of the h01 reflections in the range 2.66–2.00 Å as described by Bailey (1980) and shown in Figure 2.14. With the IIb polytype there is a group of five fairly strong h01 reflections between 2.60 and 2.25 Å, some of which, in other polytypes, are either very weak, missing or broad.

2.4.4.5 *Heat treatment.* The response of many chlorites to heating at ~ 550°C can often be used as a diagnostic feature. Such treatment results in structural rearrangement and leads to a marked intensification of the 001 peak at 14 Å, and with corresponding diminution of intensities for the higher orders of the basal reflection. Unfortunately, this simple test is not infallible and some chlorites, especially those that are iron rich, or of clay grade, may be thermally unstable at lower temperatures and be difficult to distinguish from kaolinites on this basis.

2.4.5 *Vermiculite*

With vermiculite, it is crucial to bear in mind that the basal spacings vary according to:

(i) The nature of the interlayer cation
(ii) The relative humidity (but to a lesser extent)
(iii) Treatment with organic compounds, and
(iv) Heat treatment at relatively low temperatures. Vermiculite is classed as an expansible mineral along with the smectites, and there are many obvious similarities in the behaviour of the two minerals. However, a distinction is made on the basis of vermiculite having a higher layer charge (>0.6 per formula unit), leading to a higher cation-exchange capacity and to definite differences in its response—as revealed by X-ray diffraction—to various auxiliary treatments. Vermiculite also differs from most smectites in that its

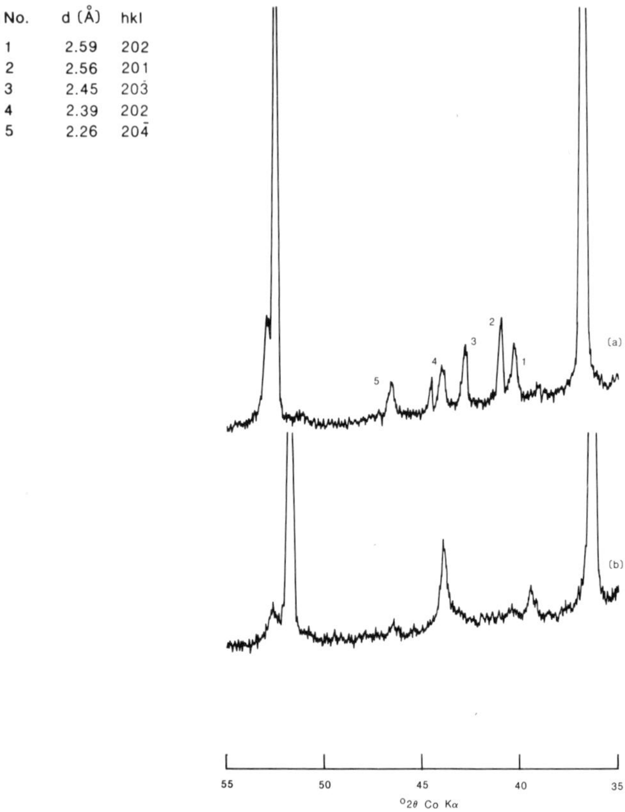

Figure 2.14 X-ray diffractometer traces of randomly oriented (a) IIb, and (b) Ia chlorite polytypes showing diagnostic reflections of the former

structure is more highly ordered, and a complete X-ray pattern of a randomly oriented powder yields many well-defined hkl reflections that are simply not present in the vast majority of smectites (Figure 2.15). It is true that some smectites occasionally show a higher degree of structural order than the norm, but this is very much the exception to the rule.

Both di- and trioctahedral vermiculite are known, but the latter has been much better characterized because it occurs in large single-crystals, usually derived from the alteration of biotite or phlogopite. In contrast, so far, dioctahedral vermiculite has been found mainly, if not exclusively, in clay fractions. These fractions are typically soil clays where the dioctahedral vermiculite may be mixed with other types of clay minerals, and may contain non-exchangeable hydroxyaluminium in the interlayer space. Di- and trioctahedral vermiculites may be distinguished on the basis of their 060 spacings, 1.49–1.50 Å compared with 1.53–1.54 Å, and their different susceptibilities to

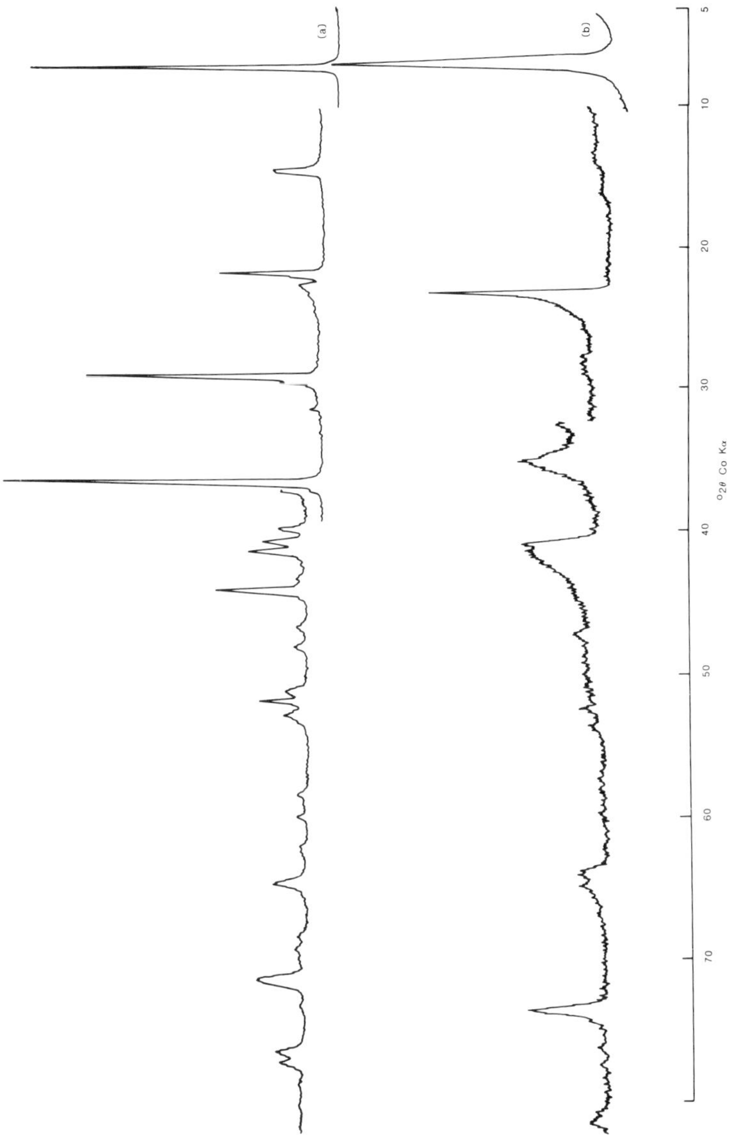

Figure 2.15 X-ray diffractometer traces of randomly oriented (a) vermiculite, Glen Urquhart, Scotland, and (b) montmorillonite, Wyoming, USA. Both samples were Mg saturated

Table 2.5 Basal spacings (Å) of vermiculites saturated with various cations at different relative humidities (after MacEwan and Wilson, 1980).

Interlayer cation	Kenya vermiculite[1] 32% RH[+]	52% RH	79% RH	Water	Loch Scye vermiculite[2] 32% RH	52% RH	79 RH	Water
Ca	14.8	14.9	14.9	14.9	14.9	14.9	14.9	15.1
Mg	14.3	14.4	14.4	14.6	14.4	14.4	14.9	15.1
Na	12.3	12.6*	14.6	14.9	12.1	14.7	14.7	14.9
K	10.4*	10.4*	10.5*	10.5–14.3*	11.0	11.0	11.0	11.0
NH_4	10.8*	10.8*	11.0*	12.9*	10.8	10.8	10.8	11.0

*broad or asymmetrical reflection.
[1] Mathieson and Walker (1954).
[2] Smith-Aitken (1965).
[+] RH Relative humidity.

acid treatment, as has been described earlier for other di- and trioctahedral minerals.

2.4.5.1 *Effect of water and inorganic cations.* The X-ray diffraction pattern of an oriented aggregate of vermiculite is usually characterized by an intense basal reflection between 12 and 15 Å, depending on the saturating cation, with much weaker higher-order reflections. Natural vermiculites would most likely be in the Ca-, Mg- or Na-saturated form, but for diagnostic purposes it is also useful to saturate vermiculite with K^+ or ammonium ions. The basal spacings for each of these cations under various relative humidities are shown in Table 2.5. The divalent cations yield basal spacings between 14 and 15 Å, which remain virtually constant at all relative humidities likely to be encountered under natural conditions. The 14–15 Å spacing is indicative of a two-layer hydrate in the interlayer space. With Na-saturation a two-layer hydrate is stable at a relative humidity of > 50%, and a one-layer hydrate becomes the stable form under drier conditions, yielding a spacing of 12–12.5 Å. Even lower (and broader) basal reflections at 10.5–11.0 Å are yielded by K^+- and NH_4^+-saturated vermiculites, because of the weak hydration of these cations. These results have been confirmed only for trioctahedral vermiculites, but are also likely to be applicable to dioctahedral forms.

2.4.5.2 *Complexes with ethylene glycol and glycerol.* As shown in Table 2.6, the exact spacing of the vermiculite–ethylene glycol/glycerol complex depends on the nature of the interlayer cation and on the layer charge of the mineral, a point that is perhaps often overlooked. Only Mg-saturated vermiculite with a charge of > 1.4 per formula unit fails to expand with ethylene glycol. In contrast, Mg-saturated vermiculite with a charge of < 1.4 per formula unit expands to 15–16 Å, while Ca- and Na-saturated vermiculite may show an expanded spacing of ~ 16 Å over a range of layer

Table 2.6 Basal spacing (Å) of ethylene glycol and glycerol complexes of vermiculites saturated with various cations (after Brindley, 1966).

Source	Layer charge	Cation	Ethylene glycol	Glycerol
Young River, W Australia	1.2	Ca	16.2	17.6
		Mg	16.3	14.3
		Na	16.3	14.8
West Chester, Pennsylvania	1.4	Ca	16.1	17.6
		Mg	14.3	14.3
		Na	16.3	14.8
Macon County, N Carolina	1.6	Ca	15.6*	14.3
		Mg	14.3	14.3
		Na	14.8*	14.8

*Irregular layer sequences

Table 2.7 Basal spacings (Å) of vermiculite (Macon County, North Carolina, saturated with various cations and heated at different temperatures, after Barshad, 1952).

Temperature (°C)	Ca	Mg	Na	K
Room	15.1	14.3	12.6	10.4
170	11.8	11.8	10.3	10.4
250	10.1	10.1	10.1	10.4
610	9.6	9.4	10.0	10.4

charges. The glycerol complex shows rather more stable behaviour, a spacing of 14.3 Å being found in Mg-saturated vermiculites, whatever the layer charge, and a spacing of 14.8 Å in most Na-saturated vermiculites. On the other hand, glycerol solvation of low charge, Ca-saturated vermiculites still leads to expanded spacings of 17.6 Å. This information suggests that the vermiculite/smectite differentiation in natural clay materials is still best accomplished by a procedure involving Mg-saturation and glycerol treatment, as advocated by Walker (1961).

2.4.5.3 *Dehydration and rehydration.* The dehydration of vermiculite by heating shows some variation, again depending on the nature of the interlayer cation (Table 2.7). With the Mg- and Ca-saturated forms, heating at 170°C results in a spacing at 11.8 Å, corresponding with a single layer of water molecules. Further heating to 250°C results in the decomposition of this water complex, and in further structural collapse to 10.1 Å. The last remaining traces of water are, however, tenaciously held, and complete contraction to the anhydrous spacing of 9.4–9.6 Å occurs only after heating to 610°C. As already shown, K-saturated vermiculite tends not to form interlayer hydrates and shows only minor changes in basal spacings on heating.

Vermiculite saturated with strongly hydrateable interlayer cations will rehydrate on exposure to moist air. For example, Mg-vermiculite heated at 450°C will rehydrate to yield a spacing of 14 Å. Such rehydration can take place very rapidly, and Walker (1961) has pointed out that clay-size Mg-vermiculite can absorb moisture from the air almost instantaneously if the sample is allowed to cool. After heating at 600°C, however, Mg-vermiculite is reported to require hydrothermal treatment before it will recover its expanding properties (Farmer, 1978).

2.4.6 *Smectites*

Like vermiculites, smectites yield X-ray diffraction patterns characterized by basal spacings that vary with exchangeable cation, relative humidity, association with certain organic molecules, and heat treatment. Unlike vermiculites, smectites generally yield X-ray diffraction patterns, characterized by broad and diffuse non-basal reflections or bands that arise from random layer stacking. These bands are typically asymmetrical towards the higher angles and indicate unambiguously the lack of three-dimensional order; for this reason they are assigned hk indices. The most prominent hk bands occur at 4.5 Å (02, 11), 2.61 Å (13, 20), 1.70 Å (15, 24, 31) and 1.50 Å (06, 33) (Figure 2.16). A higher degree of structural order is occasionally encountered in beidellite and saponite, but such specimens are rather rare.

Although there are a large number of different smectite species, the most commonly encountered are montmorillonite, beidellite, nontronite and

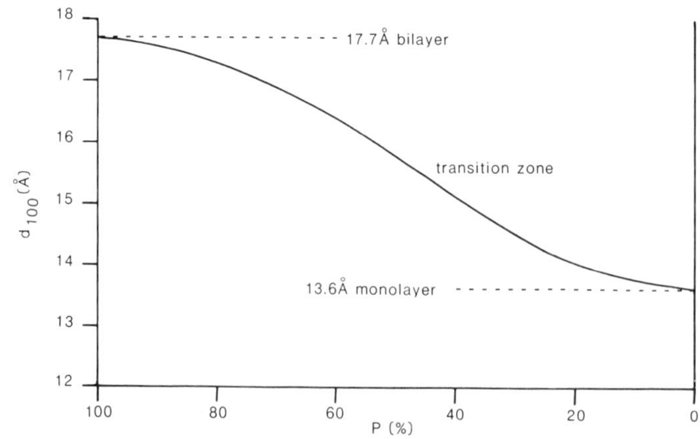

Figure 2.16 Peak-migration curve of the 001 reflections of alkylammonium chloride-smectite complexes used to determine charge-distribution histograms. P is the proportion of mono- to bilayer complexes. (After Lagaly *et al.*, 1976)

saponite, the first three being dioctahedral and the last being trioctahedral. The 06 diffraction band is sufficiently sharp for it to be measured to about 0.25% accuracy (Brindley, 1981), so that di- and trioctahedral smectites can be generally distinguished on this basis. Thus, montmorillonite and beidellite yield an 06 spacing at about 1.50 Å, whereas the same reflection for saponite usually occurs at 1.53–1.54 Å. However, the 06 reflection for nontronite often occurs at an intermediate spacing of ~ 1.52 Å, or at an even higher spacing, making differentiation uncertain on this basis. Both montmorillonite and beidellite tend to resist acid treatment, whereas nontronite decomposes easily, and in this respect, resembles the trioctahedral saponite.

2.4.6.1 *Effect of water and inorganic cations (Table 2.8).* When saturated with Ca- or Mg-ions, smectite of whatever layer charge yields a strong basal

Table 2.8 Basal spacings (Å) of smectites saturated with various cations at different relative humidities (after MacEwan and Wilson, 1980).

Wyoming montmorillonite

Interlayer cation	32% RH[1]	52% RH	79 RH	In water
Ca	15.2	15.1	15.5	19.0
Mg	15.1	15.1	15.2	19.5
Na	12.5	12.5	14.8	Ind. Sw.
K	11.9	11.9	12.1	Ind. Sw.
NH_4	11.7	11.7*	11.9	12.4*

Unterrupsroth beidellite

Interlayer cation	32% RH[1]	52% RH	79% RH	In water
Ca	15.1	15.1	15.1	15.1
Mg	14.7	14.7	15.0	15.9
Na	12.5	12.5	12.5	14.8
K	11.8	11.9	11.9	12.1*
NH_4	11.8	11.9	11.9	12.0

St Cyrus saponite

Interlayer cation	32% RH[1]	52% RH	79% RH	In water
Ca	15.1	15.2	15.2	15.8
Mg	14.7	14.7	14.7	19.0
Na	12.5	12.5	12.8*	15.5*
K	12.4	12.4	12.5	15.4*
NH_4	12.2	12.7	12.7	13.5*

*broad, probable indication of interstratification.
Ind. Sw. Indefinite swelling.
[1]RH Relative humidity.

reflection at ~ 15 Å, at all the relative humidities likely to be encountered in the laboratory. In water, this two-layer hydrate expands to a three-layer hydrate yielding a spacing of 19 Å in the case of montmorillonite. Such an expansion may not necessarily be observed in beidellite and saponite. The behaviour of different smectites in the Na form is also variable. Both Na-beidellite and saponite tend to yield a stable single-layer water complex, with a spacing of ~ 12.5 Å, even at high relative humidities, whereas Na-montmorillonite develops a two-layer hydrate at 50–60% relative humidity. In the wet state, Na-montmorillonite may show complete dissociation of the silicate layers, leading to highly swollen preparations with very high (> 40 Å) spacings, or diffuse scattering at low angles. This phenomenon may or may not occur with beidellite and saponite. Most smectites in the K- or NH_4-saturated forms yield a basal spacing at ~ 12 Å whatever the humidity. This spacing persists in water for NH_4-smectites, but K-smectites sometimes behave in much the same way as their Na-saturated counterparts.

2.4.6.2 *Ethylene glycol/glycerol complexes.* The various smectite species show a rather uniform response to ethylene-glycol or glycerol solvation, whatever the saturating cation (Table 2.9). Small differences in basal spacing may perhaps be detected, but it is doubtful if these can be used for diagnostic purposes. In general, ethylene glycol-treated smectites yield a basal spacing of ~ 17 Å, with a rational series of higher-order reflections, while glycerol-treated smectites show a basal spacing of ~ 17.8 Å. Complexes of both beidellite and saponite may show slightly lower spacings; in the K-saturated form some, but not all, beidellites only form a one-layer complex, with a spacing of ~ 14 Å.

Table 2.9 Basal spacing (Å) of ethylene-glycol- and glycerol complexes of smectites saturated with various cations (after Brindley, 1966 and Suquet *et al.*, 1975).

Mineral	Cation	Ethylene glycol	Glycerol
Wyoming montmorillonite	Ca	16.9–17.1	17.8
	Mg	16.9–17.1	17.8
	Na	16.9–17.1	17.8
Unterrupsroth beidellite	Ca	16.7	17.7
	Mg	17.1	17.9
	Na	16.9	18.0
Kozakov saponite	Ca	16.6	17.9s 14.4vw
	Mg	15.7	18.1s 14.3vw
	Na	16.7	18.1

s = strong
vw = very weak

Table 2.10 Basal spacings (Å) of montmorillonite saturated with different cations and heated at different temperatures (after Martin-Vivaldi *et al.*, 1963).

Temperature (°C)	Ca	Mg	Na	K
Room	15.1	15.1	12.5	11.9
110	10.9*	10.4	9.6	10.0
150	9.8	10.4	9.6	10.0
200	9.8	10.1	—	10.0
250	9.6*	10.1*	—	10.0

* broad peak indicating interstratification.

2.4.6.3 *Dehydration/rehydration.* All smectites lose interlayer water between 100 and 200°C, giving a 'collapsed' structure, but the extent of the dehydration depends on the nature of the exchange cation and the smectite itself. Montmorillonite contracts to ~10 Å after heating at 150°C whatever the cation, although the water associated with interlayer Ca and Mg is retained more strongly than that associated with Na and K (Table 2.10). Saponite behaves in the same way as vermiculite, and single-layer water complexes with interlayer Mg and Ca are stable at 170°C, yielding a basal spacing of ~11.5 Å. This spacing contracts to 10 Å after heating at 250°C.

The rehydration behaviour of montmorillonite and saponite is also different, with the latter being able to rehydrate much more readily. Thus, none of the cations is able to rehydrate in montmorillonite after heating to 400°C, and Mg fails to rehydrate significantly after heating to 250°C (Farmer, 1978). In contrast, Na- and K-saponite rehydrate readily after heating up to 600°C, and the Mg form is fully rehydrateable after heating to ~400°C.

2.4.6.4 *Distinction between montmorillonite and beidellite.* It has been suggested that such a distinction can be made by using the Hofmann–Klemen effect (or Greene–Kelly test), which is based on the response to successive Li-saturation (using 3 M LiCl), heating at ~250–300°C overnight, and treatment with glycerol. Hofmann and Klemen (1950) observed that heating Li-saturated smectites induces migration of Li from the interlayer positions to vacancies in the octahedral sheet. In montmorillonite, the Li ions neutralize the charge arising from octahedral substitution, thus rendering the structure non-expansible, whereas in beidellite, because the layer charge arises from tetrahedral substitution, expansibility is unaffected. It is necessary to add some cautionary comments about the uncritical use of this test. In the first place, anomalous results may be obtained if the Li clay is mounted on ordinary sodium hydroxide/calcium hydroxide glass slides, because heating of these slides is thought to induce exchange of Na from the glass for Li from the clay. This would result in an incomplete neutralization of the octahedral charge, and an overestimation of the amount of tetrahedral substitution (Bystrom–Brusewitz, 1975). In addition, Bühmann *et al.* (1985) point out that the test does not distinguish between all montmorillonites and beidellites. Thus, as much as 49% of the layer charge of montmorillonite may arise from tetrahedral substitution, and even as little as 20% would suffice to retain expansibility and lead to an identification of the beidellite. This observation is important because tetrahedrally charged smectites seem to be common, particularly in soils (Wilson, 1987).

2.4.6.5 *Determinative of layer charge characteristics by adsorption of aklyl-ammonium chloride.* Lagaly and Weiss (1970) have used alkylammonium-smectite complexes:

(i) To assess the homogeneity of the smectite charge distribution
(ii) To estimate upper and lower limits of layer charge and
(iii) To provide a charge distribution pattern for a smectite sample.

The method for making alkylammonium complexes requires some discussion. Alkylammonium chlorides ($C_nH_{2n+1}NH_3Cl$, where n ranges from 4 to 18) are prepared by dissolving the pure alkylamine in a small amount of ethanol, adding water, and adjusting the pH to 6–7 with HCl. About 10 ml of solution is added to 30 mg of freeze-dried clay and heated in a water bath at 65°C for two days, with one change of solution. The excess alkylamine–HCl is removed by washing ten times with 5 ml of 1:1 ethanol/water and twice with pure ethanol. The paste is spread over a glass slide, dried at 65°C, and then examined by X-ray diffraction. Aqueous alkylammonium complexes are used in the following concentrations: 2 M for $n = 4$–6; 0.5 M for $n = 7$–9; 0.1 M for $n = 10$–15 and 0.05 M for $n = 16$–18. The orientation of the alkylammonium ions between the silicate sheets depends on layer charge, and on the chain length of the cations. Short chains are arranged in flat-lying monolayers and yield a spacing of 13.6 Å. Double-layer complexes show a 17.7 Å spacing, and are formed when the area covered by the flat-lying cation exceeds the area available to each univalent cation.

According to Lagaly and Weiss (1970), the layer charge of smectites may vary throughout individual crystallites, with particular silicate layers or sequences of layers having different charge densities. With homogeneous charge distribution, the mono- to double-layer transition occurs sharply between two members of the homologous series. Where inhomogenous charge distribution occurs, the transition takes place over a range marked by interstratification of monolayer and bilayer complexes. Such a transition is identified by the appearance of a non-integral series of basal reflections. The complete formation of the double-layer complex is recognized by the reappearance of a rational series of basal reflections.

The procedure for determining the charge distribution of a smectite has been summarized by Bühmann *et al.* (1985), as follows:

(i) Determine the upper and lower limits of n that mark the beginning and

Table 2.11 Table for calculation of interlayer charge/$O_{10}OH_2$ from the number of carbon atoms (n) of the intercalated alkylammonium chloride molecule (after Lagaly *et al.*, 1976).

Carbon atoms in chain	Montmorillonite	Beidellite	Saponite
6	0.51	0.49	0.51
7	0.46	0.43	0.47
8	0.42	0.39	0.42
9	0.39	0.36	0.38
10	0.36	0.33	0.36
11	0.33	0.31	0.32
12	0.31	0.28	0.31
13	0.29	0.27	0.28
14	0.28	0.25	0.26
15	0.26	0.24	0.25
16	0.25	0.22	0.24
17	0.24	0.21	0.22
18	0.23	0.20	0.21

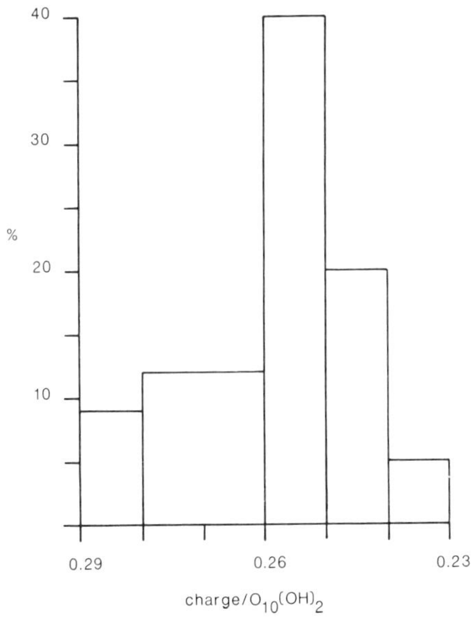

Figure 2.17 Charge distribution histogram of Pörndorf montmorillonite, South Africa (after Bühmann *et al.*, 1985)

end of the mono- to bilayer complex formation as assessed from X-ray diffraction patterns

(ii) Convert n to interlayer charge density by using Table 2.11

(iii) Obtain the charge distribution by analysing the transition zone, using a peak-migration curve such as that shown in Figure 2.16 (p. 60)

(iv) Use the information obtained to plot a histogram of charge distribution similar to that illustrated in Figure 2.17. Such diagrams have been found to be useful in predicting the effectiveness of a particular smectite for certain types of industrial application.

2.4.7 *Interstratified clay minerals*

The closely related layer structures of the clay minerals discussed so far mean that it is possible for two (or more) different clay mineral types to occur in an intimately associated stacking sequence. The sequence involves repetition of unit cells of different thicknesses, an arrangement that is referred to as 'interstratified' or 'mixed layered'. The stacking sequence is built up along the *c* axis, and the units involved are thought to form crystallites or domains, with little or no order in the *ab* directions. Consequently, the X-ray diffraction patterns obtained from these materials generally resemble those of smectites, in that they are characterized by distinct basal reflections and poorly developed hk bands. The exact type of diffraction pattern obtained depends, among other things, upon the nature, proportion and arrangement of the components making up the interstratification, differences in ordering being particularly significant.

The two extreme forms of ordering found in interstratified structures are described as regular or random. Perfect regular ordering of a two-component interstratified clay consisting of equal proportions of the two components refers to the case where the probability of an A layer being followed by a B layer is 1.0, and where the probability of a B layer being followed by an A layer is also 1.0. Thus, the clay consists of a perfectly regular structure which is built of units as follows:

ABABABAB ABABAB ABABABAB

With completely random ordering of a two-component interstratified system consisting of equal proportions of the two components, the probability of an A layer succeeding an A layer is exactly the same as the probability of a B layer following an A layer. Thus, the clay may be built up of layer sequences such as:

ABBABAABAB BAAABA BBABABBA

Regular interstratified structures (or at least structures with a strong tendency to regularity) generally yield X-ray diffraction patterns characterized by a high-spacing basal reflection which is equal to the sum of the thicknesses of the two-component units, and by a rational series of higher-order reflections. For example, hydrobiotite consists of a regular alternation of $\sim$ 10 Å biotite units and 14.7 Å vermiculite units so that, in theory at least, the 001 reflection occurs at 24.7 Å, 002 at 12.35 Å, 003 at 8.23 Å, 004 at 6.17 Å, and so on. In contrast, the X-ray diffraction patterns of randomly interstratified structures are not characterized by high-spacing basal reflections. Their fundamental basal reflection may occur at a spacing intermediate between the thickness of the component units, and the other basal-type reflections are irrational with respect to it.

Interstratification can also be recognized by the X-ray diffraction response of clay materials to auxiliary treatments like ethylene glycol solvation or heating. Such treatments generally affect only expansible components, and it is worth noting that interstratified clays nearly always contain such components. Instances of interstratification between two non-expansible components are known, but they are very much the exception to the general rule. An example of the use of such auxiliary treatments for a randomly interstratified chlorite-smectite can be considered. If the smectite is Mg-saturated, the fundamental reflection is likely to occur at ~ 14.7 Å in the air-dried state and the possibility of interstratification may not be immediately obvious. After ethylene glycol treatment, however, it may be observed that this reflection has moved to a spacing of ~ 15.7 Å, that is to a position intermediate between chlorite (14.4 Å) and fully expanded smectite (16.9 Å). Similarly, heating at 550°C may lead to an incomplete contraction, and the observed spacing at ~ 12.5 Å is intermediate between heated chlorite (~ 14 Å) and fully contracted smectite (~ 10 Å). Treatments involving saturation with particular cations (e.g. K^+ or NH_4^+) may be useful for distinguishing smectite from vermiculite as the expansible component. It is evident, therefore, that systematic study of the response of the basal reflections to auxiliary treatments provides important information about the nature of interstratified components.

2.4.7.1 *Interpretation involving calculated diffraction profiles.* So far, only the simplest types of interstratification have been considered, but there are a multitude of other possibilities that may be encountered. For example, in a two-component interstratified system, any proportion of the two components may occur. Again, it is a reasonable assumption that perfect regularity or complete randomness of an interstratified structure is the exceptional case and that we are more likely to encounter structures with variable tendencies towards randomness or regularity. Furthermore, it is possible that special types of ordering may occur, involving for example, a high degree of segregation of one component or perhaps long-range ordering of the type AAB AAB AAB. Other complications are that interstratifications may occur between three or even more components, and also that the expansible component may be variably hydrated. The potential complexity of interstratified structures is enormous, and X-ray diffraction characteristics should respond to, and reflect, this complexity. In the face of this problem, the only practicable approach is to devise methods for calculating the X-ray diffraction profiles from model interstratified structures, so that the effects of the sources of variation can be systematically explored and compared with experimentally obtained diffraction curves. Fortunately, a number of clay mineralogists have dedicated themselves to this problem and have achieved a very large measure

Table 2.12 The major interstratified clay minerals, classified according to the nature of the non-expansible component and the ordering of the structure (modified after Wilson and Nadeau, 1985).

Non-expansible component / Ordering	Mica		Chlorite		Kaolin
	Di-	Tri-	Di-	Tri-	
Regular nearest-neighbour ordering	Rectorite Mica-vermiculite	Hydrobiotite	Tosudite	Corrensite	
Long-range non-nearest neighbour ordering	Illite-smectite	Mica-vermiculite	?	?	?
Random	Illite-smectite	Mica-vermiculite (or smectite)	Chlorite-smectite (or vermiculite)	Chlorite-smectite (or vermiculite)	Kaolin-smectite

of success, for example, the work of Méring (1949), which was developed by MacEwan (1958) and further refined by Reynolds and Hower (1970). The overview of Reynolds (1980) can still by regarded as the authoritative treatment for the X-ray identification of interstratified clays, and the reader is referred to this work for further details of the methods involved. This work and others like it have led to the situation where the clay mineralogist can virtually identify an interstratified clay by comparing the experimental pattern with calculated profiles from the literature. Furthermore, these calculated profiles have led to the production of peak-migration curves, which enable a ready assessment of the relative proportion and arrangement of the component units in an interstratified structure, as will now be illustrated.

2.4.7.2 *X-ray identification of interstratified minerals.* The major types of interstratified minerals are classified in Table 2.12, according to the type of ordering and the nature of the non-expansible component, whether mica, chlorite or kaolinite.

(i) Regular interstratification: the basal reflections of the main, regularly interstratified clay minerals are illustrated in Figure 2.18 and listed in Table 2.13, together with typical responses to ethylene glycol treatments and heat treatments. As previously indicated, a high-spacing peak attributable to the sum of the thicknesses of the two-component units is the most diagnostic feature. For example, interstratifications involving mica, as in hydrobiotite and rectorite, typically show a reflection at 25 Å in the air-dried state: this being the sum of the 10 Å mica units and the 15 Å vermiculite or smectite units (Ca- or Mg-saturated). Treatment with ethylene glycol enables further differentiation, as in rectorite the 25 Å peak will shift to 27 Å, whereas in hydrobiotite the 25 Å peak remains unchanged. When the non-expansible component is chlorite, as in corrensite and tosudite, the high-spacing reflection occurs at 29–30 Å in the air-dried state, this being the sum of two components with thicknesses of 14.5–15 Å. Ethylene glycol treatment brings about a shift in this spacing to ~32 Å. Differentiation between mica- and chlorite-based structures can also be made on the basis of response to heating to 500–600°C. Thus, mica-smectite will yield a peak at ~20 Å, whereas chlorite-smectite shows a reflection at ~24 Å. Further, it may be noted that trioctahedral corrensite and hydrobiotite are easily decomposable by HCl treatment, whereas rectorite and tosudite are not.

(ii) Random interstratification: the most extensively studied system is that of illite-smectite, and excellent fits between calculated and experimental curves have been obtained. The main feature to bear in mind is that no intermediate reflection occurs between the illite 001 and the smectite 001 in the ethylene glycol-treated state. Rather, the effect of interstratifying increasing amounts of illite with smectite, is to gradually decrease the intensity of the smectite peak at ~ 17 Å. Indeed, the 17 Å peak is still discernible when there is up to 60% illite, although typically it is weak, broad, and asymmetrical, with

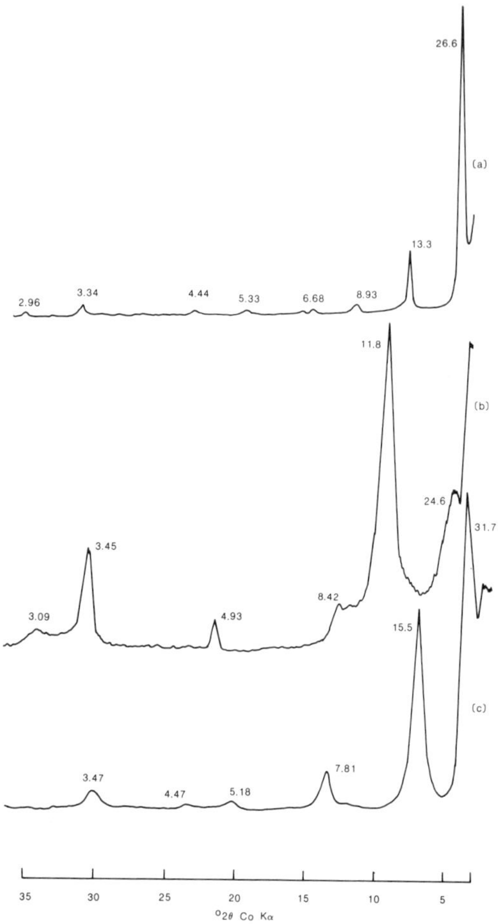

Figure 2.18 X-ray diffractometer traces of well-oriented, regularly interstratified (a) illite-smectite (rectorite), (b) biotite-vermiculite (hydrobiotite), and (c) chlorite-smectite (corrensite), all ethylene glycol salvated

Table 2.13 High-spacing basal reflections (Å) of regularly interstratified minerals before and after ethylene glycol, glycerol and heat treatments.

Treatment	Rectorite[1]	Hydrobiotite[2]	Tosudite[3]	Corrensite[4]
Air-drying	25.0	24.7	28.9	29.1
Ethylene glycol	26.5	24.7	31.1	31.0
Glycerol	27.4	24.6	32.2	32.0
Heating at 500/600°C	19.3	10.0	23.4	23.7

[1] Brown and Weir (1963).
[2] Veniale and van der Marel (1969).
[3] Brown *et al.* (1974).
[4] Earley *et al.* (1956).

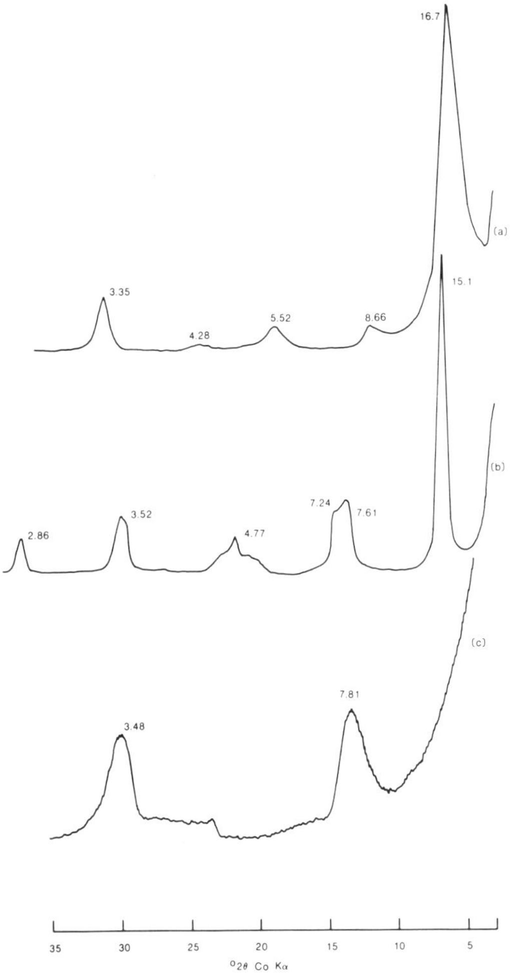

Figure 2.19 X-ray diffractometer traces of well-oriented, randomly interstratified (a) illite-smectite (~25% illite), (b) chlorite-smectite (~75% chlorite), and (c) kaolinite-smectite (~75% kaolinite), all ethylene glycol solvated

marked scattering towards the high-angle side (Figure 2.19). Fortunately, however, there are systematic spacing changes associated with other reflections, in particular the 001 illite/002 smectite and the 002 illite/003 smectite. These changes can be plotted in the form of a peak-migration curve, which can be used to ascertain the proportions of the two components fairly precisely (Figure 2.20). Even higher precision can be achieved using refinements developed by Srodon (1980), which the reader should consult for details.

For randomly interstratified mica-vermiculite, the major 001 peak is readily detectable and shows a steady shift in spacing between 10 and 14.3 Å, which is

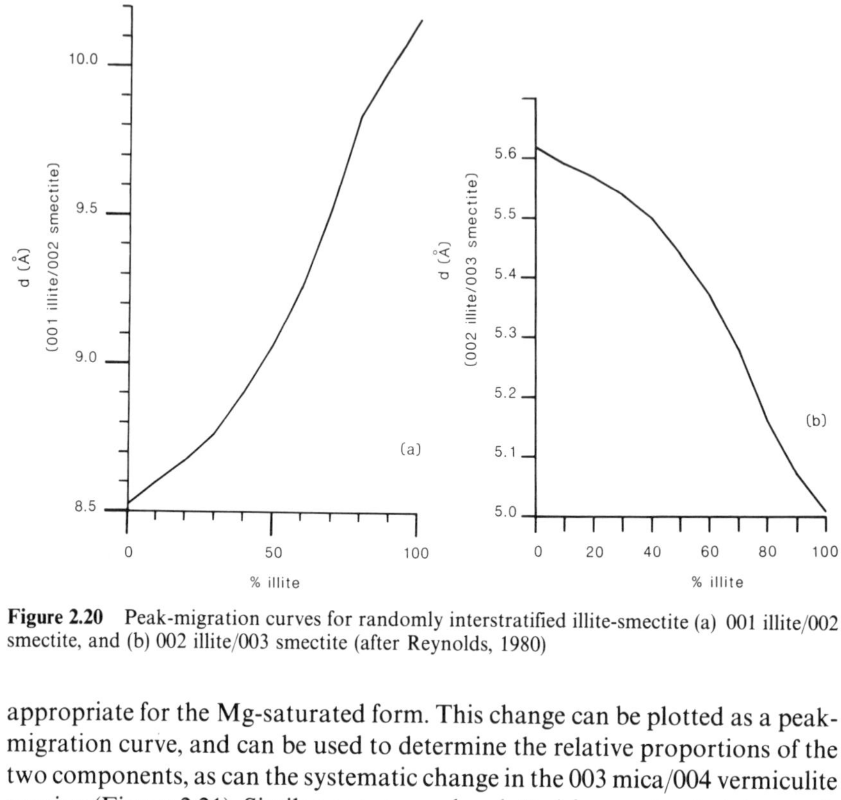

Figure 2.20 Peak-migration curves for randomly interstratified illite-smectite (a) 001 illite/002 smectite, and (b) 002 illite/003 smectite (after Reynolds, 1980)

appropriate for the Mg-saturated form. This change can be plotted as a peak-migration curve, and can be used to determine the relative proportions of the two components, as can the systematic change in the 003 mica/004 vermiculite spacing (Figure 2.21). Similar curves can be plotted for randomly interstratified chlorite-smectite (Figure 2.22(a)). Finally, mention should be made of randomly interstratified kaolinite-smectite, which is now becoming more widely recognized, especially in soils. This type of interstratification yields a reflection between 7.2 and 8.5 Å in the glycolated state, resulting from the combined effect of 001 kaolinite 002 smectite. Heating to 300°C seems to result in an increase in spacing, because the combined reflection now results from 001 kaolinite/001 'collapsed' smectite, spanning a range of 7.2 to 9.6 Å. An appropriate peak migration curve is shown in Figure 2.22(b).

(iii) Long-range ordering: this type of ordering is found particularly in interstratified illite-smectite, although, in principle, there seems to be no reason why it should not occur in other interstratifications. Such ordering becomes evident when the proportion of illite approaches 65% and is thought to correspond with structures where there are, for example, IIS and IIIS superlattices. Distinctive diffraction patterns are obtained, as shown in Figure 2.23. Thus, ethylene glycol-solvated IIS-ordered clays show a strong reflection at ~ 12.5 Å, with a somewhat weaker peak at ~ 9.6 Å. In the air-dried state only a single peak at ~ 11.0 Å is observed. The IIIS-ordered clays correspond with higher illite contents (> 80%), and yield an X-ray diffraction

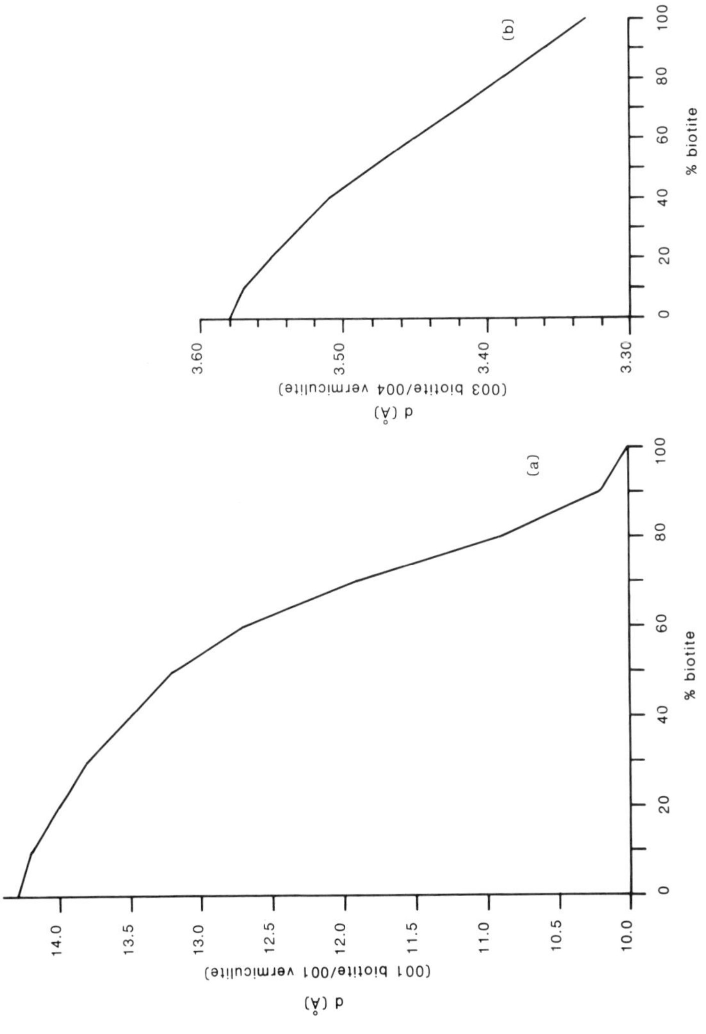

Figure 2.21 Peak-migration curves for randomly interstratified biotite–Mg vermiculite for (a) 001 biotite/001 vermiculite, and (b) 003 biotite/004 vermiculite (after Reynolds, 1980)

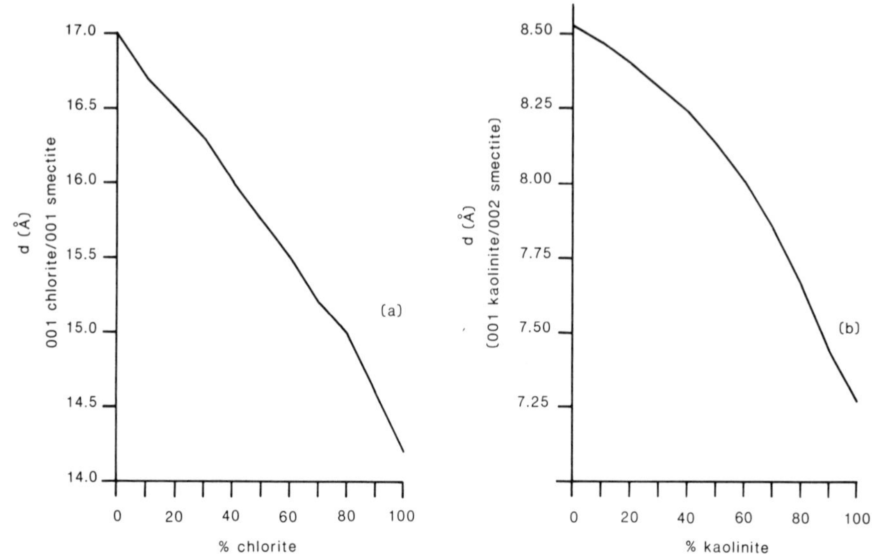

Figure 2.22 Peak-migration curves for randomly interstratified (a) chlorite-smectite–001 chlorite/001 smectite, and (b) kaolinite-smectite–001 kaolinite/002 smectite (after Reynolds, 1980)

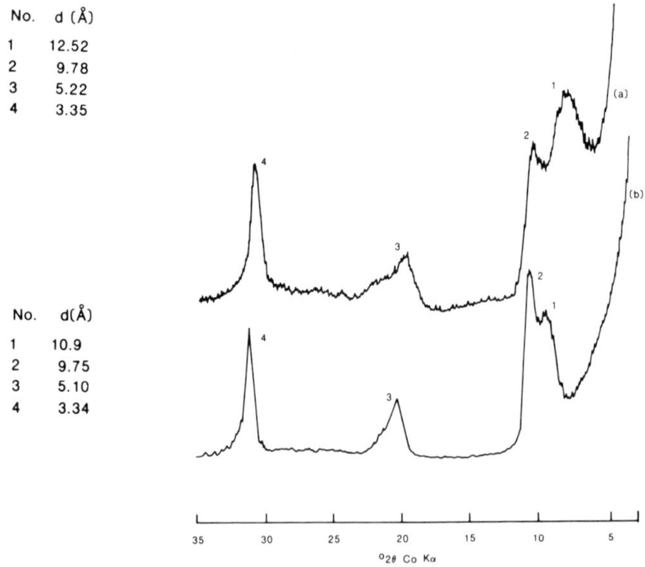

Figure 2.23 X-ray diffractometer traces of long-range ordered illite-smectite with (a) IIS, and (b) IIIS structures

pattern characterized by a strong peak at ~9.8 Å, with a subsidiary peak or shoulder at ~11 Å. A recent interpretation as to the actual physical significance of this type of ordering, as well as of the whole family of interstratified minerals, has been proposed by Nadeau *et al.* (1984).

2.4.8 *Hydroxy-interlayered clays*

Some environments, particularly soils, are characterized by clay minerals that show X-ray diffraction responses to ethylene glycol/glycerol or heat treatments intermediate between those of smectite/vermiculite and chlorite. The basal reflections of these materials tend to be broad and diffuse, becoming even more so after heating, and the complete diffraction pattern is often rather poorly developed. These soil clays appear to have been expansible minerals that have sorbed variable amounts of non-exchangeable interlayer material, frequently in the form of polymerized hydroxyhydrates of Al ions, and are described as 'intergrade' or 'intergradient'. The surest way to identify this material would be to remove the contaminating interlayer material by chemical methods, so restoring the expansible mineral to its original state. Various methods have been suggested to accomplish this, such as the use of KOH and KCl (Brown, 1953), Na-dithionite citrate (Tamura, 1956), and heating at 400°C followed by boiling in NaOH (Dixon and Jackson, 1959), but none can be regarded as completely effective. For Scottish soil clays the writer has found hot $NaCO_3$ treatment (Follett *et al.*, 1965) to be useful. Typical diffraction patterns for some Al-interlayered soil clays from Scotland, before and after treatment, are shown in Figure 2.24.

2.4.9 *Magnesium-rich clays*

In some environments, which are characterized by high magnesium activities—for example: weathering profiles and accumulations derived from ultrabasic rocks; soils in arid and semi-arid environments; alkali-rich lacustrine sediments, etc.—a variety of unusual clay minerals may occur that are usefully considered in a group. Some of the more common Mg-rich minerals have already been discussed—serpentine minerals, saponite and clinochlore—but there are other minerals like talc, kerolite, stevensite, sepiolite and palygorskite that remain to be considered.

2.4.9.1 *Talc* yields an intense 001 reflection at 9.3 Å, and a strong 003 reflection at 3.11 Å. These peaks are unaffected by heat treatment, organic solvation or cation saturation, and occur in such a unique position on the diffractogram as to enable unambiguous identification. Pyrophyllite is the only other layer-silicate mineral with a similar pattern, but its main basal reflections occur at the somewhat lower spacings of 9.20 and 3.07 Å

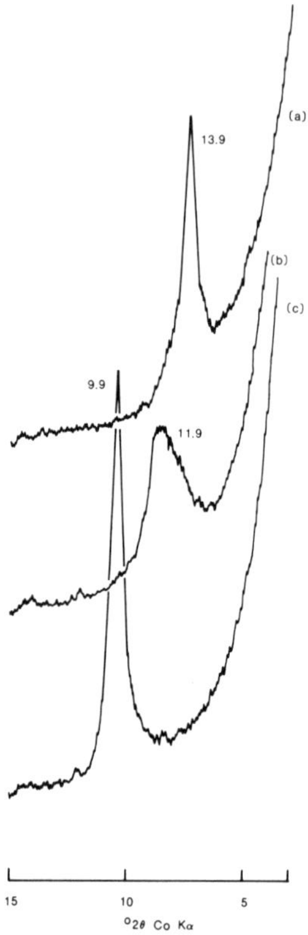

Figure 2.24 X-ray diffractometer traces of K-saturated hydroxy-interlayered soil clay (a) air-dry, (b) after heating at 300°C for 2 h, and (c) after $NaCO_3$ treatment and heating at 300°C for 2 h

respectively. In addition, talc is decomposable by HCl treatment, whereas pyrophyllite is not.

2.4.9.2 *Kerolite* is chemically and structurally similar to talc, but is extremely fine grained, contains some water and yields a basal spacing near 10 Å. The full X-ray pattern shows features typical of turbostratic disorder, being characterized by a small number of broad, diffuse and often asymmetrical non-basal reflections, which occur, however, at approximately similar spacings to the more highly crystalline talc (Brindley, 1980). Kerolite cannot be expanded in water, and expands only partially, and with difficulty, with ethylene glycol.

2.4.9.3 *Stevensite* is a trioctahedral smectite mineral, the exact nature of which is still open to question. According to Faust *et al.* (1959), it has a talc-like structure, but with defects (i.e. deficiencies) in the octahedral sheet, which give rise to a layer charge and explain its smectitic structure. On the other hand, there are also a small number of talc-like domains with no defects, and which therefore possess the non-swelling attributes of talc. However, Brindley (1955) has interpreted stevensite as an interstratified talc-saponite consisting of distinct units of ~10 and ~14 Å in thickness. Whatever its structure, stevensite yields a smectite-like diffraction pattern, characterized by an intense basal reflection at ~12 Å, a weak high-spacing reflections at ~24 Å and strong hk bands at 4.54 and 2.54 Å. The mineral swells in glycerol and ethylene glycol, and after heating to 400°C collapses to yield a strong 9.6 Å peak with a weak high-spacing reflection at ~19 Å. The occurrence of a high-spacing reflection would appear to support an interstratified structure, as suggested by Brindley (1955).

2.4.9.4 *Aliettite* is a regularly interstratified talc-saponite, and yields a fundamental basal spacing of 24.8 Å, with many higher orders. The basal spacing increases to ~26 Å on glycol treatment (Figure 2.25), and contracts to 9.6 Å after heating at 400°C (Alietti, 1958; Veniale and van der Marel, 1968).

2.4.9.5 *Swelling chlorite* characteristically yields a strong 14 Å reflection in the air-dried state, which expands to 17 or 17.8 Å on being treated with ethylene glycol or glycerol, respectively. Unlike smectite, however, the 14 Å spacing is heat-stable, perhaps not to the degree of well-crystallized chlorites, but at least up to 430°C. Stephen and MacEwan (1951) described a pure

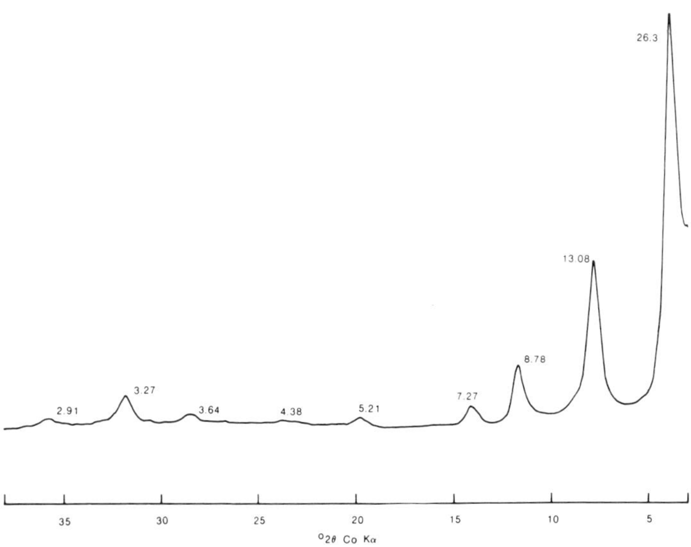

Figure 2.25 X-ray diffractometer trace of ethylene glycol-solvated aliettite–regularly interstratified talc–saponite

Table 2.14 Basal spacings (Å) of swelling chlorite and related minerals before and after glycerol- and heat treatments.

Treatment	1	2	3
Air-dry	14.0	13.6	14.9
Glycerol	17.8	14.8	18.0
Heating at 500°C	14.0	13.7	10–14 (peaking at 13.9 Å)

1 Stephen and MacEwan (1951).
2 Martin-Vivaldi and MacEwan (1957).
3 Bain and Russell (1981).

swelling chlorite, but it is more common to find this component as part of an interstratified structure. Thus, interstratified chlorite-swelling chlorite is characterized by incomplete swelling and essentially no contraction on heating (Alietti, 1958). Again, interstratified saponite-swelling chlorite shows full expansion with glycol/glycerol, but incomplete contraction after heating (Veniale and van der Marel, 1963). Table 2.14 shows the diagnostic spacings of the range of the swelling chlorite minerals just described.

2.4.9.6 *Sepiolite and palygorskite* have related fibrous structures, but somewhat different chemical compositions: the essential difference being that sepiolite is a magnesium silicate with only minor aluminium, whereas palygorskite may be regarded as an alumino-magnesium silicate. The fibre axis of both minerals may be conveniently taken as the *a*-axis and, although a fibre axis along *c* has been preferred, the former beings out the relationship with the phyllosilicate structures more clearly, with the *c* axis being normal to the 2:1 type ribbon structure. The fibrous nature of the minerals precludes the production of an oriented aggregate to enhance the 001 planes. The most intense and diagnostic reflections are from the 011 planes, yielding a strong maximum at 12.2 Å in sepiolite and at 10.5 Å in palygorskite (Figure 2.26). Other moderate intensity reflections for sepiolite are at 4.5 Å (060), 4.3 Å (131), 3.75 (260) and 3.20 Å (331), and for palygorskite 6.36 Å (200) 4.47 Å (040), 4.26 Å (121) and 3.68 Å (221).

The X-ray diffraction pattern does not change with variations in relative humidity, or with the addition of ethylene glycol/glycerol, but changes do occur after heating. Thus, sepiolite heated at 250°C for 1 h is characterized by a decreased intensity of the main reflections and the appearance of new reflections at 10.4 and 8.2 Å (Hayashi *et al.*, 1969). These peaks intensify on further heating and persist on heating up to 600°C for 120 h (Nagata *et al.*, 1974). With palygorskite, heating at 350°C decreases the intensity of the main reflections, but generates new maxima at 9.2 and 4.7 Å. The main 10.5 Å reflection disappears completely after heating at 600°C. Sepiolite decomposes more rapidly than palygorskite in acid conditions, the reaction rate-constant

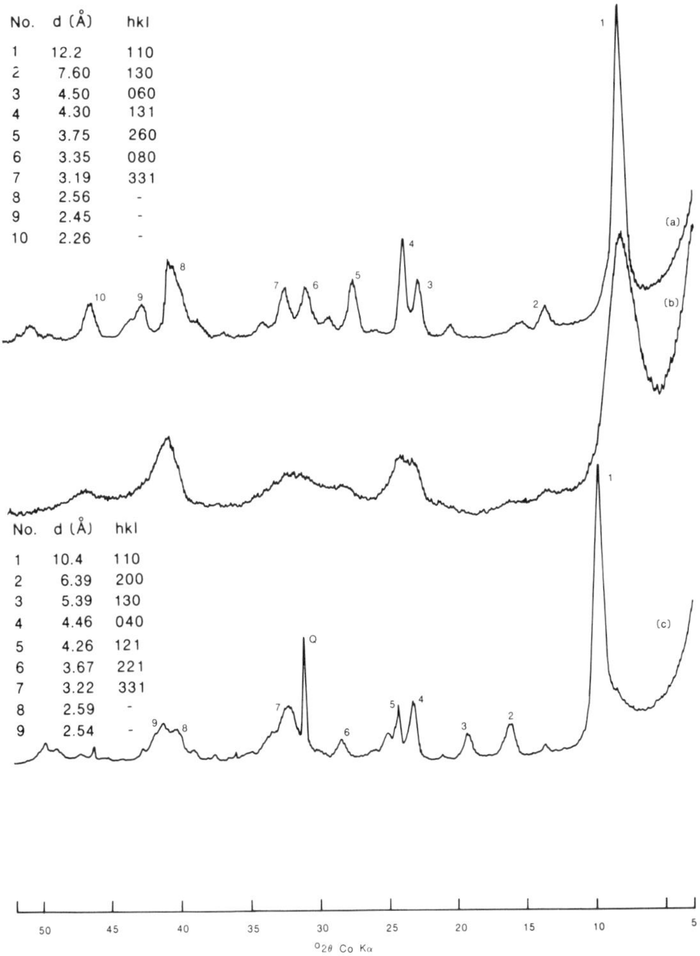

Figure 2.26 X-ray diffractometer traces of randomly oriented (a) highly crystalline sepiolite, (b) poorly ordered sepiolite, and (c) palygorskite, with quartz impurity. Partial diffraction data showing some diagnostic reflections are indicated

being about 240 times greater (Abdul-Latif and Weaver, 1969).

Finally, it may be noted that sepiolite/palygorskite clays can yield diffraction patterns that suggest considerable variation in their degree of structural order. For example, some sepiolites as in Figure 2.26(b), are characterized by diffraction patterns consisting of relatively few weak and broad peaks (Brindley, 1959) and some materials yield patterns characteristic of neither mineral. The main feature is that there is a broad reflection in the 10–12 Å region, with a central value of 11 Å, while the remainder of the pattern does not agree well with either palygorskite or sepiolite. Martin-Vivaldi and

Linares Gonzalez (1962) interpret this in terms of a fine-scale, random intergrowth between the two minerals, which on structural grounds would appear to be quite possible.

2.4.10 *Imogolite and allophane*

Both these minerals are very poorly crystalline and are sometimes regarded as being X-ray amorphous. However, they do, in fact, yield X-ray diffraction patterns which can be characteristic in the case of imogolite. This mineral consists of thin, tubular particles, and in a randomly oriented form yields a diffraction pattern with broad, quite intense reflections at 12–20, 7.8, 5.6 and 3.3 Å. Heating at 100–300°C induces the replacement of the broad high-spacing reflection by a sharper and more intense peak at 17.7 Å and by other minor changes in the rest of the pattern (Figure 2.27(a) and (b)). Brown (1980)

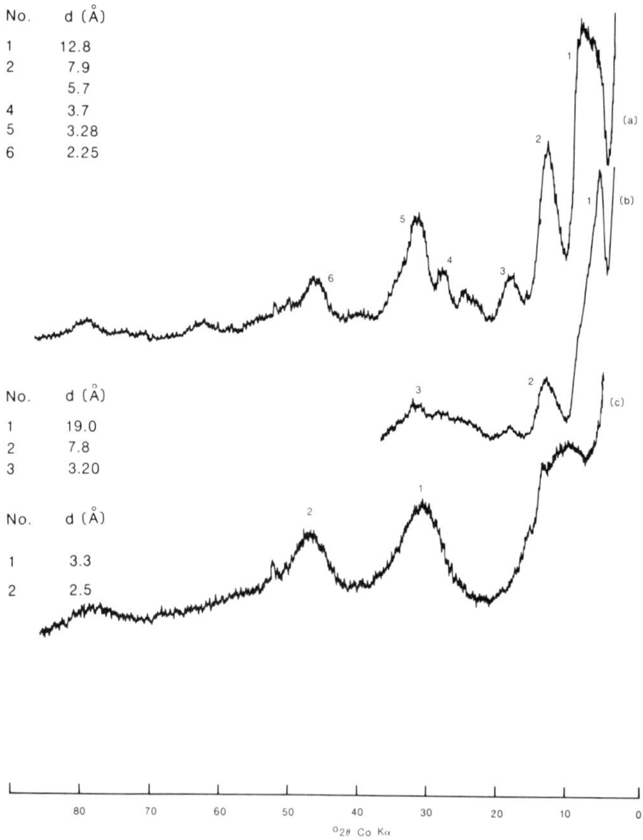

Figure 2.27 X-ray diffractometer traces of randomly oriented (a) imogolite, air-dried, (b) imogolite, heated at 200°C, and (c) allophane, air-dried

points out that this change after heating is slowly reversible in moist air, and that this may be a useful indicator of imogolite. Allophane generally shows little evidence of structural order when examined by X-ray diffraction, and shows a pattern consisting of continuous scattering, superimposed on which are a few broad, poorly defined maxima, especially at ~3.3 and 2.25 Å (Figure 2.27(c)).

It can only be concluded that neither of these minerals can be confidently identified by X-ray diffraction criteria alone, particularly when the clay sample is impure and contains a preponderance of more crystalline material, as is often the case. In these circumstances, examination by other techniques, such as electron microscopy, infrared spectroscopy and selective dissolution analysis becomes essential.

2.4.11 *Iron oxide minerals*

The X-ray identification of many iron oxide minerals is straightforward when they are present in concentrations of > 10%, but in many natural materials such as soils and sediments, much lower concentrations are the rule, and difficulties do arise. These difficulties may be compounded by the fact that some iron oxide minerals are extremely poorly ordered. It is therefore prudent to take especial care in X-ray technique, preparation procedures and auxiliary treatments when dealing with these materials. In general, it is preferable to use Co or Fe radiation in X-ray studies of iron oxides, because this avoids the high background fluorescence effect that occurs when the widely used Cu radiation is employed. However, if a monochromator is placed in the diffracted beam, then this high background fluorescence can be avoided. The use of thin, oriented preparations is not generally recommended for iron oxides, and better patterns can be obtained by utilizing randomly oriented mounts. If only a small amount of sample is available, something approaching random orientation can be achieved by drying a suspension in acetone onto a glass slide (Paterson *et al*., 1986) and excellent diffraction patterns can be obtained. Preconcentration of iron oxides is often necessary, and here it is helpful to fractionate to a fine particle size ($< 1.0\,\mu m$) and to use chemical dissolution treatments to remove kaolinite for example, which in soils is often the main non-iron-oxide contaminant. Useful procedures involve treatment with 5 N NaOH (Norrish and Taylor, 1961) or HF (Norrish, 1968). Information can also be gained by visually comparing the X-ray diffraction pattern of a sample before and after the removal of iron oxides by dithionite treatment. The advent of computer-controlled diffractometers means that a differential X-ray diffraction pattern can be obtained that even enables easier identification of poorly crystalline iron oxides at low concentrations. Differential X-ray diffractometry, as applied to iron oxides, has been illustrated by Brown and Wood (1985) and Campbell and Schwertmann (1985). For general identification purposes, the response of iron oxyhydroxide and hydrous oxide minerals to heating at 600°C is also helpful. This always results in transformation to the anhydrous oxide, hematite.

The main diagnostic characteristics of the various iron oxide minerals are now described.

2.4.11.1 *Hematite* (α-Fe_2O_3) yields sharp, strong reflections at 2.69 and 2.51 Å, and medium-intensity reflections at 3.67, 2.20, 1.83 and 1.69 Å. The hematite structure can accommodate Al and Ti ions as isomorphous substituents, but this occurs only to a small extent at low temperatures, with minimal effects on the X-ray diffraction pattern.

2.4.11.2 *Maghemite* (γ-Fe_2O_3)/*magnetite* (Fe_3O_4) are magnetic minerals and can often be concentrated with a hand magnet. Their X-ray diffraction patterns are similar, but readily distinguishable when pure materials are examined. Magnetite yields a strong reflection at 2.53 Å, and medium-intensity lines at 2.97, 2.10, 1.62 and 1.48 Å. The most intense reflections of maghemite occur at slightly lower spacings, the strongest reflection being at 2.51 Å, and medium-intensity peaks at 2.95, 2.09, 1.60 and 1.47 Å. Furthermore, pure maghemite yields many additional weak reflections not seen in the pattern of magnetite. In must be stressed, however, that in impure materials it may be difficult, if not impossible, to distinguish unambiguously between magnetite and maghemite on the basis of X-ray diffraction criteria alone.

2.4.11.3 *Goethite* (α-FeOOH) is one of the most widely occurring iron oxide minerals and is readily identifiable by an intense reflection at 4.18 Å, and by medium-intensity reflections at 2.69 and 2.45 Å. However, the exact nature of the diffraction pattern varies according to the degree of Al substitution and particle size, which affects line position and line width respectively. An approximate assessment of the amount of Al-substitution in goethite can be made by using the graph in Figure 2.28, which plots a linear shift in spacing of the 111 reflection, from 2.448 Å in the case of pure goethite, to 2.40 Å for the maximum Al-substitution corresponding with $(Al_{0.33}Fe_{0.67})OOH$. It is advisable, when making precise measurements of this kind, to add a compound that yields a strong reflection close to that being measured and that can be used as an internal reference. A suitable compound is lead nitrate, which yields a peak at 2.37 Å. An assessment of the particle size of pure goethite can of course be made using experimentally determined peak breadths in conjunction with the Scherrer equation.

2.4.11.4 *Akaganeite* (β-FeOOH) does not occur in soils or geological deposits, but is a common constituent of fine-grained corrosion products. It is readily synthesized from the hydrolysis of ferric chloride. It yields a characteristic X-ray diffraction pattern, the most diagnostic feature of which is a strong reflection at 7.40 Å. When akaganeite is heated in the temperature

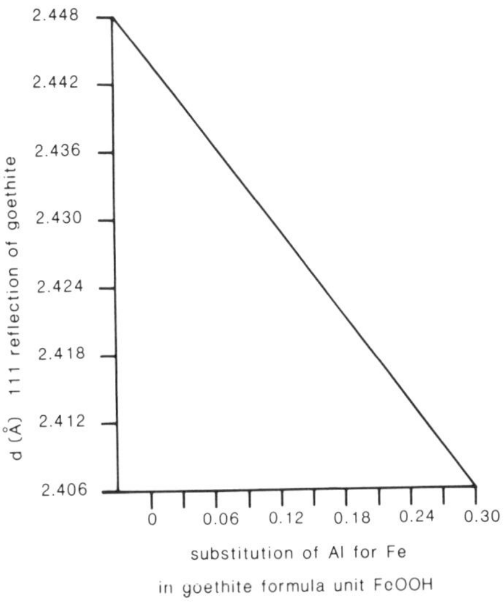

Figure 2.28 Graph for assessment of Al-substitution in goethite

range 200–400°C it gradually decomposes, and well-crystallized hematite appears above ~420°C (Brown, 1980).

2.4.11.5 *Lepidocrocite* (γ-*FeOOH*) can be identified by a strong reflection at 6.27 Å and by medium-intensity reflections at 3.29 and 2.47 Å. The only other mineral with which it may be confused is boehmite. The response to dithionite treatment could be used to provide further confirmatory evidence.

2.4.11.6 *Delta* (δ) *FeOOH* was first obtained as a synthetic product from the rapid oxidation of iron, in sulphate or chloride solutions, in strongly alkaline media. The product is rather poorly crystalline and is characterized by an X-ray powder pattern of ten lines. A naturally occurring material, which is even more poorly crystalline, yields a pattern of only four broad lines, at 2.56, 2.15, 1.71 and 1.48 Å (Figure 2.29(b)). It has been termed δ' FeOOH or feroxyhyte by Chukhrov *et al.* (1976) and is found in deep-sea Fe-Mn nodules and in some gley soils.

2.4.11.7 *Ferrihydrite* ($Fe_2O_3nH_2O$) is another poorly crystallized iron oxide mineral with an ill-defined diffraction pattern. In its most completely developed form, ferrihydrite shows a five-line pattern with maxima at 2.5, 2.2, 1.97, 1.71 and 1.50 Å (Figure 2.29(a)). All these lines are broad and weak, so that the mineral can only be detected in high concentration. Although there is an obvious similarity with the feroxyhyte pattern, the reflection at 1.97 Å is thought to be characteristic of ferrihydrite (Chukhrov *et al.*, 1976).

No.	d (Å)
1	2.54
2	2.35
3	1.97
4	1.72
5	1.51
6	1.47

No.	d (Å)
1	2.56
2	2.15
3	1.71
4	1.48

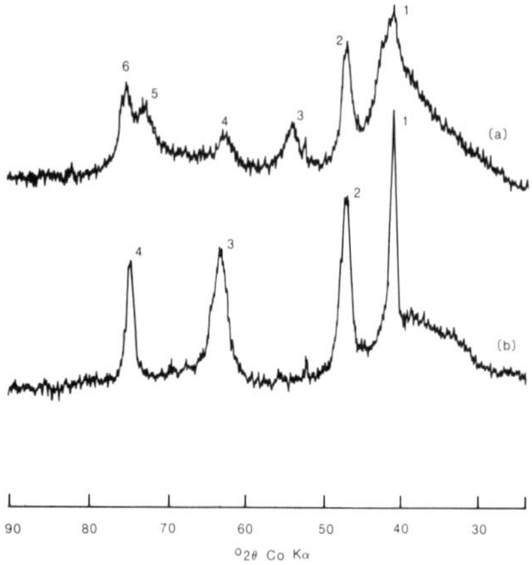

Figure 2.29 X-ray diffractometer traces of (a) ferrihydrite, and (b) feroxyhyte, both prepared experimentally

2.4.12 *Aluminium hydroxides/oxyhydroxides*

Of the three known polymorphic forms of $Al(OH)_3$—gibbsite, bayerite and nordstrandite—only the first is of common natural occurrence. Little difficulty should be experienced in identifying gibbsite. It yields a strong reflection at 4.85 Å and medium-intensity reflections at 4.37, 4.31, 2.45 and 2.38 Å. The 4.85 Å peak is the most diagnostic and is separate from the major clay-mineral reflections; furthermore, it is a basal reflection (002), and can be enhanced by preparing an oriented aggregate. Nevertheless, it may be difficult to detect gibbsite by X-ray diffraction when it is present at about the $< 5\%$ level, and in these circumstances it is better to use a more sensitive technique like differential thermal analysis (DTA).

The two oxyhydroxide polymorphs—boehmite and diaspore—are also readily identifiable by X-ray diffraction. Boehmite is characterized by an intense reflection at 6.11 Å and strong reflections at 3.16 and 2.34 Å. The strongest reflection of diaspore occurs at 3.99 Å, and there is a diagnostic group of medium-intensity reflections at 2.32, 2.13 and 2.08 Å.

2.4.13 *Manganese oxide minerals*

These minerals often yield diffuse and poorly defined diffraction patterns, and are difficult to identify except when they occur in the form of relatively pure segregations. Apart from their poorly ordered nature, another difficulty is the strong secondary fluorescence that they excite when examined by Co or Cu radiation. The most common manganese-oxide mineral in clay materials is probably birnessite, which has a layer structure, and which yields a pattern with a prominent reflection at 7.2 Å, and others at 3.6, 2.46, 2.33 and 2.04 Å. Other manganese minerals yielding a strong reflection in the 7 Å region include cryptomelane (KMn_8O_{16}) and hollandite ($BaMn_8O_{16}$), both of which occur in soils. It is useful to remember that manganese minerals can be decomposed by reducing agents such as hydroxylamine hydrochloride, and when these minerals contain Mn^{4+} they may be dissolved by acidified hydrogen peroxide treatment.

2.4.14 *Silica minerals*

Although there are a large number of silica polymorphs, those most likely to occur in clay materials are quartz, cristobalite and the so-called opal CT.

2.4.14.1 *Quartz* is, of course, of almost universal occurrence in clay, at least in trace amounts. It is readily recognized by an intense reflection at 3.34 Å, and by a weaker, but still strong reflection at 4.26 Å. Although quartz has no cleavage, Eslinger *et al.* (1973) have pointed out that the mineral still shows an orientation effect when prepared as a sedimented aggregate. This effect takes the form of an enhanced intensity of the 3.34 Å peak relative to that at 4.26 Å, and is attributed to orientation along well-developed prism faces.

2.4.14.2 *Cristobalite* although not common in clays, may occur in soils and other deposits derived from volcanic materials. It yields a strong reflection at 4.05 Å, and may be suspected particularly when a reflection in this position cannot be attributed to feldspar, which also yields a strong peak at 4.04 Å.

2.4.14.3 *Opal CT* which is found in bentonites and deep-sea oozes, is thought to consist of disordered sequences of cristobalite and tridymite-like units. However, electron-diffraction and infrared evidence suggest a predominance of tridymite-like stacking (Wilson *et al.*, 1974). Typically this material yields a broad peak at $\sim$4.10 Å (Figure 2.30).

2.4.15 *Other minerals*

Clay materials may contain subordinate amounts of a wide variety of minerals not so far discussed. Often, only the strongest line of these minerals appears on the X-ray diffraction pattern, and in these circumstances, identification is greatly assisted by awareness of the likely possibilities. Minerals belonging to the feldspar, carbonate, and titanium oxide groups are relatively common in clays. Zeolites, sulphates, sulphides and phosphates are rather less abundant. In Table 2.15, the strongest line of particular minerals within the above groups is listed in order of decreasing *d* spacing. With some minerals, however, the most easily observed line is not necessarily the strongest one, but rather, one that occurs at a relatively high spacing. These reflections are marked with an asterisk.

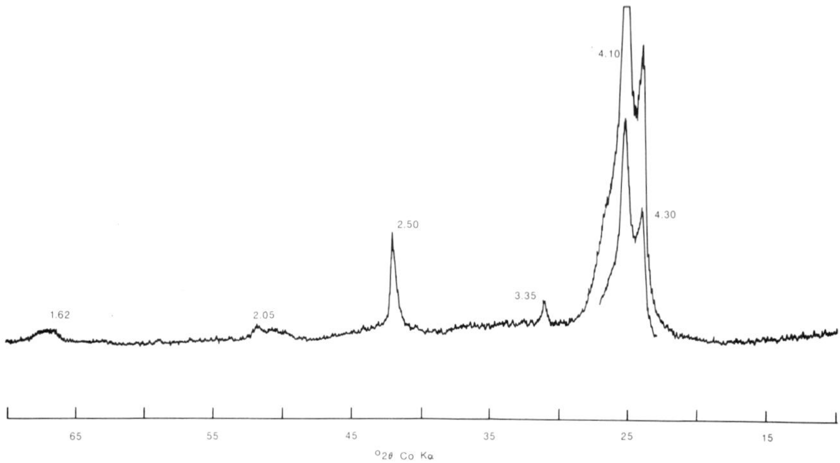

Figure 2.30 X-ray diffractometer trace of opal CT from deep-sea chert

2.5 Quantitative analysis

Undoubtedly, quantitative X-ray diffraction analysis of clay minerals must lead to a more precise understanding of the numerous properties of the soils, sediments and rocks that these minerals control or influence. However, such an analysis is a difficult undertaking. Many factors need to be taken into account, not least the highly variable nature of the clay minerals themselves. Nevertheless, in principle, clay materials should be susceptible to quantitative X-ray diffraction analysis, despite the problems involved.

The basic equation governing the intensity of a diffracted X-ray beam from a selected plane of a mineral, in a multicomponent mixture, was developed by Alexander and Klug (1948). This equation is:

$$Ii = \frac{KiXi}{\rho i \mu s}$$

where Ii is the intensity of a line from component i; Xi is the weight fraction of component i; ρi is the density of component i; μs is the mass-absorption coefficient of the whole sample; and Ki is a constant depending on the nature of the component and the geometry of the apparatus. Rearrangement of the preceding equation gives:

$$Xi = \frac{\rho i Ii \mu s}{Ki}$$

In the general case, where the mass-absorption coefficient of component i differs from that of the sample, and the latter is unknown, then the addition of an internal standard and the construction of calibration curves for component

Table 2.15 Strongest or most easily observed reflections (Å) of accessory minerals that may occur in clay materials.

Mineral group	Mineral	Formula	Diagnostic reflection
Feldspars	Sanidine	$KAlSi_3O_8$	3.31
	Microcline	$KAlSi_3O_8$	3.25
	Albite	$NaAlSi_3O_8$	3.19
	Anorthite	$CaAl_2Si_2O_8$	3.18
Carbonates	Aragonite	$CaCO_3$	3.40
	Calcite	$CaCO_3$	3.03
	Ankerite	$Ca(Mg, Fe)(CO_3)_2$	2.90
	Dolomite	$CaMg(CO_3)_2$	2.89
	Siderite	$FeCO_3$	2.79
	Magnesite	$MgCO_3$	2.74
Titanium oxides	Anatase	TiO_2	3.52
	Rutile	TiO_2	3.25
Zeolites	Laumontite	$Ca_4(Al_8Si_{16}O_{48})\cdot 16H_2O$	9.43*
	Chabazite	$Ca_2(Al_4Si_8O_{24})\cdot 12H_2O$	9.40*
	Stilbite	$NaCa_4(Al_9Si_{27}O_{72})\cdot 30H_2O$	9.12
	Mordenite	$Na_3KCa_2(Al_8Si_{40}O_{96})\cdot 28H_2O$	9.06*
	Heulandite	$(Na, K)Ca_4(Al_9Si_{27}O_{72})\cdot 24H_2O$	8.96
	Clinoptilolite	$(Na, K)_6(Al_6Si_{30}O_{72})\cdot 20H_2O$	8.96
	Phillipsite	$K_2(Ca_{0.5}Na)_4(Al_6Si_{10}O_{32})\cdot 12H_2O$	7.14*
	Thomsonite	$Na_4Ca_8(Al_{20}Si_{20}O_{80})\cdot 24H_2O$	6.63
	Mesolite	$Na_{16}Ca_{16}(Al_{48}Si_{72}O_{240})\cdot 64H_2O$	6.59
	Analcite	$Na_{16}(Al_{16}Si_{32}O_{96})\cdot 16H_2O$	5.61*
Sulphates	Gypsum	$CaSO_4\cdot 2H_2O$	7.61
	Bassanite	$Ca_2(SO_4)_2\cdot \frac{1}{2}H_2O$	6.01
	Alunite	$(K, Na)Al_3(SO_4)_2\cdot (OH)_6$	5.77*
	Mirabilite	$Na_2SO_4\cdot 10H_2O$	5.49
	Hexahydrate	$MgSO_4\cdot 6H_2O$	5.45*
	Jarosite	$KFe_3(SO_4)_2(OH)_6$	5.09*
	Thenardite	Na_2SO_4	4.66*
	Epsomite	$MgSO_4\cdot 7H_2O$	4.21
	Baryte	$BaSO_4$	3.45
Sulphides	Mackinawite	$(Fe, Ni)S$	5.03
	Marcasite	FeS_2	3.44
	Pyrite	FeS_2	2.71
Phosphates	Taranakite	$H_6K_3Al_5(PO_4)_8\cdot 18H_2O$	15.9
	Brushite	$CaHPO_4\cdot 2H_2O$	7.57
	Vivianite	$Fe_3(PO_4)_2\cdot 8H_2O$	6.80
	Leucophosphite	$K_2(Fe^{3+}, Al)_4(PO_4)_4(OH)_2\cdot 4H_2O$	6.79
	Plumbogummite	$PbAl_3(PO_4)_2(OH)_5\cdot H_2O$	5.73*
	Crandallite	$CaAl_3(PO_4)_2(OH)_5\cdot H_2O$	5.68*
	Strengite	$FePO_4\cdot 2H_2O$	5.50
	Variscite	$AlPO_4\cdot 2H_2O$	5.36*

i are required. This method has been successfully applied to the analysis of clay minerals (Gibbs, 1967), as have other schemes which have been developed from the same basic principles. For example, Norrish and Taylor (1962) demonstrated that if the mass-absorption coefficient of the sample to the radiation used was directly measured, then quantitative analysis was possible without internal standards or calibration charts. Again, Moore (1968) developed a method based upon the evaluation of so-called linear interaction factors for mineral pairs in a mixture. However, despite the theoretical soundness of each of these methods, they have been relatively little used by clay mineralogists, and certainly not on a routine basis, for large numbers of samples. The reason for this probably lies in the apparently time-consuming nature of preliminary work, the preparation and examination of internal/external standard mixtures, the determination of mass-absorption coefficients, etc. Coupled with this is the realization that, despite the utmost care at all stages, the accuracy of the final results ultimately depends upon obtaining standard minerals which are similar in their diffracting power to those being sought in the mixture. Gibbs (1967) circumvented this problem by using clay-mineral standards separated from the samples to be analysed, but this is a time-consuming procedure and cannot always be achieved in practice.

Many clay mineralogists have adopted compromise procedures which sacrifice accuracy, but enable a rapid, semiquantitative analysis of a clay to be made directly from an X-ray diffraction trace (Biscaye, 1965; Griffiths, 1967). Such procedures usually involve the use of weighting factors. For example, Johns *et al.* (1954) suggested that a weighting factor of four times the 10 Å peak area is necessary for the comparison of illite with montmorillonite. Again, Weaver (1960) suggested that for a comparison of illite with kaolinite, a factor of 2.5 times the 10 Å peak is necessary. In this type of approach, no account is taken of the presence of other minerals, or of the variable diffracting powder of individual mineral species which results from differences in composition, degree of crystallinity, etc. More accurate determination of relative abundances can be achieved by the addition of an internal standard (Heath and Pisias, 1979), or alternatively, the minerals analysed may be assigned intensity factors relative to a standard mineral such as quartz.

The author has applied the quantitative X-ray analysis methods of Chung (1974a, b) to clay minerals and to the commonly occurring subordinate minerals, because, in principle, the methods are capable of high precision and accuracy, even with poorly crystallized materials and are also simple to use. In addition, the construction of calibration curves is not required, and the intensity/concentration equations are free from the matrix effect. Needless to say, there are many special factors which need to be taken into consideration when these methods are applied to clays.

2.5.1 *The methods of Chung*

Chung developed two methods for the quantitative interpretation of multi-component mixtures. The first was described as the matrix-flushing method (Chung, 1974a). Absorption factors complicate the X-ray diffraction intensity–concentration equation, and therefore the best approach is to remove these factors by adding a flushing agent to the sample. This agent may be any pure compound not present in the sample. Chung arrived at the following equation, all the preceding steps being set out in the original paper:

$$Xi = \left(\frac{X_f}{ki}\right)\left(\frac{Ii}{I_f}\right)$$

where subscript f stands for flushing agent and k stands for Ii/Ic which is the reference intensity ratio relative to corrundum in a one-to-one weight ratio mixture. This is the basic working equation for quantitative multicomponent analysis. It is simple, free from matrix effects, requires no previous information regarding the approximate concentration ranges of the various components sought, and uses the intensity ratios derived from the same scan as the experimental data. In principle, it can also be used for the determination of amorphous material.

The second method, which is described as using the adiabatic principle (Chung, 1974b), eliminates the need for a flushing agent, and enables quantitative interpretation directly from the X-ray diffraction trace of a mixture. For quantitative analysis of a mixture of n components the basic working equation has the following form:

$$Xi = \left(\frac{ki}{Ii}\sum_{i=1}^{n}\frac{Ii}{ki}\right)^{-1} \tag{1}$$

The essential characteristics of the two methods of analysis can be summarized as follows. The matrix flushing-method involves the addition of a flushing agent, and can be used to analyse for specific components of interest, even in samples containing unidentified components or amorphous materials. The adiabatic method does not require the addition of a flushing agent and can only be applied to samples where all the components are identified. All components must be determined and the analysis is self-normalizing. In practice, the author has found that the adiabatic method is more usefully applied to clay materials.

2.5.2 *Quantitative analysis of non-clay minerals*

The use of Chung's methods for mineralogical analysis can be illustrated with reference to the highly crystalline non-clay minerals: quartz, orthoclase feldspar, labradorite, calcite and corundum (used as flushing agent). The minerals were ground to $< 10\,\mu m$, mixtures were homogenized and prepared for X-ray analysis by filling the sample-holders through a 100 mesh sieve, thus avoiding preferred orientation of the mineral grains. The *ki* factors for the various minerals were determined after taking the mean of three 10-second counts, less the background, on selected diffraction peaks in a 50/50 mixture with corundum. The diagnostic peaks used were quartz 3.33 Å, orthoclase feldspar 3.24 Å, labradorite 3.18 Å and calcite 3.03 Å, using divergent, receiving and antiscatter slits of 1°, 0.2 mm and 1° respectively. For corundum, the

Table 2.16 Quantitative determination of mineral mixtures using the matrix-flushing (1) and adiabatic (2) principles of Chung (1974a, b).

Sample	Composition	I (cps)[1]	*Ki*	% known	% found (1)	% found (2)
1	Quartz	7917	2.89	49.8	48.7	48.5
	Orthoclase	812	0.45	30.5	32.1	31.9
	Corundum	1109	1.00	19.7	—	19.6
					100.5	
2	Quartz	5031	2.89	30.3	32.5	30.3
	Labradorite	1039	0.56	31.3	34.6	32.3
	Orthoclase	653	0.45	25.2	27.1	25.2
	Corundum	707	1.00	13.2	—	12.3
					107.1	
3	Quartz	1591	2.89	29.9	30.9	31.7
	Orthoclase	213	0.45	30.0	26.6	27.2
	Calcite	883	1.65	30.0	30.1	30.8
	Corundum	178	1.00	10.0	—	10.2

[1] cps = counts per second.

reflection used was at 2.09 Å with slits at 2°, 0.2 mm and 2°. The results for three different mineral mixtures, determined by both the adiabatic and matrix flushing principles are shown in Table 2.16. It can be seen that the results are reasonably good. Sample 1 is used to present a typical calculation using the adiabatic principle and the previous equation.

$$\% \text{ quartz} = \frac{1}{1 + \dfrac{2.89}{7917}\left(\dfrac{812}{0.45} + \dfrac{1109}{1.00}\right)} = 48.5\%$$

$$\% \text{ orthoclase} = \frac{1}{1 + \dfrac{0.45}{812}\left(\dfrac{7917}{2.89} + \dfrac{1109}{1.00}\right)} = 31.9\%$$

$$\% \text{ corundum} = \frac{1}{1 + \dfrac{1.00}{1109}\left(\dfrac{7917}{2.89} + \dfrac{812}{0.45}\right)} = 19.6\%$$

Using the matrix-flushing method and equation (1) the calculation is as follows:

$$\% \text{quartz} = \frac{19.7}{2.89} \times \frac{7917}{1109} = 48.7\%$$

$$\% \text{orthoclase} = \frac{19.7}{0.45} \times \frac{812}{1109} = 32.1\%$$

2.5.3 *Quantitative analysis of clay minerals*

There are many difficulties associated with quantitative X-ray diffraction analysis of clay minerals, the most important of which is their inherent variability. In conjunction with this basic difficulty are secondary problems involving experimental technique, especially specimen preparation.

2.5.3.1 *Sample preparation.* Quantitative X-ray analysis of clays requires a simple and reproducible method of specimen preparation. Many authors have developed techniques for ensuring that the minerals in a sample are randomly oriented. However, the disadvantage of these procedures is that the reduction in the intensity of the diagnostic basal reflections of the clay minerals is so great that detection limits for these minerals can become impracticably high and even their identification is made difficult. The technique of drying clay/water suspensions onto glass slides takes advantage of the enhanced intensities of the basal reflections that result from preferred orientation, the only proviso being that such a procedure should be reasonably reproducible. Rapid drying on a hot-plate at 50–60°C is recommended to minimize preferential concentration of fine-grained minerals, such as montmorillonite, at the slide surface.

2.5.3.2 *Choice of reference intensity material.* As pointed out by Chung (1974b), ideally, the reference intensity material should have the same level of crystallinity as the components being sought for in the sample. For clay-mineral analysis it is impossible for any single reference material to be always appropriate, because clays frequently consist of mixtures of highly ordered and poorly ordered minerals. All that can be done here is to adopt a compromise. The reference material used in this study was synthetic boehmite, which has a platy morphology similar to that of the clay minerals and yields an intense, conveniently measured reflection at 6.2 Å.

2.5.3.3 *Clay-mineral variability.* The inherent variability of clay minerals means that chemical composition, crystallinity, grain size, morphology, interstratification, etc., may differ greatly within clay-mineral groups, and even within individual species of the groups. Consequently, the diffracting power of a particular clay mineral also varies greatly. In terms of Chung's reference-intensity method this implies that the use of a single Ki value for a clay mineral, irrespective of geological source, cannot be regarded as valid. Consideration of Table 2.17 emphasizes this point. It can be seen that the Ki value of the 7 Å reflection of kaolinite ranges from 0.07 to 0.45, the 10 Å reflection of illite/mica from 0.05 to 0.43, the 14 Å reflection of chlorite from 0.39 to 0.99 and the 15 Å reflection of Ca-montmorillonite from 0.17 to 0.55. These values are based upon peak-height measurements of the above reflections, but a similar spread was found using peak areas. The samples were prepared by packing the powder into an aluminium holder, using a hydraulic press with a pre-set

Table 2.17 *ki* factors for kaolinite, illite, Ca-montmorillonite and chlorite, using the 7.14, 10.0, 15.1 and 14.4 Å reflections respectively, with reference to the 6.11 Å reflection of boehmite. Samples were pressed into an aluminium holder.

Clay mineral	Source/nature		*ki*
Kaolinite	Carboniferous fireclay, Scotland, poorly ordered		0.07
Kaolinite	Sedlec, Czechoslovakia, moderately ordered		0.25
Kaolinite	Carboniferous sandstone, Scotland, well crystallized		0.45
Illite	Le Puy en Velay, France	$2M_1$ (90%)	0.04
Illite	Fithian, Illinois	1Md	0.04
Illite	Old Red Sandstone, Scotland	1M	0.06
Illite	Ballater, Scotland. Hydrothermal		0.10
Muscovite		$2M_1$	0.43
Montmorillonite	Mississippi. Bentonite		0.17
Montmorillonite	Arizona. Bentonite		0.19
Montmorillonite	Dordrecht, South Africa. Bentonite		0.29
Montmorillonite	Texas. Bentonite		0.31
Montmorillonite	Wyoming. Bentonite		0.55
Chlorite	Wanibuchi, Japan		0.23
Chlorite	Penninite		0.53
Chlorite	Salida		0.72
Chlorite	Prochlorite		0.99

Table 2.18 *ki* factors used for prepared clay-mineral mixtures made up as oriented aggregates on glass slides and with boehmite as the reference mineral.

Mineral	Source	Reflection used (Å)	*ki* factor
Kaolinite	Unknown	7.15	2.66
Illite	Fithian	10.0	0.41
Montmorillonite	Mississippi	17.8	0.52
Chlorite	Unknown	14.3	0.25
Boehmite	Synthetic	6.1	1.00

pressure, in an attempt to induce a reproducible degree of preferred orientation.

2.5.3.4 *Analysis of prepared sedimented aggregates.* The utility of reference-intensity methods for clay-mineral analysis was assessed by examining known mixtures made up in the form of sedimented aggregates. The *ki* factors of the minerals were first determined on Mg-saturated, glycerol-solvated 50/50 mixtures with boehmite (Table 2.18). Divergent, receiving and antiscatter slits of 1/2°, 0.2 mm and 1/2°, respectively, were used for kaolinite, illite and boehmite, and 1/4°, 0.2 mm, 1/4° for chlorite and montmorillonite. Duplicate aggregates were scanned at the rate of 1/2° 2Θ/min, the diffraction patterns recorded and peak heights measured directly from the chart above a visually estimated background. Where kaolinite and chlorite occur together there can be a problem with peak overlap at 7 Å. Here, a proportionation procedure was

Table 2.19 Quantitative determination of clay-mineral mixtures in oriented aggregates on glass slides, using the matrix-flushing (1) and adiabatic (2) principles of Chung (1974a, b). Intensities measured from a visually estimated background.

Sample	Composition	I (mm)	% known	% found (1)	% found (2)
1	Kaolinite	480	28.7	30.7	28.8
	Illite	92	30.4	38.1	35.8
	Montmorillonite	79	28.9	25.8	24.2
	Boehmite	70	11.9	—	11.2
				106.5	
2	Kaolinite	376	28.2	27.1	29.4
	Illite	48	31.0	22.5	24.4
	Chlorite	40	29.0	30.7	33.3
	Boehmite	62	11.9	—	12.9
				92.2	
3	Kaolinite	514	27.5	35.8	31.0
	Montmorillonite	97	29.2	34.6	29.9
	Chlorite	42	29.2	31.2	26.9
	Boehmite	76	14.1	—	12.2
				115.7	
4	Illite	54	29.8	24.3	25.5
	Montmorillonite	65	26.6	23.1	24.2
	Chlorite	47	30.4	34.7	36.5
	Boehmite	71	13.1	—	13.7
				95.2	

used, based on the relative intensities of the more easily separable 002 kaolinite and 004 chlorite reflections. The values in Table 2.19 show that reasonable results can be obtained using reference intensity methods and the simple procedure above, provided that the *ki* factors are known. Even better results might be obtained by correcting for transparency and matrix effects according to the iterative procedure of Davis (1980).

2.5.3.5 *Quantitative analysis of unknown clay materials.* The normal situation is, of course, one where the *ki* factors of the clay minerals to be analysed are unknown. What can be done in these circumstances? Realistically, true quantitative analysis, on a routine basis by reference intensity methods, is a practical impossibility, largely because of the problem of clay-mineral variability. This problem also affects all the other methods of quantitative analysis by X-ray diffraction. So far we have been dealing with well-characterized 'standard' minerals, and have not taken into account the effect of complicating factors such as interstratification between different types of clay minerals, or interlayer intercalation of non-exchangeable hydroxy species. Such complications are all too common in soils and sediments. Probably the best tactic is to adopt a compromise procedure which enables a semiquantitative analysis that is not based entirely upon arbitrary assumptions. The

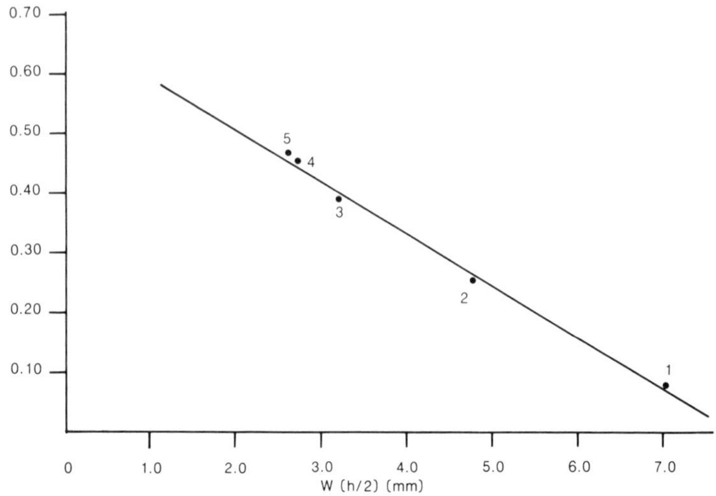

Figure 2.31 Plot of *ki* factor for different kaolinites against peak width at half height, $W(h/2)$ for 001 reflection. 1. Fireclay kaolinite, Scotland, 2. Sedlec kaolinite, Czechoslovakia, 3. Osmo kaolinite, 4. highly crystalline kaolinite, Scotland, 5. highly crystalline kaolinite, Pugu, Tanzania

author has attempted to make a more meaningful assessment of the *ki* factors of the clay minerals to be analysed by taking into account factors such as crystallinity and interstratification. For example, Figure 2.31 shows that the *ki* value for the 001 reflection of different kaolinites varies inversely with peak width at half height, the latter being broadly correlated with crystallinity. Again, the *ki* values of illite vary inversely with peak width at half height, which may correlate roughly with interstratified expansible layers. Certainly, a different *ki* factor must be used for a muscovite-type mica with a sharp, intense 10 Å reflection compared with an illite with marked asymmetry of its 10 Å peak. Such observations remain to be systematized, but they do indicate how the choice of a particular *ki* value might be rationalized.

References

Abdul-Latif, N. and Weaver, C.E. (1969) Kinetics of acid-dissolution of palygorskite (attapulgite) and sepiolite. *Clays & Clay Miner.*, **17**, 169–178.

Alexander, L. and Klug, H.P. (1948) Basic aspects of X-ray absorption in quantitative diffraction analysis of powder mixtures. *Anal. Chem.*, **20**, 886–889.

Alietti, A. (1958) Some interstratified clay minerals of the Taro Valley. *Clay Miner. Bull.*, **3**, 207–211.

Bailey, S.W. (1980) 'Structures of layer silicates,' in *Crystal Structures of Clay Minerals and their X-ray Identification*, (G.W. Brindley and G. Brown eds.), Mineralogical Society, London, 1–123.

Bailey, S.W. and Brown, B.E. (1962) Chlorite polytypism. I. Regular and semi-random one layer structures. *Am. Mineral.*, **47**, 819–850.

Bain, D.C. and Russell, J.D. (1981) Swelling minerals in a basalt and its weathering products from Morvern, Scotland. II. Swelling chlorite. *Clay Miner.*, **16**, 203–212.

Barshad, I. (1948) Vermiculite and its relation to biotite as revealed by base exchange reactions. X-ray analyses, differential thermal curves and water content. *Am. Mineral.*, **33**, 655–678.
Barshad, I. (1950) The effect of interlayer cations on the expansion of mica-type crystal lattices. *Am. Mineral.*, **35**, 225–238.
Barshad, I. (1952) Factors affecting the interlayer expansion of vermiculite and montmorillonite with organic substances. *Proc. Soil Sci. Soc. Am.*, **16**, 176–182.
Biscaye, P.E. (1965) Mineralogy and sedimentation of recent deep-sea clay in the Atlantic Ocean and adjacent seas and oceans. *Geol. Soc. Am. Bull.*, **76**, 803–831.
Brindley, G.W. (1955) Stevensite, a montmorillonite-type mineral showing mixed-layer characteristics. *Am. Mineral.*, **40**, 239–247.
Brindley, G.W. (1956) Allevardite, a swelling double layer mica mineral. *Am. Mineral.*, **41**, 91–103.
Brindley, G.W. (1959) X-ray and electron diffraction data for sepiolite. *Am. Mineral.*, **44**, 495–500.
Brindley, G.W. (1961) 'Kaolin, serpentine and kindred minerals', in *The X-ray Identification and Crystal Structures of Clay Minerals*, (G. Brown, ed.), Mineralogical Society, London, 51–131.
Brindley, G.W. (1966) Ethylene glycol and glycerol complexes of smectites and vermiculites. *Clay Miner.*, **6**, 237–260.
Brindley, G.W. (1980) 'Quantitative X-ray analysis of clays,' in *Crystal Structures of Clay Minerals and their X-ray Identification*, (G.W. Brindley and G. Brown eds.), Mineralogical Society, London, 411–438.
Brindley, G.W. and Brown, G. (eds.) (1980) *Crystal Structures of Clay Minerals and their X-ray Identification*, Mineralogical Society, London.
Brindley, G.W. and Kurtossy, S.S. (1961) Quantitative determination of kaolinite by X-ray diffraction. *Am. Mineral.*, **46**, 1205–1215.
Brindley, G.W. and Porter, A.R.D. (1978) Occurrence of dickite in Jamaica—ordered and disordered varieties. *Am. Mineral.*, **63**, 554–562.
Brown, G. (1953) The dioctahedral analogue of vermiculite. *Clay Miner. Bull.*, **2**, 64–69.
Brown, G. (1980) 'Associated minerals', in *Crystal Structures of Clay Minerals and their X-ray Identification*, (G.W. Brindley and G. Brown eds.) Mineralogical Society, London, 361–410.
Brown, G. and Weir, A.H. (1963) 'The identity of rectorite and allevardite', in *Proc. Intern. Clay Conf., Stockholm.* Vol. 1, Pergamon Press, Oxford 27–35.
Brown, G., Edwards, B., Omerod, E.C. and Weir, A.H. (1972) A simple diffractometer heating stage. *Clay Miner.*, **9**, 407–413.
Brown, G. and Brindley, G.W. (1980) 'X-ray diffraction procedures for clay mineral identification,' in *Crystal Structures of Clay Minerals and their X-ray Identification*, (G.W. Brindley and G. Brown eds.) Mineralogical Society, London, 305–359.
Brown, G. and Wood, I.G. (1985) Estimation of iron oxides in soil clays by profile refinement combined with differential X-ray diffraction. *Clay Miner.*, **20**, 15–27.
Brown, G., Bourgignon, P. and Thorez, J. (1974) A lithium bearing aluminium regular mixed layer montmorillonite-chlorite from Huy, Belgium. *Clay Miner.*, **10**, 135–144.
Buckley, H.A., Bevan, J.C., Brown, K.M., Johnson, L.R. and Farmer, V.C. (1978) Glauconite and celadonite: two separate mineral species. *Miner. Mag.*, **42**, 373–382.
Buckley, H.A., Easton, A.J. and Johnson, L.R. (1984) Compositional variation in glauconite. *Miner. Mag.*, **48**, 119–126.
Bühmann, C., Fey, M.V. and de Villiers, J.M. (1985) Aspects of the X-ray identification of swelling clay minerals in soils and sediments. *S. African. J. Sci.*, **81**, 505–509.
Bystrom-Brusewitz, A.M. (1975) Studies of the Li-test to distinguish beidellite and montmorillonite. *Proc. Intern. Clay Conf., 1972, Mexico City.* Applied Publishing Ltd., Wilmette, Illinois, USA, 419–428.
Campbell, A.S. and Schwertmann, U. (1985) Evaluation of selected dissolution extractants in soil chemistry and mineralogy by differential X-ray diffraction. *Clay Miner.*, **20**, 515–519.
Chukhrov, F.V., Zvyagin, B.B., Gorshkov, A.I., Yermilova, L.P., Korovushkin, V.V., Rudnitskaya, Ye. S. and Yakubovskaya, N.Yu. (1976) Feroxyhyte, a new modification of FeOOH. *Intern. Geology. Rev.*, **19**, 873–890.
Chung, F.H. (1974a) Quantitative interpretation of X-ray diffraction patterns of mixtures. I. Matrix flushing method for quantitative multi-component analysis. *J. Appl. Crystallog.*, **7**, 519–525.
Chung, F.H. (1974b) Quantitative interpretation of X-ray diffraction patterns of mixtures. II. Adiabatic principle of X-ray diffraction analysis of mixtures. *J. Appl. Crystallog.*, **7**, 526–531.

Churchman, G.J., Whitton, J.S., Claridge, G.G.C. and Theng, B.K.G. (1984) Intercalation method using formamide for differentiating halloysite from kaolinite. *Clays & Clay Miner.*, **32**, 241–248.

Cowking, A., Wilson, M.J., Tait, J.M. and Robertson, R.H.S. (1983) Structure and swelling of fibrous and granular saponitic clay from Orrock Ouarry, Fife, Scotland. *Clay Miner.*, **18**, 49–64.

Cullity, B.D. (1956) *Elements of X-ray Diffraction.* Addison-Wesley Publishing Company., Reading, Massachussetts, USA.

Davis, B.L. (1980) 'Standardless' X-ray diffraction quantitative analysis. *Atmospheric Environment*, **14**, 217–220.

Dixon, J.B. and Jackson, M.L. (1959) Dissolution of interlayers from intergradient soil clays after pre-heating at 400°C. *Science*, **129**, 1616–1617.

Earley, J.W., Brindley, G.W., McVeagh, W.J. and Vanden Heuvel, R.C. (1956) A regularly interstratified montmorillonite-chlorite. *Am. Mineral.*, **41**, 258–267.

Eslinger, E.V., Mayer, L.M., Durst, T.L., Hower, J. and Savin, S.M. (1973) An X-ray technique for distinguishing between detrital and secondary quartz in the fine-grained fraction of sedimentary rocks. *J. Sedim. Petrol.*, **43**, 540–543.

Farmer, V.C. (1978) 'Water on particle surfaces,' in *The Chemistry of Soil Constituents.* (D.J. Greenland and M.H.B. Hayes eds.). Wiley, New York, 405–448.

Faust, G.T., Hathaway, J.C. and Millot, G. (1959) A restudy of stevensite and allied minerals. *Am. Mineral.*, **44**, 342–370.

Follett, E.A.C., McHardy, W.J., Mitchell, B.D. and Smith, B.F.L. (1965) Chemical dissolution techniques in the study of clays. Part I. *Clay Miner.*, **6**, 23–24.

Gibbs, R.J. (1965) Error due to separation in quantitative chemical mineral X-ray diffraction mounting techniques. *Am. Mineral.*, **50**, 741–751.

Gibbs, R.J. (1967) Quantitative X-ray diffraction analysis using clay mineral standards extracted from the samples to be analyzed. *Clay Miner.*, **7**, 79–90.

Griffiths, J.C. (1967) *Scientific Method in the Analysis of Sediments.* McGraw-Hill, New York, 508 pp.

Grim, R.E., Bray, R.H. and Bradley, W.F. (1937) The mica in argillaceous sediments. *Am. Mineral.*, **22**, 813–829.

Hauff, P.L., Pevear, D.R., Blackmon, P.D., Reynolds, R.C. and Phillips, R.E. (1982) *Oriented Sample Mounts for the Analysis of Clay Minerals by X-ray Diffraction.* Proposed U.S. Geol. Surv. Open File Rep.

Hayashi, H., Otsuka, R. and Imai, N. (1969) Infrared study of sepiolite and palygorskite on heating. *Am. Mineral.*, **53**, 1613–1624.

Heath, G.R. and Pisias, N.G. (1979) A method for the quantitative estimation of clay minerals in North Pacific deep sea sediments. *Clays & Clay Miner.*, **27**, 175–184.

Hinckley, D.N. (1963) Variability in 'crystallinity' values among the kaolin deposits of the coastal plain of Georgia and South Carolina. *Clays & Clay Miner.*, **11**, 229–235.

Hofmann, U. and Klemen, R. (1950) Verlust der Austauschfahigeit von Lithiumionen an Bentonit durch Erhitzung. *Zeit. Anorganische Chem.*, **262**, 95–99.

Hughes, R. and Bohor, B. (1970) Random clay powders prepared by spray drying. *Am. Mineral.*, **55**, 1780–1786.

Hughes, J.C. and Brown, G. (1979) A crystallinity index for soil kaolins and its relation to parent rock, climate and soil maturity. *J. Soil Sci.*, **30**, 557–563.

Johns, E.D., Grim, R.E., and Bradley, W.F. (1954) Quantitative estimations of clay minerals by diffraction methods. *J. Sediment. Petrol.*, **24**, 242–251.

Jonas, E.C. and Kuykendall (1966) Preparation of montmorillonite for random powder diffraction. *Clay Miner.*, **6**, 232–235.

Kinter, E.B. and Diamond, S. (1956) A new method for preparation and treatment of oriented aggregate specimens of soil clays for X-ray diffraction analysis. *Soil Sci.*, **81**, 111–120.

Klug, H.P. and Alexander, L.E. (1974) *X-ray Diffraction Procedures.* Wiley, New York.

Kubler, B. (1964) Les argiles, indicateurs de métamorphisme. *Rev. Inst. Franc. Pétrole*, **19**, 1093–1112.

Lagaly, G. and Weiss, A. (1970) 'Inhomogeneous charge distributions in mica-type layer silicates,' in *Proc. Réunion Hispano-Belge Minerales de la Arcilla.* Consejo Superior de Investigaciones Cientificos, Madrid, 179–287.

Lagaly, G., Fernandez-Gonzalez, M. and Weiss, A. (1976) Problems in layer charge determination of montmorillonite. *Clay Miner.*, **11**, 173–187.

MacEwan, D.M.C. (1958) Fourier transform methods for studying X-ray scattering from lamellar systems. II. The calculation of X-ray diffraction effects for various types of interstratification. *Kolloidzeitschrift*, **156**, 61–67.

MacEwan, D.M.C. and Wilson, M.J. (1980) 'Interlayer and intercalation complexes in clay minerals,' in *Crystal Structures of Clay Minerals and their X-ray Identification*, (G.W. Brindley and G. Brown eds.) Mineralogical Society, London, 197–248.

Martin-Vivaldi, J.L. and Linares Gonzalez, J. (1962) A random intergrowth of sepiolite and attapulgite. *Clays & Clay Miner.*, **9**, 592–602.

Martin-Vivaldi, J.L. and MacEwan, D.M.C. (1957) Triassic chlorite from the Jura and Catalan coastal range. *Clay Miner. Bull.*, **3**, 177–183.

Martin-Vivaldi, J.L., MacEwan, D.M.C. and Rodriguez-Gallego, M. (1963) *Proc. Intern. Clay Conf., Stockholm.* Vol. I, Pergamon Press, Oxford, 45–51.

Mathieson, A.McL. and Walker, G.F. (1954) Crystal structure of Mg-vermiculite. *Am. Mineral.*, **39**, 231–255.

Maxwell, D.T. and Hower, J. (1967) High grade diagenesis and low grade metamorphism of illite in the Precambrian Belt series. *Am. Mineral.*, **52**, 843–857.

Méring, J. (1949) L'interference des rayons X dans les systèmes à stratification desordonées. *Acta Crystallog.*, **2**, 371–377.

Milne, I.H. and Warshaw, C.M. (1956) Methods of preparation and control of clay mineral specimens in X-rays diffraction analysis. *Clays & Clay Miner.*, **4**, 22–30.

Moore, C.A. (1968) Quantitative analysis of naturally occurring multicomponent mineral systems by X-ray diffraction. *Clays & Clay Miner.*, **16**, 325–336.

Nadeau, P.H. and Bain, D.C. (1986) Composition of some smectites and diagenetic illitic clays and implications for their origin. *Clays & Clay Miner.*, (in press).

Nadeau, P.H., Wilson, M.J., McHardy, W.J. and Tait, J.M. (1984) Interparticle diffraction: a new concept for interstratified clays. *Clay Miner.*, **19**, 757–769.

Nagata, H., Shimoda, S. and Sudo, T. (1974) On dehydration of bound water of sepiolite. *Clays & Clay Miner.*, **22**, 285–293.

Nemecz, E. (1981) *Clay Minerals*, Akademiai Kiado, Budapest, Hungary.

Norrish, K. (1968) Some phosphate minerals of soils. *Trans. 9th Intern. Congr. Soil Sci.*, **III**, 713–723.

Norrish, K. and Taylor, R.M. (1961) Isomorphous replacement of iron by aluminium in soil goethites. *J. Soil Sci.*, **12**, 294–306.

Norrish, K. and Taylor, R.M. (1962) Quantitative analysis by X-ray diffraction. *Clay Miner. Bull.*, **5**, 98–109.

Novich, B.E. and Martin, R.T. (1983) Solvation methods for expandable layers. *Clays & Clay Miner.*, **31**, 235–238.

Nuffield, E.W. (1966) *X-ray Diffraction Methods*, Wiley, New York.

Odom, I.E. (1984) 'Glauconite and celadonite minerals,' in *Micas*, (*Rev. Mineral.*, **13**) (S.W. Bailey ed.) Mineralogical Society of America, 545–572.

Parrish, W. and Taylor, J.A. (1956) Beta filters for X-ray diffractometry. *Norelco Reporter*, **3**, 105–106.

Paterson, E., Bunch, J.L., and Duthie, D.M.L. (1986) Preparation of randomly oriented samples for X-ray diffractometry. *Clay Miner.*, **21**, 101–106.

Plançon, A. and Tchoubar, C. (1977) Part II. Nature and proportion of defects in natural kaolinites. *Clays & Clay Miner.*, **25**, 436–450.

Range, K.J., Range, A. and Weiss, A. (1969) Fireclay type kaolinite or fire-clay mineral? Experimental classification of kaolinite-halloysite minerals. *Proc. Intern. Clays Conf.*, 1969, Tokyo. **1**, Israel Universities Press, Jerusalem, 3–13.

Reynolds, R.C. (1967) Interstratified clay systems: calculation of the total one-dimensional diffraction function. *Am. Mineral.*, **50**, 990–1001.

Reynolds, R.C. (1980) 'Interstratified clay minerals,' in *Crystal Structures of Clay Minerals and their X-ray Identification*, (G.W. Brindley and G. Brown eds.). Mineralogical Society, London, 249–303.

Reynolds, R.C., and Hower, J. (1970) The nature of interlayering in mixed layer illite-montmorillonite. *Clays & Clay Miner.*, **18**, 25–36.

Rich, C.I. (1969) Suction apparatus for mounting clay specimens on ceramic tile for X-ray diffraction. *Proc. Soil Sci. Soc. Am.*, **33**, 815–816.

Shaw, H.F. (1972) The preparation of oriented clay mineral specimens for X-ray diffraction analysis by a suction-onto-ceramic tile method. *Clay Miner.*, **9**, 349–350.

Smith-Aitken, W.W. (1965) An occurrence of phlogopite and its transformation to vermiculite by weathering. *Miner. Mag.*, **35**, 151–164.

Srodon, J. (1980) Precise identification of illite/smectite interstratifications by X-ray powder diffraction. *Clays & Clay Miner.*, **28**, 401–411.

Srodon, J. (1984) X-ray identification of illitic materials. *Clays & Clay Miner.*, **32**, 337–349.

Srodon, J. and Eberl, D.D. (1984) 'Illite', in *Micas*, (*Rev. Miner.*, **13**) (S.W. Bailey ed.) Mineralogical Society of America, 495–544.

Stephen, I. and MacEwan, D.M.C. (1951) Some chloritic minerals of unusual type. *Clay Miner. Bull.*, **1**, 157.

Sterne, E.J., Reynolds, R.C. and Zantop, H. (1982) Natural ammonium illites from black shales hosting a stratiform base metal deposit. Delong Mountains, northern Alaska. *Clays & Clay Miner.*, **30**, 161–166.

Suquet, H., de la Calle, C. and Pezerat, H. (1975) Swelling and structural organization of saponite. *Clays & Clay Miner.*, **23**, 1–9.

Tamura, T. (1956) Weathering of mixed-layer clays in soils. *Clays & Clay Miner.*, **4**, 413–422.

Thomson, A.P., Duthie, D.M.L. and Wilson, M.J. (1972) Randomly oriented powders for quantitative determination of clay minerals. *Clay Miner.*, **9**, 345–348.

Thompson, G.R. and Hower, J. (1975) The mineralogy of glauconite. *Clays & Clay Miner.*, **23**, 289–300.

Tien, Pei-lin (1974) A simple device for smearing clay-on-glass slides for quantitative X-ray diffraction studies. *Clays & Clay Miner.*, **22**, 367–368.

Veniale, F. and van der Marel, H.W. (1963) An interstratified saponite-swelling chlorite mineral as a weathering product of lizardite rock from S. Margherita Staffora (Pavia Province), Italy. *Beitr. Mineral. u. Petrogr.* **9**, 198–245.

Veniale, F. and van der Marel, H.W. (1968) A regular talc-saponite mixed layer mineral from Ferriere, Nure Valley (Piacenza Province, Italy). *Contrib. Miner. Petrol.*, **17**, 237–254.

Veniale, F. and van der Marel, H.W. (1969) Identification of some 1:1 regular interstratified trioctahedral clay minerals. *Proc. Intern. Clay Conf., Tokyo*, 233–244.

Wada, K. (1965) Intercalation of water in kaolin minerals. *Am. Mineral.*, **50**, 924–941.

Walker, G.F. (1950) Trioctahedral minerals in soil clays of north-east Scotland. *Miner. Mag.*, **29**, 72–84.

Walker, G.F. (1961) 'Vermiculite minerals,' in *The X-ray Identification and Crystal Structures of Clay Minerals*, (G.W. Brindley and G. Brown eds.) Mineralogical Society, London, 297–324.

Weaver, C.E. (1960) Possible uses of clay minerals in the search for oil. *A.A.P.G. Bull.*, **44**, 1505–1518.

Wilson, M.J. (1987) Soil smectites and associated interstratified minerals: recent developments. *Proc. 1985 Intern. Clay Conf. Denver, USA* (in press).

Wilson, M.J., Russell, J.D. and Tait, J.M. (1974) A new interpretation of the structure of disordered α-cristobalite. *Contr. Miner. Petrol.*, **47**, 1–6.

Wilson, M.J. and Nadeau, P.H. (1985) 'Interstratified clay minerals and weathering processes,' in *The Chemistry of Weathering*, (J.I. Drever ed.) D. Reidel Publishing Co., Dordrecht, 97–118.

3 Thermal analysis

E. PATERSON AND R. SWAFFIELD

3.1 Introduction

Of all the techniques currently in use for the characterization of clays, thermal analysis is one of the oldest. Interest in the effect of heat on minerals stems from the realization by early man that the physical and chemical properties of minerals could be readily modified using controlled amounts of heat. Thus, smelting of various ores was practised some 5000 years ago (Mackenzie, 1984a). At the turn of the present century, some of the links between mineralogy and thermal analysis were being forged through the work of Roberts-Austen, Le Chatelier and others. As Mackenzie (1984a) has pointed out, the earlier work of Josiah Wedgwood on temperature measurement provides an often-overlooked link between mineralogy, ceramics and thermal analysis. Clearly, the driving force in all these early works was empiricism, a factor that still plays an important role in current investigations.

The primary use of thermoanalytical methods in present-day mineralogy lies in the qualitative and quantitative analysis of mineral mixtures. In this role, thermoanalytical methods are complementary to the principal investigative tool, X-ray diffraction (XRD). The complementary nature of the two techniques reflects their contrasting backgrounds; XRD derives its strengths from the absolute nature of the measurements, and the direct representation of mineral structure and identity, whereas in thermal analysis many physical, chemical and structural factors contribute to the results. In consequence, no simple, straightforward relationship exists between the thermoanalytical properties of a mineral and any single physical or chemical property. Given an appreciation of this fact, and consequent appropriate usage, thermal analysis can be of considerable value in mineralogy. Indeed, it may be argued that XRD tends to oversimplify the physical description of minerals. In general, while XRD is particularly useful in establishing similarities between samples, thermoanalytical methods tend to emphasize the differences. In this regard, heat treatments are widely used in XRD to establish differences in mineralogical composition and properties.

This chapter attempts to briefly outline the multiplicity of available thermoanalytical methods, to describe the basic principles of instrumentation and methodology, and to indicate their value in various aspects of mineralogical investigation.

3.2 Thermoanalytical methods

Thermal analysis is actually a group of techniques, in which a physical property of a substance and/or its reaction products is measured as a function of temperature, while the substance is subjected to a controlled temperature programme (Lombardi, 1980). Clearly, this definition covers a very wide range of techniques, ranging from very widely used and well-established methods, such as differential thermal analysis, to those at a very early stage of their development, and as yet, untested in mineralogical investigations, for example thermosonimetry. This chapter concentrates on the more commonly encountered thermoanalytical techniques and their use in characterization of clays, concentrating on methods involving measurement of the thermal properties, *per se*, of the sample, and on other methods concerned, for example, with decomposition reactions involving the loss of volatile products, or with dimensional changes.

3.2.1 *Methods measuring thermal properties*

3.2.1.1 *Differential thermal analysis (DTA).* The physical parameter measured in DTA is ΔT, the difference in temperature between the sample and reference material, while both are subjected to a controlled temperature programme. The experimental record is the DTA curve, in which, by convention, ΔT is plotted on the ordinate with endothermic peaks pointing downward, while on the abscissa, time or temperature, either that of the reference material or of the sample, is recorded increasing from left to right (Figure 3.1).

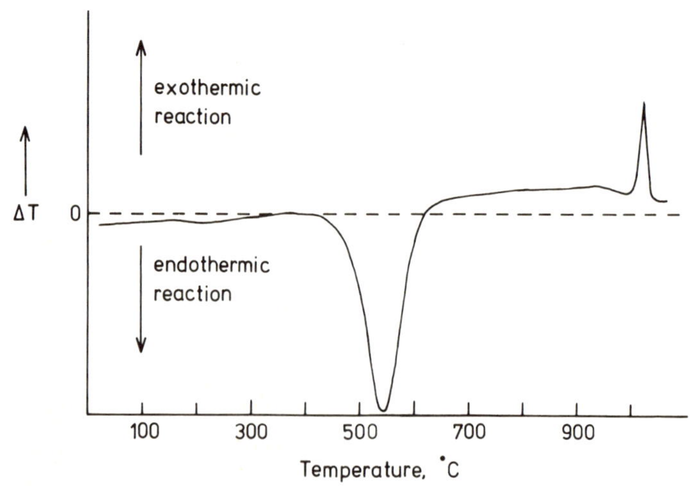

Figure 3.1 Recommended presentation for DTA curve (Lombardi, 1980)

3.2.1.2 *Differential scanning calorimetry (DSC).* The physical parameter measured in DSC is the difference in energy inputs into the sample and reference material, when both are subjected to a controlled temperature programme. The development of two distinct methods, by which measurement is carried out, has resulted in a subdivision of DSC into heat-flux and power-compensation methods. These two distinct approaches use different methods for reporting their curves. For the heat-flow instruments, the same convention is followed as that used in DTA, with endothermic peaks pointing downward and temperature increasing from left to right (Figure 3.2(a)).

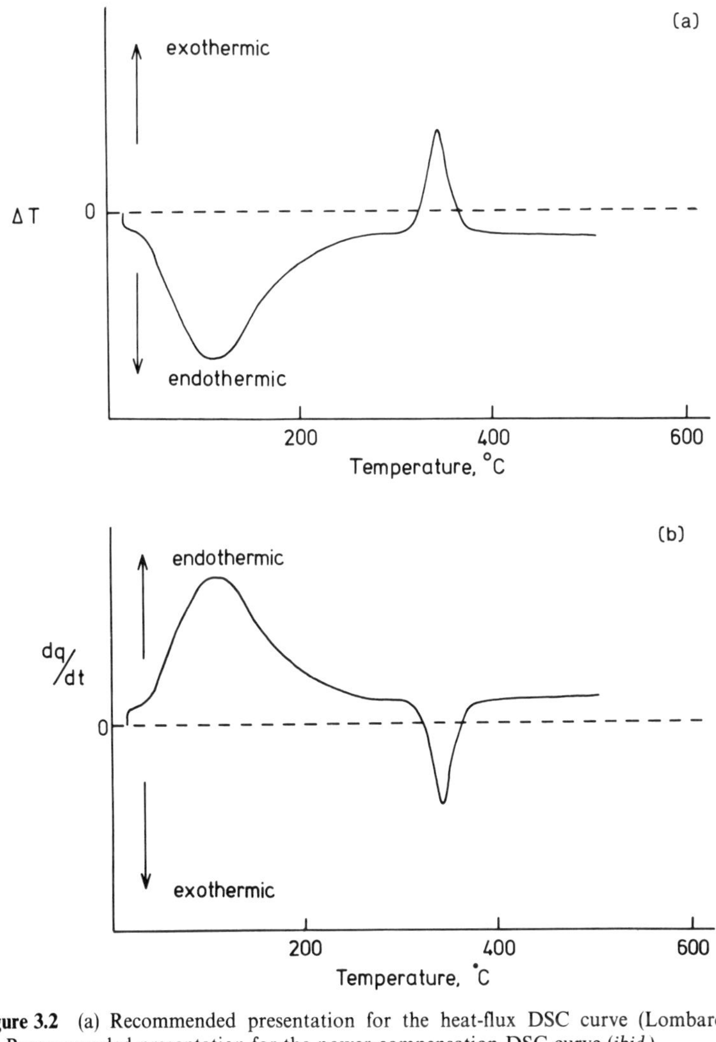

Figure 3.2 (a) Recommended presentation for the heat-flux DSC curve (Lombardi, 1980) (b) Recommended presentation for the power-compensation DSC curve (*ibid.*)

However, in power-compensation DSC curves, a calorimetric convention is followed, where endothermic effects result in an upward displacement of the ordinate, while on the abscissa, temperature or time increases from left to right (Figure 3.2(b)).

3.2.2 *Other methods*

3.2.2.1 *Thermogravimetry (TG).* The measured parameter in TG is the mass of the sample, and in reporting the curve, the mass of the sample should be recorded on the ordinate decreasing downwards, while on the abscissa, the temperature or time should increase from left to right (Figure 3.3(a)). In contrast to DTA and DSC, TG curves are integral curves and often it is useful to recast the data in a derivative form. The curve thus obtained, whether the differentiation is performed manually, by analogue electronics, or by digital computer, is the derivative thermogravimetric (DTG) curve (Figure 3.3(b)).

3.2.2.2 *Evolved gas detection/analysis.* In evolved gas *detection* (EGD), the evolution of gas is detected, whereas in evolved gas *analysis* (EGA), the nature and/or amount of gas is measured as a function of temperature, while the substance is subjected to a controlled temperature programme.

3.2.2.3 *Thermodilatometry.* In this technique, a dimension of a substance is measured, under negligible load, as a function of temperature, while the substance is subjected to a controlled temperature programme.

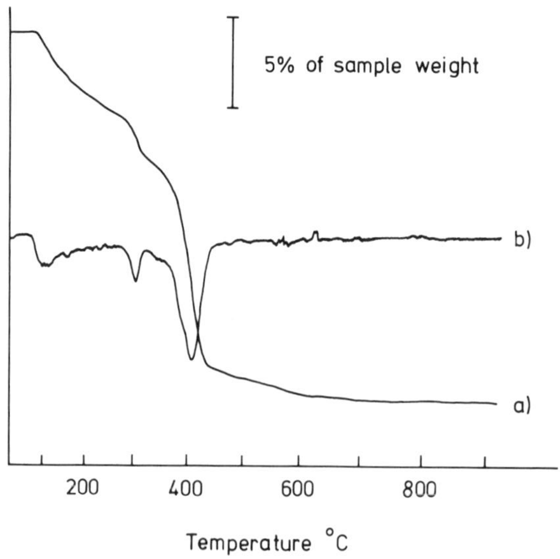

Figure 3.3 (a) TG and (b) DTG curves in recommended format

3.2.3 *Combined/simultaneous techniques*

Frequently, it is found that in order to elucidate fully the thermal behaviour of a substance, the results from different thermoanalytical methods have to be compared. One method of achieving this is to combine the facilities for monitoring more than one property of the heated sample into one apparatus. Of the various possible permutations, most include DTA as one member and, although combination with evolved gas detection or analysis is in some ways the simplest to achieve, the most popular simultaneous methods are DTA/TG/DTG.

3.2.4 *Conclusions*

Of the techniques summarized in section 3.2, those in the first group are generally the most versatile. In fact, DTA occupies a pre-eminent role in the study of clays. There are a number of reasons for this. In DTA and DSC the curves reflect changes in the thermal properties of the sample, and can indicate the occurrence of a very wide range of reaction types, including solid-state phase changes, decomposition reactions and removal of sorbed species. On this basis, DSC is potentially just as versatile as DTA, but the temperature range is generally more restricted and this introduces some limitations in its use. The other methods mentioned cannot characterize such a wide range of reaction types and are of more restricted use. This applies particularly to thermodilatometry, which is included mainly because of its empirical use in the ceramics industry.

3.3 Instrumentation

From the general definition of thermal analysis it is clear that all thermoanalytical instrumentation must meet two main requirements. The first is that a system must be provided to monitor and record the chosen physical parameter, and the second, that the environment of the sample is controlled during examination. The primary objective of this controlled environment is to regulate the temperature regime, but other factors, such as the atmosphere surrounding the sample, must also be considered. Although the monitoring systems obviously vary from one technique to another, the control systems for the sample temperature and the atmosphere show less variation.

The heat source for almost all modern thermoanalytical instruments is a resistively heated electric furnace. The exact details of furnace design and control vary from technique to technique, and indeed, from apparatus to apparatus. Temperature control is normally provided using a null-point system, in which the output from a temperature sensor within the furnace is compared with a ramp signal corresponding with the predetermined temperature programme. This control signal is provided either by an electromechanical system of cams and servo-motors or, in more recently developed instruments, by solid-state circuitry. The temperature range required for most mineralogical work is about 1200°C, although for studies of the firing of ceramics higher temperatures may be required.

It is now accepted that control of the composition of the atmosphere surrounding the sample is highly desirable. Atmosphere control is often equated with gas flow through the furnace, and hence, over and around the sample, but this does not necessarily control the atmosphere in the immediate vicinity of the sample. Ideally, gas flow through the sample is required, and in one DTA cell, this is achieved using a porous ceramic base and lid. In EGA instuments, particularly those using mass spectrometers, the atmosphere in the heating cell is determined by the requirements of the gas-analysis technique.

3.3.1 *Differential thermal analysis*

In DTA, measurement of the physical parameter used, ΔT, is normally carried out using a differential thermocouple assembly, one leg of which is sometimes used to indicate the sample or reference temperature. The most commonly encountered materials used for thermocouples are base-metal combinations, such as Chromel *v.* Alumel, or noble-metal ones, such as platinum *v.* platinum/rhodium. The output from the former, for a given temperature difference, is an order of magnitude greater than that obtained from the noble metals, thus enabling higher sensitivity to be attained. This is a significant advantage in mineralogy, where some of the peaks to be monitored are small. But this advantage may, in some cases, be outweighed by the more limited temperature range of base metals, and a tendency towards easier contamination. The μV output from the differential thermocouple pair is then amplified, and either recorded directly on a potentiometric recorder, or transferred via an analogue-to-digital converter to a microcomputer for data treatment and/or storage on disc, tape or even erasable solid-state devices.

The link between the monitoring system, the sample and the controlled environment is provided by the sample-holder assembly. Despite extensive considerations of the practical (Mackenzie and Mitchell, 1970), and theoretical aspects (Wilburn, 1980), of sample-holder design and performance, this remains an area where considerable differences exist between different instruments.

The most important single factor distinguishing the various sample-holder configurations is the distinction between qualitative and quantitative cells. As Wilburn (1980) has shown, when the sample and reference are contained in cylindrical wells within a highly conducting metal block, with the thermocouples centrally placed in the sample and reference material, the shape of the DTA peak due to a particular reaction closely reflects the progress of that reaction. The relationship between peak area and heat of reaction for a cell of this type is dependent on the thermal conductivity of the sample, and in view of the difficulty involved in calibration, the cell is considered to be qualitative. On the other hand, if the sample and reference materials are contained within two crucibles separated by an air-gap, with the sensing thermocouples placed at the base of the crucibles, then the relationship between peak area and heat of reaction does not include the thermal conductivity of the sample, but instead incorporates the conductivity of the dividing air-gap. Thus, the area is independent of the physical properties of the sample and the cell may be considered quantitative. Unfortunately, however, the peak shape suffers considerable distortion compared with the actual course of the reaction. For studies aimed at elucidating the kinetics of a particular reaction, the qualitative sample-holder is obviously preferable, whereas if the heat of reaction is the focus of attention, then the quantitative cell should be chosen. However, it is important to appreciate that, given suitable calibration, both types of cell can be quantitative—in the sense of determining the amount of a particular mineral in a mixture.

3.3.2 *Differential scanning calorimetry*

The detection system in DSC is more variable than for DTA. As already discussed, there are two extreme types of DSC cell—the heat-flow and power-compensation types—but there are a number of intermediates (Mackenzie, 1980).

In the case of the heat-flow system, produced by Du Pont, there exists a well-defined heat-exchange path between the sample, the reference material and the heat source. Thus, any imbalance in the system, caused by the absorption or evolution of heat by the sample, is countered by conductive heat flow through this well-defined path. The parameter measured is differential temperature, as in DTA, but the closer approach to equilibrium conditions enables a calorimetric response to be registered. In the power-compensation instruments—such as the Perkin-Elmer DSC—the sample and reference are heated separately, and when an imbalance is detected the appropriate cell is heated differentially. The rate of power consumption required to return to the equilibrium position is then recorded against time. One distinct advantage of this method is that isothermal work may be carried out.

Despite the theoretical superiority of DSC in a calorimetric sense, the limited temperature range has restricted its application in mineralogical studies. This is unfortunate, since experience in the authors' laboratory has suggested that for quantitative determinations of gibbsite, goethite and kaolinite, the increased sensitivity, improved baseline performance, and improved atmosphere control of a heat-flow DSC (Figure 3.4(b)) yields better results than those from a conventional DTA produced by the same company (Figure 3.4(a)).

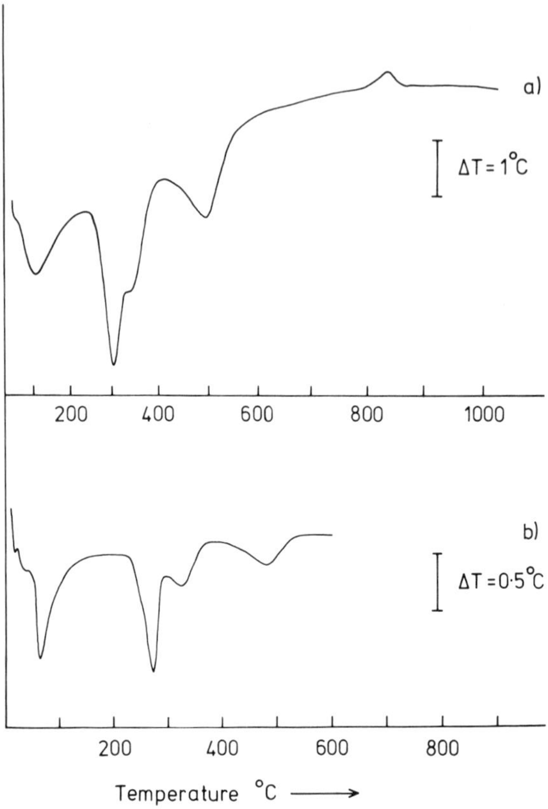

Figure 3.4 (a) DTA curve for tropical soil clay containing gibbsite, goethite and halloysite (b) Heat-flux DSC curve for the same sample showing improved resolution and baseline

3.3.3 *Thermogravimetry*

For TG, where the measured parameter is the mass of the sample, the detection system will obviously be based on a balance assembly. Early instruments were designed around analytical balances, with a variety of methods for the detection of beam movement (more or less continuously), and hence, mass changes. An excellent summary of the principles and requirements of a modern thermobalance has been given by Redfern (1970).

Many recent designs are based on electrobalances, which are null-point instruments, i.e. any change in sample mass, and hence on the force acting on the balance beam, is detected and a restoring force applied. By monitoring this restoring force, the mass changes in the sample may be recorded. This system has many advantages, for example, that there is no requirement for transducers to convert mechanical movement into an electrical signal. More importantly, perhaps, the sample position within the furnace does not alter with change in mass, and the requirement for a large, uniform temperature zone within the furnace is removed, allowing a dramatic reduction in furnace size. This has reduced the cooling time to a matter of minutes in some cases, thus decreasing the instrument time per sample by a factor of at least two.

For a more detailed discussion of the instrumentation briefly described in this section, as well as of other techniques currently in use, the reader is referred to the books by Mackenzie (1970a) and Daniels (1973).

3.4 Experimental technique

Thermoanalytical curves are influenced by many factors pertaining to the sample, the apparatus and the experimental conditions. The object of good experimental technique is to control the number of variables to be considered, and hence achieve the ultimate goal: reproducibility. In this way, meaningful comparisons between various samples can be made, and a major strength of thermal analysis can be exploited. The analyst can exercise control over a number of factors relating to sample preparation and instrument operation, and this section will indicate the relative importance of the various factors.

3.4.1 *Sample preparation*

(i) *Physical factors.* The examination of clays generally deals with finely particulate material, that is aggregated to a greater or lesser extent. The degree of aggregation can be important in controlling many of the reactions that minerals undergo during heating, and clearly some standardization of aggregate size is necessary. This may be readily achieved by light grinding in a mortar, followed by dry sieving. Additional benefits of a standardized comminution procedure include easier representative sampling and more uniform packing within the heating cell of the instrument. The thermal characteristics of a powdered sample in a DTA cell are a function of the properties of the whole sample. This includes both the solid phase and the gaseous phase between the constituent particles, and, in order to reduce variability arising from this source, the particle-size distribution and packing density of the samples should be maintained as constant as possible. By extending this criterion to the reference material, baseline performance can also be enhanced.

Another widely used technique which can improve baseline performance in DTA is dilution of the sample with the thermally inert reference material. This practice should be viewed with some caution, because of the possibility of secondary reactions between the diluent and the sample, and the influence of dilution on peak temperatures (Warne and Mackenzie, 1971). There is also the more philosophical view that the baseline itself reflects the thermal properties of the sample and should not be obscured. As is so often the case in thermal analysis, a dogmatic approach is best avoided and the analyst should use his discretion.

In many mineral systems, surface-controlled reactions, such as the loss of adsorbed water, are important diagnostic criteria. In attempting to standardize pretreatment, it should be noted that the fundamental parameter in the sorption of water is the relative humidity, i.e. the vapour pressure divided by the saturated vapour pressure of pure water at the same temperature. For this reason, it is convenient to equilibrate the sample at a known relative humidity in a desiccator over a saturated salt solution. A good choice in temperate climates is saturated $Mg(NO_3)_2 \cdot 6H_2O$, which produces a relative humidity of 56% over a fairly wide temperature range. The small differential between the ambient humidity and that chosen for equilibration enables sample transfer to be made without seriously altering the amount of sorbed water. However, it may be desirable to remove sorbed water in order to reduce peak overlap and, in these cases, dehydration by evacuation over phosphorus pentoxide is often effective.

(ii) *Chemical pretreatment.* Chemical pretreatments can be of considerable value in interpreting thermoanalytical data, but it is important to obtain the characteristics of the untreated sample for comparison.

One of the most important chemical pretreatments used in the examination of clays is cation saturation. This is particularly important for minerals in the smectite and vermiculite groups. In Figure 3.5, a series of DSC curves for different cation forms of montmorillonite shows that variation of the exchangeable cation causes dramatic changes in the course of the dehydration reaction. Indeed, the effect is now so well documented that the influence of cation saturation on the DTA curve can be used to discriminate between, for example, montmorillonite and halloysite, which both show large dehydration peaks in the low-temperature region.

Another of the commonly employed pretreatments for soil clays is peroxidation to remove organic matter; it vividly illustrates the dangers associated with chemical treatments. In the examination of soil clays in static-air atmospheres, any organic material in the sample can undergo oxidative decomposition in the temperature range 200–500°C. This can result in large exothermic peaks which can obscure the intrinsically weaker endothermic mineral peaks. Although peroxidation can reduce the organic content, insoluble oxalates can be produced, that are in themselves thermally active (Mitchell and Farmer, 1963). A modified procedure, involving presaturation with ammonium ion, minimizes the problem because any oxalate formed can be removed by washing with hot water. This problem is largely circumvented by the use of inert atmospheres, such as nitrogen, that can suppress the oxidation reaction.

Finally, the selective chemical dissolution methods discussed in Chapter 8 can also aid in interpretation of the thermoanalytical behaviour of a soil clay. In a recent study of tropical soil clays in the authors' laboratory, it was found

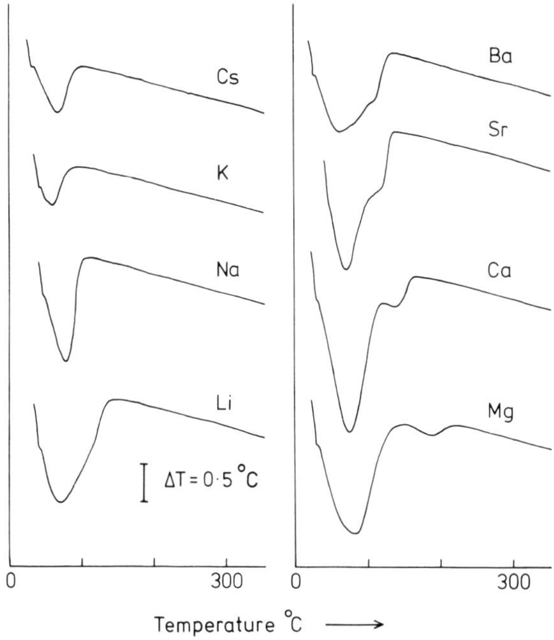

Figure 3.5 Heat-flux DSC curves for Wyoming bentonite, (containing montmorillonite) saturated with indicated cation and equilibrated at 56% RH (relative humidity)

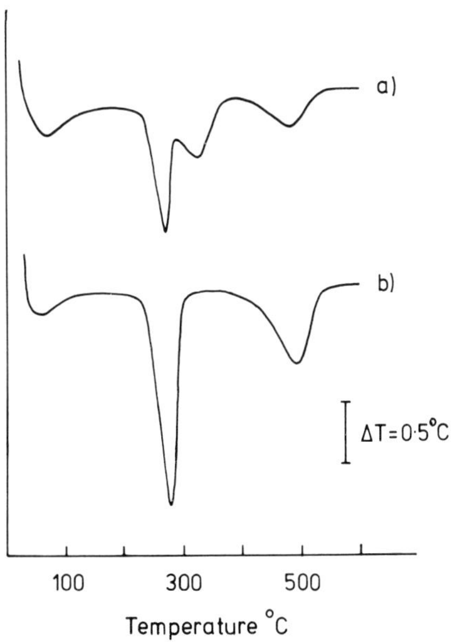

Figure 3.6 Heat-flux DSC curves for tropical soil clay containing gibbsite, goethite and halloysite, (a) before and (b) after treatment with sodium dithionite

that halloysite was the dominant component, with minor amounts of gibbsite and goethite. Two overlapping endothermic peaks at around 300°C suggested the presence of both gibbsite and goethite (Figure 3.6(a)). Treatment of the sample with sodium dithionite removed the peak at 325°C due to goethite, and enabled a quantitative estimation of both minerals (Figure 3.6). A similar approach can be envisaged for other soil-clay components, such as poorly ordered aluminosilicate materials that are removable by alkali dissolution.

3.4.2 *Instrumental variables*

In this section, attention will be concentrated on those instrumental variables that can be controlled by the analyst, rather than those intrinsic to a specific instrument. Thus, no attempt will be made to recommend a particular instrument for mineralogical investigations. For advice on the applicability of the many instruments commercially available, other mineralogically orientated users are the best source. The International Confederation for Thermal Analysis can assist in contacting national user groups in many countries of the world.

One of the most operator-dependent procedures in thermal analysis, and also a potential source of considerable variability, is the packing of the sample in the reaction cell of the instrument. Even placing a powdered sample on the flat Al pan used in heat-flux DSC measurements, can cause considerable variability (Figure 3.7). Good thermal contact between the sample and holder is crucial; this may be achieved by judicious tamping of the sample, using a simple tool constructed by driving a suitably sized nail into a piece of dowel. Another commonly used method is to lightly tap the sides of the cell, although with some of the more fragile DTA cells, caution is recommended. In the deep wells found in many DTA instruments, small amounts of sample should be added, and gently tamped after each addition to obtain a homogeneously packed sample. In cases where the amount of sample is limited and the cell design is suitable, DTA curves can be obtained using a sandwich-packing technique, where the bottom of the well is filled with the thermally inert reference material, and sufficient sample is added to surround the thermocouple. The cell is then levelled off with more inert material. However, both tamping and sandwich packing can have deleterious effects on atmosphere control.

The use of inert atmospheres, to suppress the oxidation of organic material in clays during heating, was recognized by Mitchell and Mackenzie (1959) to be a valuable technique for examination of any material likely to undergo oxidative degradation. This led to the wide acceptance of atmosphere control as a prerequisite for any modern DTA instrument used in mineralogical studies. Gases that may undergo dissociation within the temperature range of the investigation should not be used as the inert atmosphere. For example, carbon dioxide, although suppressing oxidation at room temperature, can

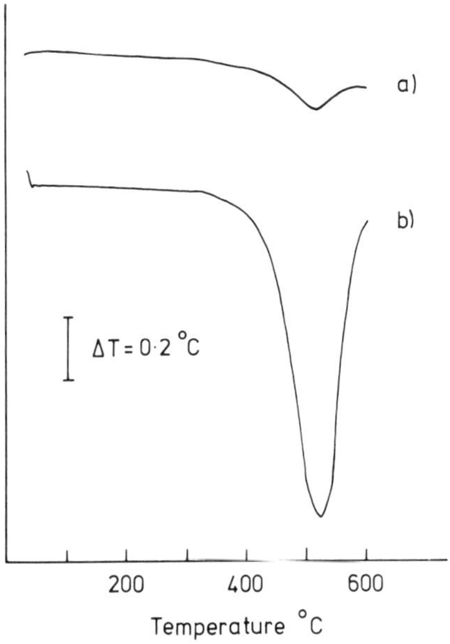

Figure 3.7 Heat-flux DSC curves for poorly crystallized Georgia kaolin, < 2 μm, showing effect of (a) poor thermal contact between sample and (b) good sample contact

undergo dissociation at higher temperatures to yield carbon monoxide and oxygen. However, nitrogen and argon are safe to use within the temperature range of mineralogical investigations.

These examples concern inert gases, but sometimes, the introduction of product gases can be instructive. This is illustrated by the decomposition of the carbonate minerals (Webb and Kruger, 1970), the change in peak shape and position for the decomposition of calcite in a CO_2 atmosphere having markedly increased the detection limit of calcite in soils (Warne and Mitchell, 1979). In addition, for mixed cation species, such as dolomite, the decomposition is split into two reactions in a pure CO_2 atmosphere (Figure 3.8). The use of TG and EGD (evolved gas detection) in the study of carbonates is well illustrated by Milodowski and Morgan (1981). Relatively little work has been reported on the influence of water, the most common product gas, and the thermoanalytical behaviour of minerals under hydrothermal conditions would seem ripe for exploration using modern instruments.

Atmosphere control is not always efficiently achieved by simply changing the composition of the atmosphere in the furnace. The design of the sample-holder must enable product gases in the immediate vicinity of the sample to be efficiently removed. Recent work has shown that the advantages of DSC over DTA in the resolution of small thermal events (associated with surface dehydroxylation of synthetic goethite) result from more effective removal of

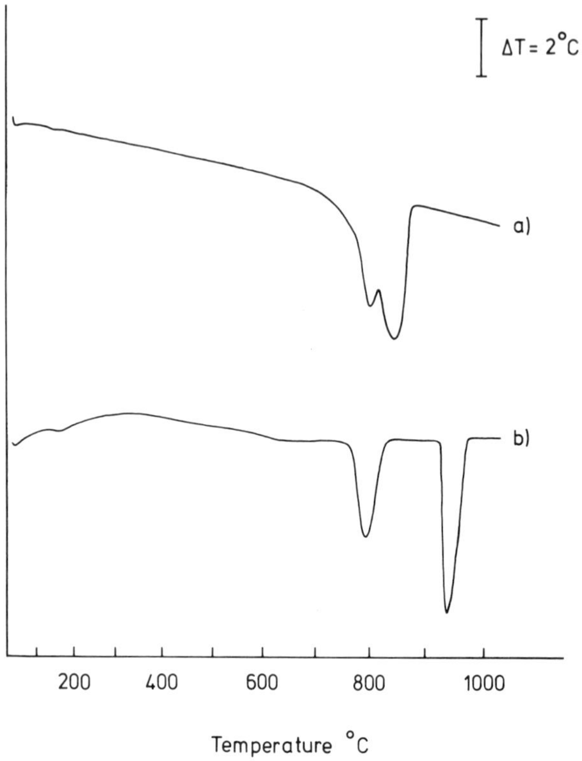

Figure 3.8 DTA curves for dolomite heated in (a) nitrogen atmosphere and (b) carbon dioxide atmosphere

the water vapour produced by the reaction, rather than any intrinsic superiority of DSC over DTA (Mackenzie *et al.*, 1981).

Finally, the influence of heating rate must be considered. Rates of ~ 10K/min have become almost standard in mineralogical investigations, but this is largely due to historical factors rather than any inherent correctness. The resolution of peaks in DTA does deteriorate with increasing heating rate, but for many mineralogical applications this is not of any great significance, because the peaks are inherently wide. In DTA, peak temperatures increase with increased heating rates, but again the importance of this effect can be overemphasized. Peak temperature in DTA is not an absolute determination in the same sense as a *d*-spacing in XRD or a band position in IR spectra. Only in power-compensation DSC is there little or no influence of the heating rate on the peak position and peak area. On the other hand, the advantages of using faster heating rates for DTA include speed of determination and increased peak size, which can be a significant factor in the analysis of minor components. However, it is important to ensure that the same heating rate is used for any samples that must be compared.

To conclude, in thermal analysis a pragmatic approach is often the most fruitful, and the analyst should show some flexibility in order to derive the most information from the particular technique employed.

3.5 Qualitative analysis

Qualitative analysis constitutes one of the most important applications of thermoanalytical methods in mineralogy. As already discussed, the thermal analysis curve for any sample is the product of an interplay between many factors, some intrinsic to the sample, some which may be controlled by the analyst, and others—related to the apparatus—which cannot be readily altered. Thus, in order to identify, isolate and compare the intrinsic properties of the material under investigation, careful attention must be paid to experimental technique and reproducibility of experimental conditions.

The sample properties that may be reflected in the thermoanalytical curve include, in addition to the principal structural features of the mineral species, other properties, such as crystallite size, chemical composition and degree of structural order. Many claims have been made concerning the ability of thermal analysis to distinguish samples using this information, However, unsubstantiated claims have sometimes been made; in particular, the distinction of polymorphic forms has frequently been asserted often without the structural disorder, particle size or chemical composition being held constant.

Despite these reservations, the thermal behaviour of the common clay and accessory minerals does enable qualitative, and even quantitative, analysis to be made. It must be stressed, however, that the interpretation of thermoanalytical data is fundamentally different from most other investigative techniques. Frequently, peak shape and size are intuitively assessed by the experienced analyst, and a 'peak temperature and correlation table' approach should not be used as an absolute indicator of mineralogical composition.

Particular care must be taken to describe the experimental conditions as fully as possible (Lombardi, 1980). For the curves reproduced here, the following conditions were used:

DTA

Instrument	*Du Pont 900 Differential Thermal Analyser* with 1200°C cell, using Pt *v.* Pt/Rh thermocouples
Heating rate	20 K/min
Atmosphere	N_2 flowing over sample at 200 ml/min
Sample weight	30 mg
Reference	20 mg calcined kaolinite
Sensitivity	marked on curve (N.B., due to thermocouple non-linearity this is an average value)

DSC

Instrument	*Du Pont 900 Differential Thermal Analyser* with heat-flux DSC cell
Heating rate	20 K/min
Atmosphere	N_2 flowing over sample at 200 ml/min
Sample weight	10 mg
Reference	empty pan
Sensitivity	marked on curve

Any variation from these standard conditions is given in the appropriate caption.

3.5.1 *Kaolinite group*

The general form of the DTA and TG curves for minerals of this group is now well established. The DTA curve for a well-crystallized kaolinite shows two main peaks, with an endothermic peak in the temperature range 500–600°C, and an exothermic one at around 900–1000°C (Figure 3.9(a)). In some cases, an additional low-temperature endothermic peak at around 100°C, due to the loss of adsorbed water, may be observed. The main endothermic peak is due to dehydroxylation of the mineral and is associated with a weight loss of around

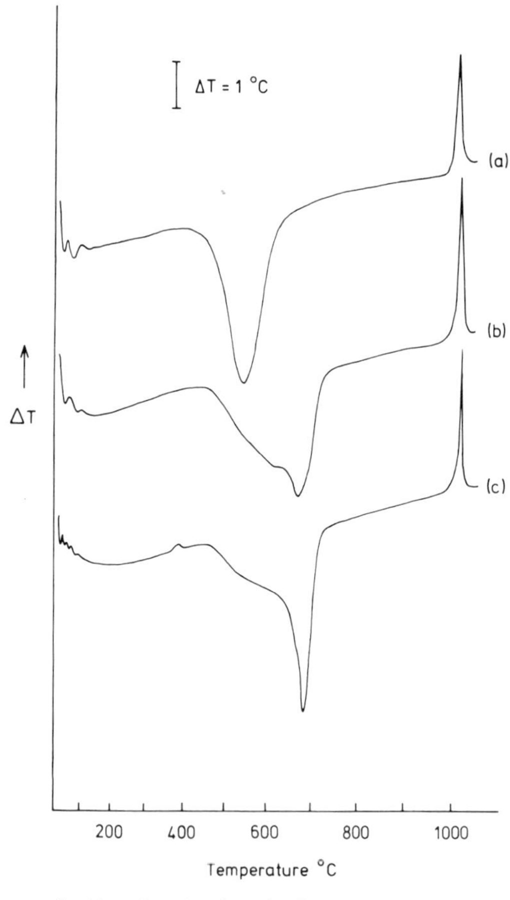

Figure 3.9 DTA curves for (a) $< 2\,\mu m$ fraction of well-crystallized kaolinite, (b) $< 2\,\mu m$ fraction of dickite and (c) well-crystallized geode kaolinite (Keller *et al.*, 1966)

5%, corresponding with the reaction:

$$Al_2Si_2O_5(OH)_4 \longrightarrow Al_2Si_2O_7 + 2H_2O$$

The exothermic peak is probably due to a reaction involving the crystallization of mullite and/or γ-Al_2O_3; this is in accord with the shape of the peak, its exothermicity and temperature. The extent to which it is influenced by mineralizers (Bulens and Delmon, 1977) and by other materials in admixture is also consistent with the suggestion that it arises from a recrystallization of the dehydroxylate. In the study of soil clays, particularly from highly weathered tropical soils, admixed Fe oxides can lead to some difficulties in interpretation (Hughes, 1979), due to alteration of the high-temperature exotherm.

The DTA and TG curves for kaolinites show less variability than those exhibited by other phyllosilicate groups. This is largely due to the compositional uniformity generally observed in kaolinite. The principal source of variability in thermoanalytical behaviour, apart from interactions with other components in mixtures, lies in physical properties such as particle size and the degree of crystalline order. There is a general tendency for peak temperatures of the main dehydroxylation reaction to decrease with decreasing structural order (Mackenzie, 1970a).

Although it is very easy to distinguish hydrated halloysite from kaolinite on the basis of a low-temperature endothermic peak (due to the loss of interlamellar water), dehydrated forms of halloysite present some difficulty. The dehydroxylation endotherm and the recrystallization exotherm are similar to those obtained for a kaolinite. Thus, if the halloysite has suffered dehydration at some stage during sampling or preparation, then the distinction cannot readily be made. Although there is some tendency towards increased asymmetry of the dehydroxylation peak with increasing disorder within the kaolin group and extending into the halloysites, this is not universal, and care must be exercised in interpreting the results from an isolated sample.

In a comparison of the curves for kaolinite and dickite, the DTA curve for a 'well-crystallized' dickite (Figure 3.9(b)) is quite distinct from that of a 'well-crystallized' kaolinite (Figure 3.9(a)), suggesting that a distinction may readily be made between the polymorphs. However, examination of the samples by scanning electron microscopy (SEM) shows that the dickite crystals are considerably thicker and better-developed than those of the kaolinite. Factors such as these can have a considerable influence on thermoanalytical behaviour. Thus, the DTA curve for a geode kaolinite (Keller *et al.*, 1966) is similar (Figure 3.9(c)) to that of dickite. The particle size and morphological development of the kaolinite is comparable with dickite, suggesting that the DTA curves reflect these features, rather than polymorphic form.

Despite these reservations, the experimenter should not be deterred from using trends in peak shape and peak temperature with care. In many investigations of a group of samples with similar origin, or displaying some common factor, relative changes in the DTA curves for the samples within the group may be of considerable value. A particularly good example of such comparative DTA is described by Lombardi and Sheppard (1977).

3.5.2 *Serpentines*

The serpentine group of minerals represents the trioctahedral analogues of kaolinite, and enables assessment of the influence of octahedral layer occupancy on thermoanalytical behaviour. The general form of the DTA curve (Figure 3.10) for chrysotile shows an increased dehydroxylation peak temperature in comparison with that observed for kaolinites. In contrast, the peak temperature of the exothermic recrystallization is lower in trioctahedral minerals, because of the lower stability of the dehydroxylate. Both these features are general for many trioctahedral minerals as compared with their dioctahedral analogues, and may often be employed successfully in qualitative analysis.

Unlike the kaolinite minerals, the serpentine group can display considerable compositional variation. As the degree of substitution increases, the peak temperature of the dehydroxylation reaction decreases; this reflects the decreased thermal stability often observed for mixed cation mineral species.

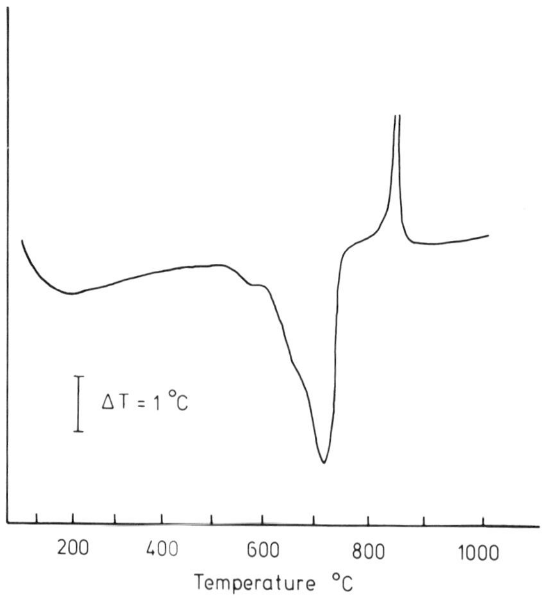

Figure 3.10 DTA curve for chrysotile

3.5.3 *Smectite group*

Smectites display a wide range of compositions, quite apart from that arising from the inclusion of both di- and trioctahedral species in the group. Isomorphous substitution is common, both in the octahedral and tetrahedral positions within the structure. In addition to the direct effect of substitution, the resultant layer charge and the presence of exchangeable interlayer cations has a significant effect on the thermoanalytical behaviour of this group of minerals.

3.5.3.1 *Dioctahedral smectites.* The most intensively studied mineral in this group is montmorillonite, and its thermoanalytical behaviour may be taken as

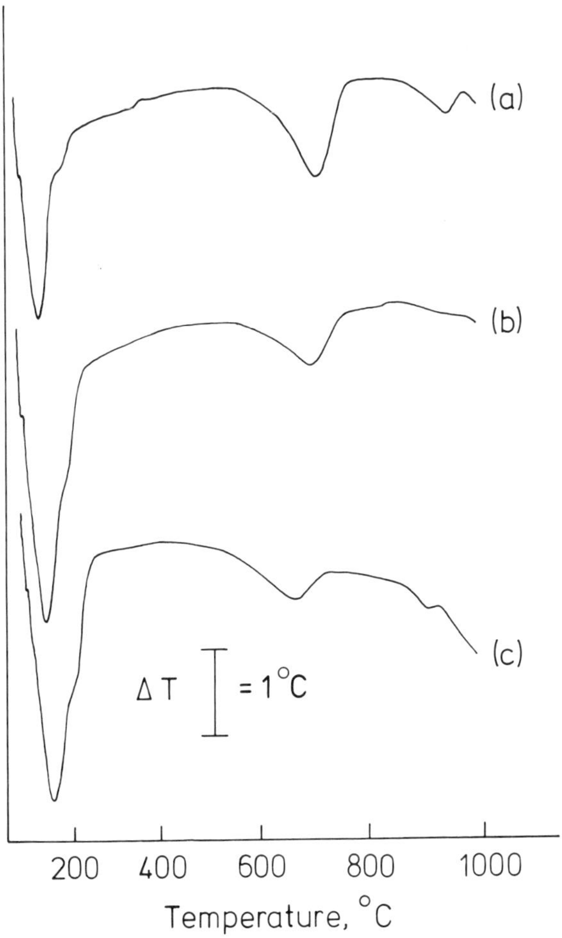

Figure 3.11 DTA curves for bulk samples, Source Clay Mineral Repository, 56% RH, (a) Wyoming, (b) Texas, and (c) Arizona, montmorillonites

a convenient yardstick for other members of the group. The general features of the DTA curve are the presence of a large, low-temperature endothermic peak system at ~100–250°C, followed by a second endotherm at ~700°C, although in some cases this peak occurs at lower temperatures of ~500°C, and finally, a shallow endothermic/exothermic effect at ~800–900°C (Figure 3.11).

The low-temperature endothermic peaks are due to the loss of adsorbed moisture, with most of the water being lost from interlayer regions. Thus, the principal two factors determining the size, shape and temperature of this peak system are the saturating cation present, and the relative humidity at which the sample has been equilibrated. Clearly, equilibration of the sample, at a known relative humidity, before examination, is crucial.

Much of the early work on the thermal behaviour of montmorillonite used DTA. Recent dehydration studies carried out in the authors' laboratory

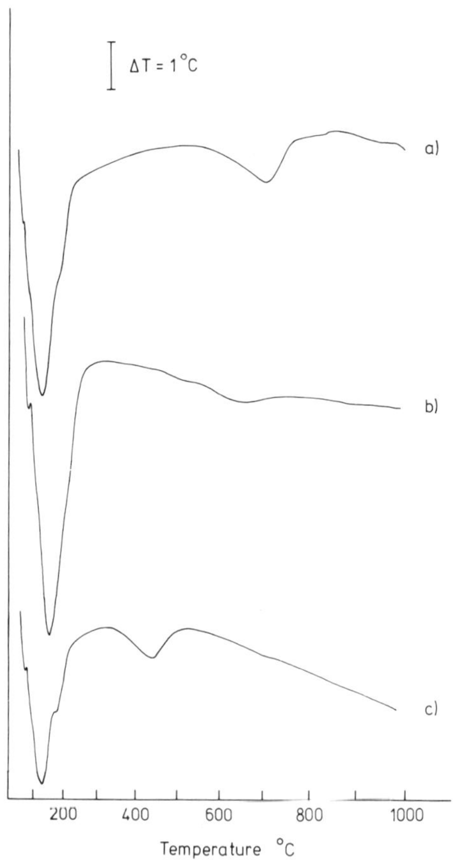

Figure 3.12 DTA curves for $<2\,\mu m$ dioctahedral smectites, Ca-saturated, 56% RH, (a) montmorillonite, (b) volkonskoite and (c) nontronite.

(Paterson, 1981) using both heat-flux and power-compensation DSC, suggest that these techniques offer significant advantages over DTA. The heat-flux DSC curves for various cation forms of Wyoming bentonite (Figure 3.5) show the paramount influence of the saturating cation, but with dehydration peak temperatures lower than those recorded in early DTA studies. This is a reflection of the more efficient removal of product gases from samples which were spread thinly on flat pans, rather than being packed into a DTA cell. The relative humidity at which the sample is equilibrated can have a considerable influence for some cation forms. From the magnitude of the changes observed, it is clear that for the smectites, cation saturation and equilibration at a known humidity are vital in qualitative analysis.

In addition to changes in the low-temperature endotherm, the saturating cation also has some effect on the dehydroxylation reaction (Mackenzie and Bishui, 1958). The reason for the observed increase in peak temperature of the dehydroxylation reaction with increasing size of the interlayer cation is not clear, but has been described by Mackenzie (1970) as having 'a steric origin'.

Methods based on differential measurements, such as DTA, suggest that the dehydration and dehydroxylation reactions are distinct. However, integral

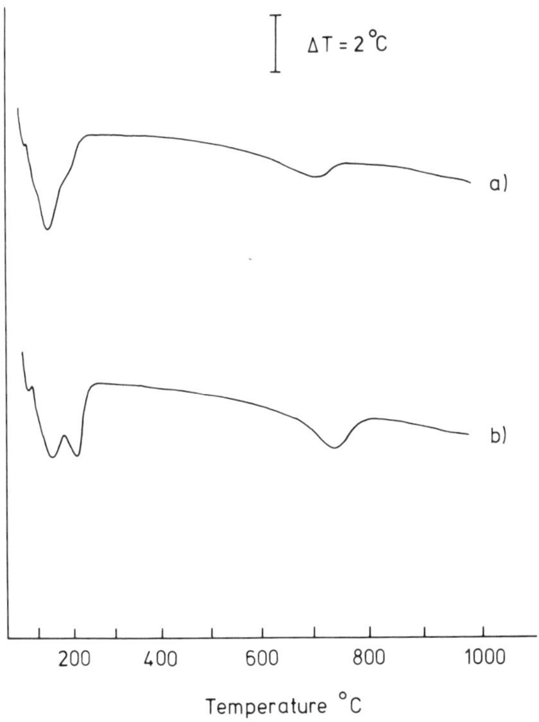

Figure 3.13 DTA curves for $< 2\,\mu m$ dioctahedral smectites, Ca-saturated, 56% RH, (a) montmorillonite and (b) beidellite

methods such as TG, show that this is not necessarily the case. Heating Mg-montmorillonite results in a continuous weight loss, from room temperature to 1000°C, with superimposed periods of accelerated weight loss. On the other hand, the monovalent cation species show a well-developed plateau between the dehydration and dehydroxylation reactions.

The effect of extreme compositional variation within the smectites can be seen in the comparison of the DTA curves for nontronite, volkonskoite and montmorillonite. Although the general features of the curves are similar, the temperature of the dehydroxylation endotherm is much lower for nontronite (Figure 3.12(c)) than that for montmorillonite (Figure 3.12(a)), with volkonskoite occupying an intermediate position with a peak temperature in the 500–600°C region (Figure 3.12(b)).

Finally, the influence of the source of the charge may be readily illustrated by comparing the montmorillonite curve (Figure 3.13(a)) with that for beidellite (Figure 3.13(b)). The principal difference is observed in the dehydration endotherm where, despite both samples having been Ca-saturated and equilibrated at 56% RH (relative humidity) the beidellite sample shows a more pronounced two-stage loss of interlayer water. This probably arises from an

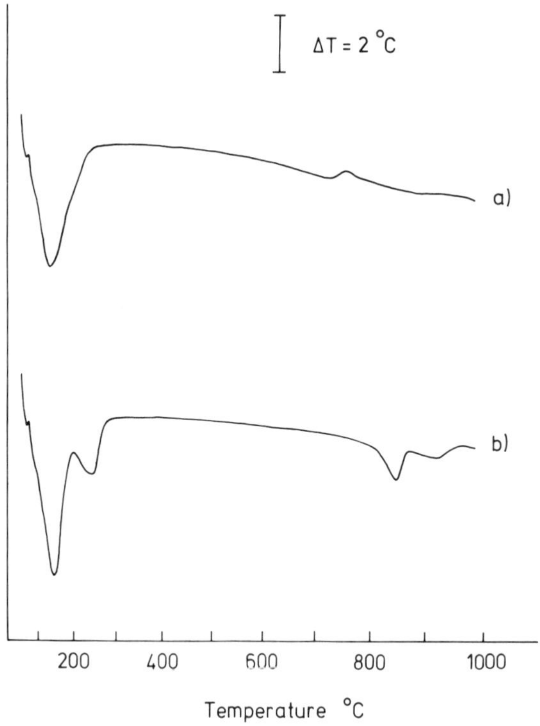

Figure 3.14 DTA curves for $< 2\,\mu m$, trioctahedral smectites, Ca-saturated, 56% RH, (a) hectorite and (b) saponite

alteration in the nature of the interactions between exchangeable cations, coordinated water molecules and the source of charge (Farmer and Russell, 1972).

3.5.3.2 *Trioctahedral smectites.* The DTA curve for saponite (Figure 3.14(b)), the most common trioctahedral smectite, shows the same general features as the dioctahedral smectites. Again, the most striking feature is the low-temperature endothermic peak system, due to the loss of sorbed water. Comparison of the curves for saponite and hectorite (Figure 3.14(a)) shows the same differences as described previously for beidellite and

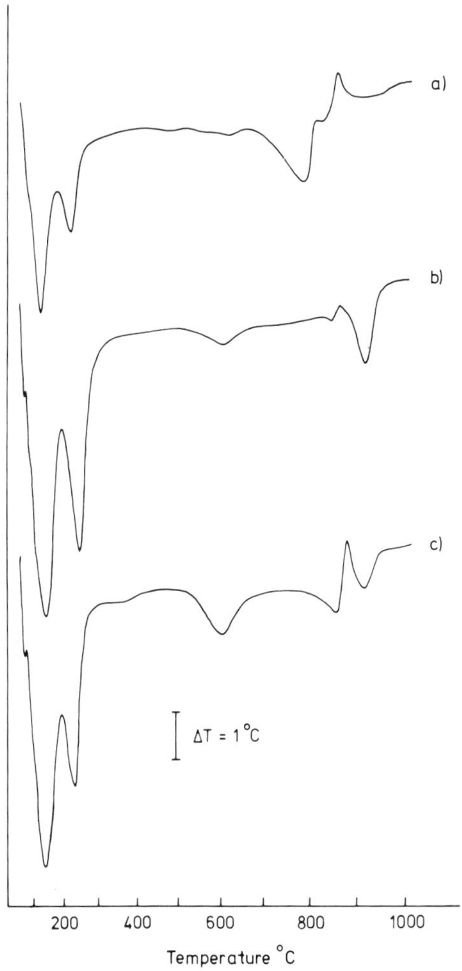

Figure 3.15 DTA curves for Ca-saturated vermiculites showing (a) reduced dehydration endotherm, (b) and (c) partial chloritization

montmorillonite. This confirms that the source of charge is reflected in the nature of the low-temperature dehydration endotherm.

Clearly, the curves for smectites are less well defined than those obtained for the kaolin group. However, this is probably a reflection of the considerable compositional and structural variations within the smectite group.

3.5.4 *Vermiculite group*

The structural similarity between the smectites and vermiculites is borne out by the similarity of their thermoanalytical behaviour. The DTA curve for vermiculite is dominated by the endothermic peak system arising from the loss of sorbed water (Figure 3.15) and, as with the smectites, the influence of the saturating cation is paramount. As the sample-temperature increases, TG curves show an almost continuous weight loss, with the dehydration reaction merging into the dehydroxylation zone. In many vermiculites, a high-temperature endo/exo peak system is observed at 800–900°C. Chloritic impurities may be readily detected by DTA (Figure 3.15(b) and (c)), and although the identification of micaceous layers is more difficult, these may be

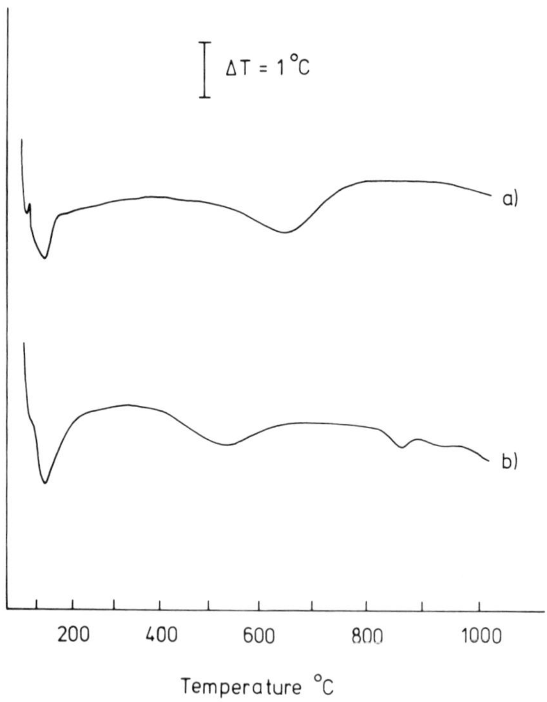

Figure 3.16 DTA curves for $<2\,\mu m$ illite, Ca-saturated, 56% RH: (a) 'normal' and (b) 'abnormal' (see Mackenzie, 1970b)

reflected in the reduced size of the dehydration endotherm (Figure 3.15(a)) or reduced weight loss on the TG curve.

3.5.5 *Illite group*

The considerable range of chemical and structural characteristics within the illite group is reflected in their thermoanalytical behaviour. For example, substantial variations in their DTA curves have been reported. A major difference noted by Mackenzie (1970) and others is that, in common with montmorillonite, two distinct forms exist that may be differentiated by their dehydroxylation peak temperatures (Figure 3.16). This differentiation cannot, however, be detected by any other technique. In many instances, the dehydroxylation reaction occurs over a wide temperature range, such that, on a differential curve, little displacement from the baseline is observed. Finally, in the high-temperature region, the endo/exo peak system noted in other 2:1

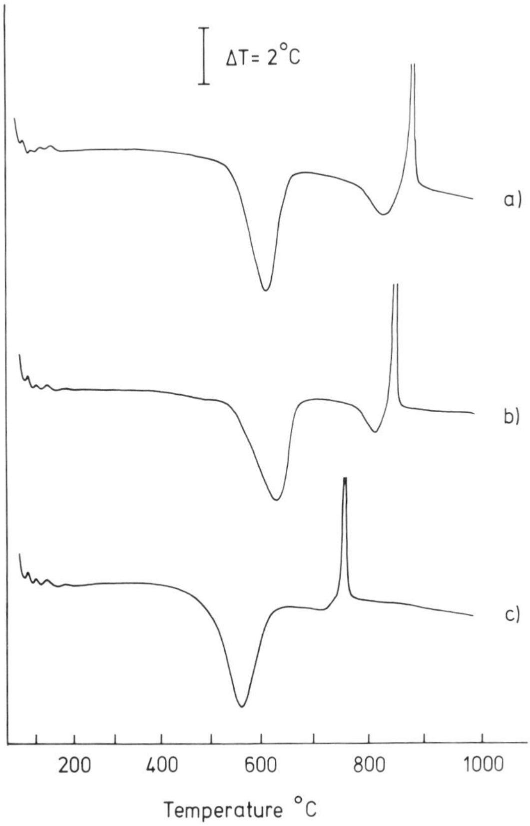

Figure 3.17 DTA curves for chlorites of increasing Fe content, (a) sample 5 (b) sample 6, (c) sample 9 (see Goodman and Bain, 1979)

layer silicates is sometimes, but not always, observed (Figure 3.16). Thus, the thermal properties of illites are very variable and it must be concluded that thermoanalytical methods are of limited value for diagnostic purposes. However, as demonstrated for other mineral groups, comparison of a series of samples can provide valuable information.

3.5.6 *Chlorite group*

In contrast to the somewhat featureless curves typical of the illites, those for chlorites are characterized by well-defined endothermic and exothermic peaks, with correspondingly distinct weight losses on the TG curves.

The general form of the DTA curves for chlorites can be seen in Figure 3.17. Essentially they consist of two distinct peak systems with a well-defined endotherm at 500–600°C, due to dehydroxylation of the interlayer hydroxide sheet, followed by a higher-temperature peak system resulting from the dehydroxylation and recrystallization of the silicate sheet. The other curves shown in Figure 3.17 illustrate the effect of increasing Fe content on the DTA characteristics, with curve (b) showing how a moderate increase in Fe results in a significant shift in the endo/exo system. In curve (c), the Fe content of the mineral is so high that it is unlikely that all of the Fe is accommodated in the talc sheet (Goodman and Bain, 1979). Some is probably located within the interlayer hydroxide sheet, this being confirmed by the decrease in peak temperature of the dehydroxylation endotherm.

Partially chloritized layer silicates may also be readily identified from thermoanalytical curves, as illustrated in the discussion of vermiculite. Cowking *et al.* (1983) compared two morphologically distinct forms of saponite, and concluded that DTA was able to detect partial chloritization more easily than could XRD or IR.

3.5.7 *Palygorskite/sepiolite*

Pure samples of these minerals have quite distinctive DTA curves. Some representative examples are shown in Figure 3.18. The major feature common to both palygorskite and sepiolite is the large endothermic peak at around 100°C, due to the loss of sorbed water. In the intermediate temperature region at around 400–450°C, palygorskite frequently shows a single, relatively large endothermic peak (Figure 3.18(c) and (d)), that has been ascribed either to dehydration of chemically bound water (Martin-Vivaldi and Cano-Ruiz, 1956) or to partial dehydroxylation and loss of residual bound water (Huggins *et al.*, 1962). For sepiolite this peak is often absent or much reduced in intensity (Figure 3.18(a) and (b)). In the high-temperature region, endothermic/exothermic reactions can be observed for both minerals, although they are often better developed in sepiolite than in palygorskite. The weaker high-temperature endotherm in palygorskite lends weight to the

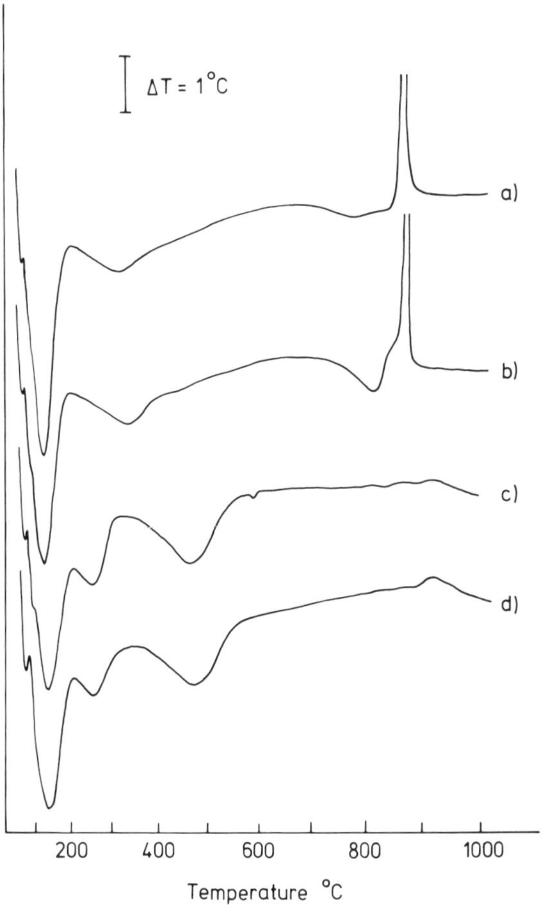

Figure 3.18 DTA curves for sepiolite, curves (a) and (b), and palygorskites, curves (c) and (d)

hypothesis that partial dehydroxylation of palygorskite occurs at intermediate temperatures.

3.5.8 *Allophane/imogolite*

Thermal analysis, in particular DTA, has been widely used to characterize poorly ordered soil components. Indeed, for some time, a verbal description of the DTA curve was used in the definition of allophane. However, it is now known that DTA is not diagnostic for allophane. The presence of this mineral generally results in a large endothermic peak in the low-temperature region, often accompanied by a high-temperature exothermic peak similar to that observed for the kaolin minerals, with no appreciable dehydroxylation

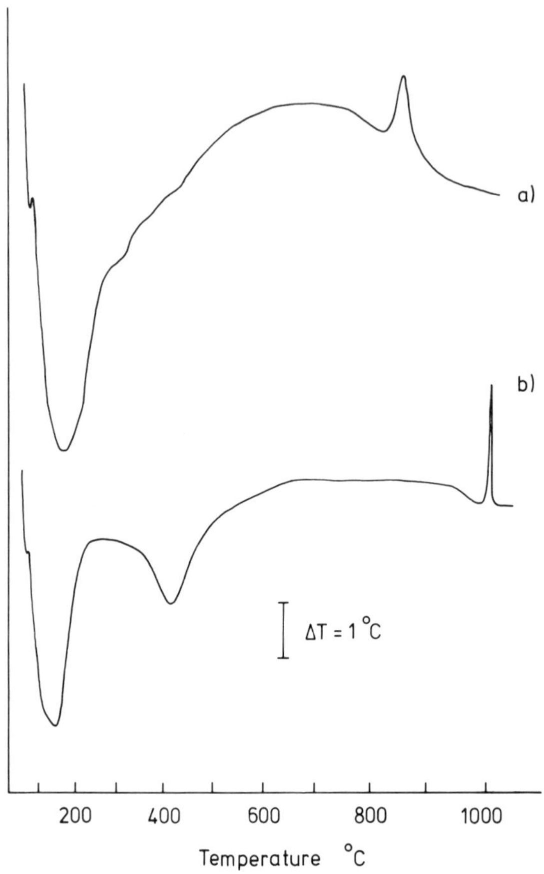

Figure 3.19 DTA curves for (a) $< 2\,\mu m$ allophanic soil clay and (b) Kitikami imogolite

reaction at intermediate temperatures (Figure 3.19(a)). However, other soil components in admixture can influence the curve considerably (Campbell *et al.*, 1968). Thus, it is advisable to confirm an identification of allophanic material by other techniques, such as selective chemical dissolution, TEM or IR.

Imogolite has a well-defined structure, but the principal feature dominating its XRD behaviour is the physical size of the constituent particles. In this case, TEM and IR spectroscopy are diagnostic, but the DTA curve of imogolite is distinctive, with a large dehydration endotherm (occasionally, two component peaks of roughly equal size may be discerned) followed by a dehydroxylation endotherm at around 400°C and a high-temperature exotherm at about 1000°C (Figure 3.19(b)).

3.5.9 *Accessory minerals*

Natural clays often contain accessory minerals in addition to the phyllosilicates referred to previously. Some of the major accessory minerals will now be discussed.

3.5.9.1 *Silica minerals.* Pure samples of the three silica polymorphs, quartz, cristobalite and tridymite, all have distinctive DTA curves (Dawson and Wilburn, 1970). The DTA curve for quartz, the most commonly encountered polymorph, exhibits a sharp endothermic peak at 573°C on heating (Figure 3.20(a)). The α–β transition which gives rise to the peak is reversible, and on cooling an exothermic peak may be observed.

3.5.9.2 *Aluminium oxide/hydroxides.* The pure oxides are relatively inert thermally although a weak exothermic effect has been reported for the conversion of γ-Al_2O_3 to the stable polymorph α-Al_2O_3. However, this is of little relevance to natural clay materials where hydroxides and oxide-hydroxides are more common.

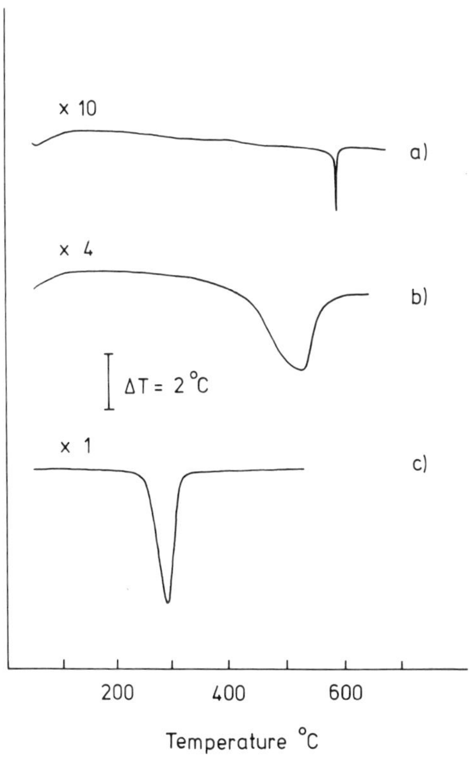

Figure 3.20 Heat-flux DSC curves for (a) quartz, (b) boehmite and (c) gibbsite

The DTA curves for the two polymorphic forms of aluminium oxide—hydroxide/boehmite and diaspore—are very similar, with a large endothermic peak at around 500°C (Figure 3.20(b)), corresponding with the reaction:

$$2AlOOH \longrightarrow Al_2O_3 + H_2O$$

Differentiation of the polymorphs by DTA alone is normally not possible.

The aluminium hydroxides, gibbsite and bayerite, give similar DTA curves with a large endothermic peak at 300–350°C (Figure 3.20(c)), corresponding with the dehydroxylation reaction,

$$2Al(OH)_3 \longrightarrow Al_2O_3 + H_2O$$

Again, differentiation of the polymorphs in natural samples is not possible.

3.5.9.3 *Iron oxide/hydroxides.* The oxides of Fe likely to be encountered in clay materials are relatively inert thermally. On the other hand, the oxide-hydroxides are very reactive and may be readily identified—as a group—through the dehydroxylation reaction that occurs at ~300°C (Figure 3.21(b) and (c)). Identification of pure γ-FeOOH may be made by observing an exothermic peak at ~550°C, due to the conversion of γ-Fe_2O_3 to the stable oxide phase, α-Fe_2O_3 (Figure 3.21(c)). This peak is unlikely to be observed in a mixture with other minerals, however. The curve for the poorly ordered mineral, ferrihydrite, is characterized by a large, low-temperature endotherm, with no indication of the sharp dehydroxylation reaction seen for the crystalline oxide-hydroxides (Figure 3.21(a)).

3.5.9.4 *Carbonate minerals.* This group of minerals is particularly amenable to study by thermoanalytical methods (Webb and Kruger, 1970) and the DTA curves are characterized by large endothermic peaks in the temperature range 500–800°C, when heated in a nitrogen atmosphere. In CO_2 the curves often show dramatic changes, with sharper peaks occurring at higher temperatures (Figure 3.8).

3.6 Quantitative analysis

It is clear from the preceding section that a number of clay minerals undergo specific reactions on heating, that give rise to clearly defined events on the experimental curve. Of the main techniques in use, those producing *differential* curves (DTA and DSC) are of particular value in quantitative determinations. It might seem that TG is more suitable for such estimations, but there are two distinct problems relating to its use. The first concerns the difficulty of determining the starting point and end-point of a specific reaction against a background of continuous weight loss. The second is that normally TG is used as an absolute determination, with an implied stoichiometry which may not be realistic for a clay mineral. On differential curves, the starting points of peaks

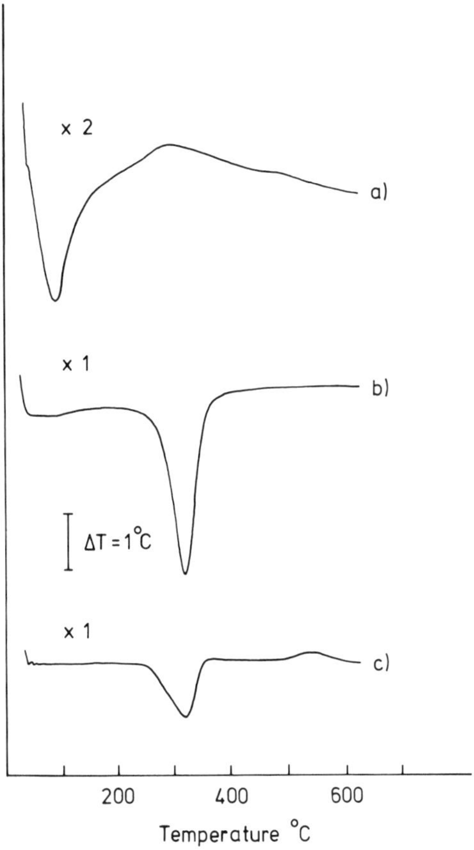

Figure 3.21 Heat-flux DSC curves for (a) poorly ordered ferrihydrite, (b) goethite, and (c) lepidocrocite

may be more readily identified, and are often sufficiently free from overlapping thermal events to enable peak-area determinations to be made.

It is assumed that, for any single set of determinations, all the experimental conditions, such as the instrument used, the heating rate, the atmosphere, sample packing, etc. have been maintained constant.

In differential curves, peak areas and not peak heights should be used to indicate the amount of any particular component. The determination of area involves the construction of a baseline, with two commonly used methods (Figure 3.22) being the tangential and the stepped construction (Wilburn, 1980). The actual area can then be determined by a variety of methods, including cutting out the peak and weighing, planimetry or tracing onto squared paper and manually counting the squares. Of these, the tracing method is recommended, although it is admittedly tedious and time-

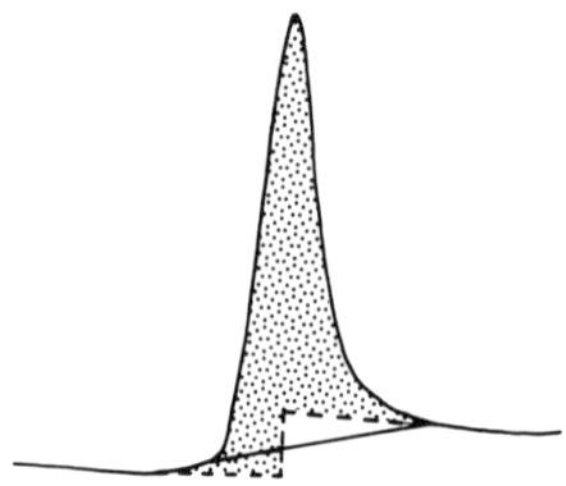

Figure 3.22 Baseline drawn across the 'base' of the peak i.e. tangential (————) compared with stepped construction (— — — —) (Wilburn, 1980)

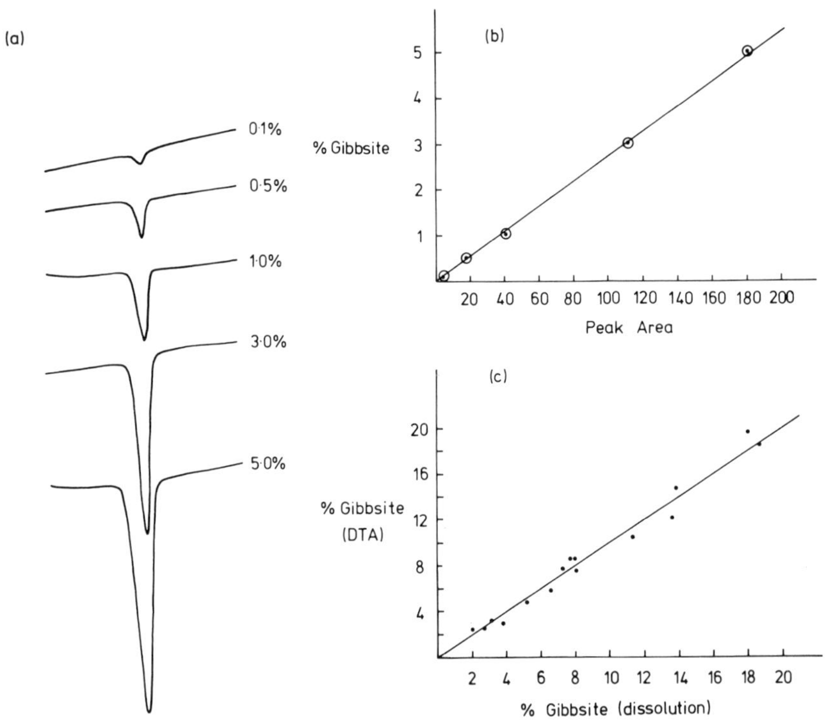

Figure 3.23 (a) Heat-flux DSC curves for indicated proportion of gibbsite in calcined kaolinite, (b) a calibration curve prepared from (a), and (c) gibbsite determination (Jorgensen *et al.*, 1970)

consuming. On many modern instruments, microcomputer-based data acquisition facilities will enable peak areas to be estimated readily. However, the baseline constructions used by many of these computer programs are far from ideal, and the analyst should be satisfied that a suitable construction is used.

The preparation of a standard calibration curve (Figure 3.23) follows normal analytical practice, in that it should encompass the actual concentration expected. When selecting a standard, the main criterion to be followed is that its thermal characteristics should match, as far as possible, those of the unknown; in this regard, dilution can be useful. In some circumstances, where only relative changes in the amount of a particular component within a series of samples are required, no standards need be used and the peak areas can be compared directly.

Quantitative mineralogical analysis, using thermoanalytical methods, is viable for many of the minerals discussed in this chapter, but particularly for the determination of gibbsite, goethite and carbonate minerals.

References

Bulens, M. and Delmon, B. (1977) The exothermic reaction of metakaolinite in the presence of mineralizers. Influence of crystallinity. *Clays & Clay Miner.*, **25**, 271–277.

Campbell, A.S., Mitchell, B.D. and Bracewell, J.M. (1968) Effects of particle size, pH and organic matter on the thermal analysis of allophane. *Clay Miner.*, **7**, 451–454.

Cowking, A., Wilson, M.J., Tait, J.M. and Robertson, R.H.S. (1983) Structure and swelling of fibrous and granular saponitic clay from Orrock Quarry, Fife, Scotland. *Clay Miner.*, **18**, 49–64.

Cunningham, A.D. and Wilburn, F.W. (1970) 'Theory', in *Differential Thermal Analysis*, Vol. I, (R.C. Mackenzie ed.), Academic Press, London, 32–62.

Daniels, T. (1973) *Thermal Analysis*, Kogan Page, London.

Dawson, J.B. and Wilburn, F.W. (1970) 'Silica minerals', in *Differential Thermal Analysis*, Vol. I, (R.C. Mackenzie ed.), Academic Press, London, 477–495.

Farmer, V.C. and Russell, J.D. (1972) Interlayer complexes in layer silicates. The structure of water in lamellar ionic solutions. *Trans. Faraday Soc.*, **67**, 2737–2749.

Goodman, B.A. and Bain, D.C. (1979) 'Mössbauer spectra of chlorites and their decomposition products', in *Proc. Intern. Clay Conf., 1978*, (M.M. Mortland and V.C. Farmer eds.), Elsevier, Amsterdam, 65–74.

Huggins, W., Denny, M.V. and Shell, H.R. (1962) Properties of palygorskite, an asbestiform mineral. *U.S. Bur. Mines, Rep. Invest.*, **RI 6071**, 1–17.

Hughes, J.C. (1979) The effects of experimental conditions on the 950°C kaolinite exotherm in some tropical soil clays. *Clay Miner.*, **14**, 21–28.

Jorgensen, S.S., Birnie, A.C., Smith, B.F.L. and Mitchell, B.D. (1970) Assessment of gibbsitic material in soil clays by differential thermal analysis and alkali dissolution methods. *J. Thermal Anal.*, **2**, 277–282.

Keller, W.D., Pickett, E.E. and Reesman, A.L. (1966) 'Elevated dehydroxylation temperature of the Keokuk geode kaolinite—a possible reference mineral', in *Proc. International Clay Conf., 1966*, Vol. I, (L. Heller and A. Weiss eds.), Israel Program for Scientific Translations, Jerusalem, 75–85.

Lombardi, G. (1980) *For Better Thermal Analysis*, International Confederation for Thermal Analysis, and the Univ. of Rome, Italy.

Lombardi, G. and Sheppard, S.M.F. (1977) Petrographic and isotopic studies of the altered acid volcanics of the Tolfa-Cerite area, Italy: the genesis of the clays. *Clay Miner.*, **12**, 147–162.

Mackenzie, R.C. (1970a) *Differential Thermal Analysis*, Vol. 1, *Fundamental Aspects*, Academic Press, London.

Mackenzie, R.C. (1970b) 'Simple phyllosilicates based on gibbsite- and brucite-like sheets', in *Differential Thermal Analysis* Vol. 1, (R.C. Mackenzie ed.), Academic Press, London, 498–537.

Mackenzie, R.C. (1980) Differential thermal analysis and differential scanning calorimetry: similarities and differences. *Anal. Proc.*, **17**, 217–220.

Mackenzie, R.C. (1984a) *De Calore*: prelude to thermal analysis. *Thermochim. Acta*, **73**, 251–306.

Mackenzie, R.C. (1984b) Origin and development of differential thermal analysis. *Thermochim. Acta*, **73**, 307–367.

Mackenzie, R.C. and Bishui, B.M. (1958) The montmorillonite DTA curve. II. Effect of exchangeable cations on the dehydroxylation of normal montmorillonite. *Clay Miner. Bull.*, **3**, 276–286.

Mackenzie, R.C. (1970) 'Simple phyllosilicates based on gibbsite- and brucite-like sheets', in *Differential Thermal Analysis*, Vol. I, (R.C. Mackenzie, ed.), Academic Press, London, 497–537.

Mackenzie, R.C. and Mitchell, B.D. (1970) 'Instrumentation', in *Differential Thermal Analysis*, Vol. I, (R.C. Mackenzie ed.), Academic Press, London, 63–99.

Mackenzie, R.C., Paterson, E. and Swaffield, R. (1981) Observation of surface characteristics by DSC and DTA. *J. Thermal Anal.*, **22**, 269–274.

Martin-Vivaldi, J.L. and Cano-Ruiz, J. (1956) Contribution to the study of sepiolite: III. The dehydration process and the types of water molecules. *Clays & Clay Miner.*, **4**, 177–180.

Milodowski, A.E. and Morgan, D.J. (1981) 'Thermal decomposition of minerals of the dolomite–ferroan dolomite–ankerite series in a carbon dioxide atmosphere', in *Proc. Second Europ. Symp. Thermal Anal.*, (D. Dollimore ed.), Heyden & Son, London, 468–471.

Mitchell, B.D. and Farmer, V.C. (1963) Occurrence of oxalates in soil clays following hydrogen peroxide treatment. *Soil Sci.*, **96**, 221–229.

Mitchell, B.D. and Mackenzie, R.C. (1959) An apparatus for DTA under controlled-atmosphere conditions. *Clay Miner. Bull.*, **4**, 31–43.

Paterson, E. (1981) 'The value of differential scanning calorimetry in assessing the physical and chemical properties of particle surfaces', in *Proc. Second Europ. Symp. Thermal Anal.*, (D. Dollimore ed.), Heyden & Son, London, xxii–xxx.

Redfern, J.P. (1970) 'Complementary methods', in *Differential Thermal Analysis*, Vol. I, (R.C. Mackenzie ed.), Academic Press, London, 123–158.

Warne, S.St.J. and Mackenzie, R.C. (1971) The thermal dissociation of some carbonate minerals. *J. Thermal Anal.*, **3**, 49–55.

Warne, S.St.J. and Mitchell, B.D. (1979) Variable atmosphere DTA in identification and determination of anhydrous carbonate minerals in soils. *J. Soil Sci.*, **30**, 111–116.

Webb, T.L. and Kruger, J.E. (1970) 'Carbonates', in *Differential Thermal Analysis*, Vol. I, (R.C. Mackenzie, ed.), Academic Press, London, 303–341.

Wilburn, F.W. (1980) Criteria for quantitative DTA and DSC. *Anal. Proc.*, **17**, 221–223.

4 Infrared methods

J.D. RUSSELL

4.1 Introduction

Infrared absorption spectroscopy is a rapid, economical and nondestructive physical method universally applicable to structural analysis. The technique is so versatile that it can be used both as a source of the physical parameters of crystal lattice determinations, and as a means of eliciting purely empirical qualitative relationships between specimens. It is an intrinsically simple technique that deserves to be more widely used in clay mineralogy and soil science, and this chapter aims to provide some basic information to help bring this about. It is proposed to indicate briefly the basic principles of IR, describe the instrumentation and methods, present some typical spectra of standard minerals, and finally to show spectra of a few representative soil clays, indicating how they may be used to assess the constituent minerals.

4.2 Principles

The absorption of infrared radiation by clay minerals depends critically on atomic masses, and the lengths, strengths and force constants of interatomic bonds in the structures of these minerals. It is also controlled by the constraints of the overall symmetry of the unit cell, and the local site symmetry of each atom within the unit cell. The total number of potentially active internal vibrations is given by $3n - 6$, where n is the number of atoms in the unit cell. Not all of these vibrations are active in the infrared, only those that undergo a change in dipole moment during the absorption process. More detailed theoretical treatments of vibration spectroscopy as applied to inorganic solids may be found in the books by Hadni (1967) and Turrell (1972). The monograph by Farmer (1974) provides one of the most appropriate texts relating specifically to clay minerals.

The absorption of infrared radiation is also strongly influenced by the degree of crystalline order (Lazarev, 1974) and the size and shape of the mineral particles (Farmer and Russell, 1966; Rendon and Serna, 1981; Serna *et al.*, 1982). It is crucial also to be aware of the effect of particle orientation relative to the intercepting beam of IR radiation. Absorption of radiation by a vibrating interatomic bond will be at a maximum when the axis of the bond is parallel to the electric vector of the beam, and will be at a minimum when the axis and vector are mutually perpendicular. Thus, the extent to which clay-mineral particles adopt a preferred orientation—for example in the sedimentation plane for platy particles of a layer silicate—will be important in the identification and structural orientation of specific interatomic groupings in the mineral. The infrared spectrum of a clay mineral is, therefore, sensitive to chemical composition, isomorphous substitution and crystallinity, and provides fundamental information not only on mineral identification, but also on surface properties and reactions of minerals with chemicals in their environment. This chapter aims to provide a representative collection of IR spectra of the most important mineral groups likely to be encountered in clay-size fractions isolated from soils, mineral deposits, etc., and to indicate methods of spectral interpretation that can identify these minerals and the types of isomorphous substitution present.

4.3 Instrumental requirements

To provide adequate characterization of a clay mineral by infrared spectroscopy, the spectrum should be recorded over the range 4000 to 250 cm^{-1}, and if possible to 200 cm^{-1}. Such coverage ensures that most of the useful vibrations active in the infrared will be included, although some minerals, such as micas, carbonates and sulphides, exhibit a few characteristic bands in the region below 200 cm^{-1}. Facilities should also be available to expand the wavenumber scale of particular spectral regions to provide good resolution of closely spaced absorption bands, for example in the hydroxyl stretching region of kaolinite.

The spectrometers commonly in use are basically of two types, dispersive and Fourier transform (FT).

4.3.1 *Dispersive instruments*

In these instruments, the detector monitors the intensity of narrow bandwidths of radiation, dispersed by diffraction gratings and/or prisms in a monochromator. The operational mode is double beam, and may be either null balance, in which a servo system equalizes the intensity of radiation in the two beams by attenuating the reference beam, or ratio recording, in which the absolute value of the radiation intensity ratio in both reference and sample beams is continuously monitored. Both types have advantages, the major one for the latter being that in regions of strong absorption the detection system can still measure with high response, whereas the former loses response with increasing absorption.

The power and versatility of a dispersive instrument can be greatly enhanced by interfacing with a microcomputer and, with appropriate programming, (Adams and Black, 1983) can acquire a potential approaching that of an FT instrument.

4.3.2 *Fourier transform instruments*

In nondispersive FTIR spectrometers, the detector continuously monitors the full wavenumber range of radiation emitted by the IR source, providing an inherently more sensitive system than that in a dispersive instrument. FT instruments are interferometers, and they require a dedicated computer to transform their output—an interferogram—into an absorption spectrum. FTIR spectrometers may be either single beam, in which the sample spectrum must be ratioed against a background, or double beam, in which ratioing against background is carried out continuously. Martin (1980) and Durig (1980) have provided excellent texts on the details of FTIR spectrometers.

4.3.3 *Spectral format adopted*

Samples were dispersed, usually in KBr, or occasionally CsI, pressed discs and the spectra illustrated in this chapter were recorded in percent transmission (%T) over the range 4000 to 250 (KBr) or 200 cm^{-1} (CsI) on a *Perkin-Elmer 580B* spectrometer. Three types of spectra are shown, but not for every sample:

(i) High sample concentration, 1–2 mg (Figure 4.1(a)) usually showing total absorption of radiation in the 1000–1100 and 430–530 cm^{-1} silicate regions

(ii) High concentration, 1–2 mg, after heating the disc to 150°C for 16 h to minimize the absorption of adsorbed water, with the wavenumber scale expanded by a factor of 5 to provide higher resolution (Figure 4.1(b))

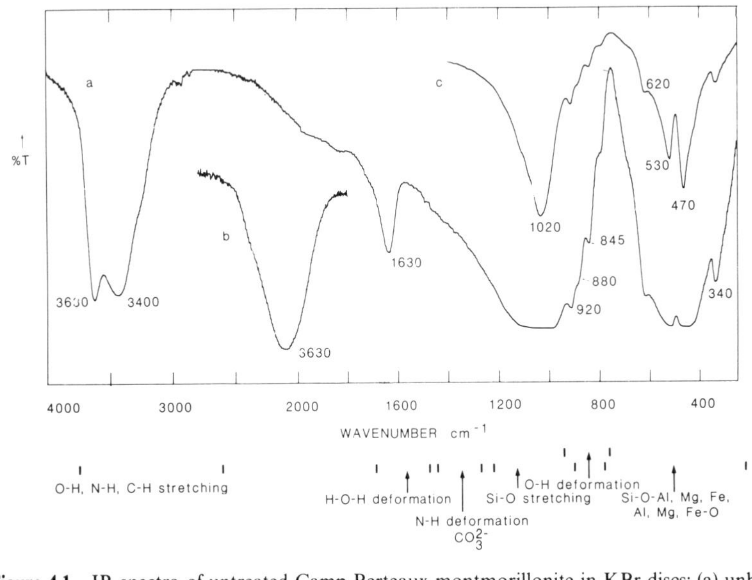

Figure 4.1 IR spectra of untreated Camp Berteaux montmorillonite in KBr discs: (a) unheated, 3 mg, (b) heated 16 h, 150°C, (c) 0.25 mg, T = transmission

(iii) Low concentration, 0.25–0.5 mg (Figure 4.1(c)), showing the spectral detail in the regions of intense silicate absorption near 1000 cm^{-1}.

Absorption bands most characteristic of each particular mineral will be presented, as shown in Figure 4.1 for montmorillonite (Wyoming type, see later for fuller description). Absorption bands of adsorbed molecular water appearing at 3400 and 1630 cm^{-1} will not normally be identified, and neither will C–H absorption bands at 2850–2960 cm^{-1} which arise from organic contaminants.

Also shown on Figure 4.1 are wavenumber ranges for some of the characteristic absorption bands likely to be encountered in spectra of soil clays.

4.4 Preparative methods and pretreatments

4.4.1 *Particle size*

By definition, soil clay minerals are less than 2 μm equivalent spherical diameter, a size that is ideal for examination by IR spectroscopy. Samples whose particle size is much greater than 2 μm, must be preground thoroughly to minimize scattering of incident IR radiation and distortion and broadening

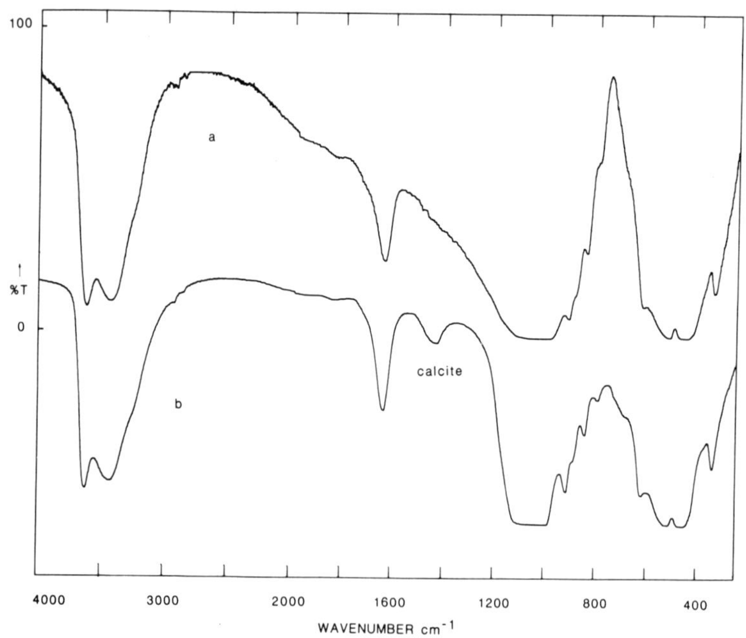

Figure 4.2 IR spectra of Camp Berteaux montmorillonite in KBr discs. (a) untreated, 3 mg; (b) preground, 2 mg

of absorption bands. Pregrinding is most effectively achieved with minimal damage to the mineral structure by maintaining the sample moist with a volatile, inert and organic liquid such as isopropyl alcohol. Grinding may be done mechanically in a stainless-steel capsule with ball-bearings, or by hand with a conventional agate mortar and pestle, usually for 1 to 5 min. The alcohol must be allowed to evaporate completely before proceeding further.

The effect of pregrinding a coarse-grained montmorillonite (Camp Berteaux) can be clearly seen in Figure 4.2, where the broad distorted bands of the original sample (a) become narrower, on a flatter background, and show greatly improved resolution (b). A small amount of coarse-grained calcite absorbing near 1430 cm^{-1} is scarcely detectable before pregrinding, but is very obvious after.

4.4.2 *Removal of organic matter*

If a soil clay contains significant organic matter it may be advantageous to remove it, so as to minimize complications in the IR spectrum. This may be done by boiling with H_2O_2, or by treatment with alkaline hypobromite. While these methods are generally effective in oxidizing organic matter, the former has the disadvantage of producing complex oxalates which may not be completely removed by water washing (Farmer and Mitchell, 1963). The effect of such contamination will be shown later.

4.4.3 *Alkali halide pressed discs*

The required weight of clay mineral (1–2 mg) and 170 mg alkali halide are thoroughly mixed for 1–2 min in a vibrating stainless-steel capsule containing three ball-bearings (2–3.5 mm diameter). The mixture is then pressed in an evacuated die to give a 13 mm diameter disc. The discs may be heated overnight at 150°C to remove adsorbed water, whose absorption bands overlap some of the spectral regions that are important for clay-mineral identification. Heated discs must be allowed to cool to room temperature in a desiccator before recording the spectra. For routine survey work on soil clays, the KBr pressed disc is the most appropriate method, and is used throughout this chapter to illustrate the utility of IR. For more demanding investigations, such as sorption of organic molecules by smectites or determination of intrinsic structural properties of minerals, other methods of sample preparation may be used.

4.4.4 *Sedimented films and deposits*

In this method, which is analogous to the oriented aggregate method used in X-ray diffraction studies, the clay mineral is sedimented, usually from 0.5–1% w/v aqueous suspension, onto either an IR-transparent plate (AgCl, CaF_2,

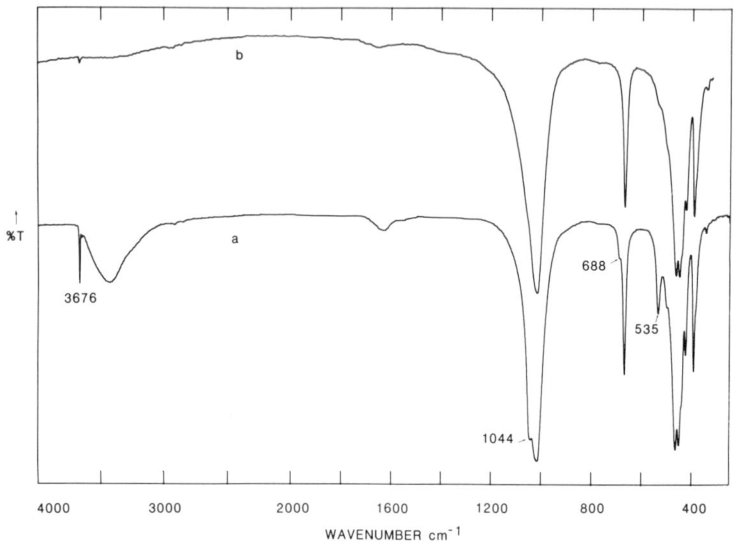

Figure 4.3 IR spectra of talc: (a) random particles (KBr disc), (b) oriented particles with *c** parallel to the IR beam, i = 0°.

Irtran 2, Irtran 4, Ge, Si) to give a uniform deposit, or onto a thin plastic sheet from which a self-supporting film of the clay mineral may be peeled after evaporation to dryness. Further details on other aspects of sample preparation for IR analysis have been described by Russell (1974, 1986) and White (1981). For the platy layer silicates, such preparations behave like quasi-two-dimensional crystals and have important properties that help to identify certain clay minerals. To illustrate this concept, a comparison is made between the spectra of talc, from randomly oriented particles in a KBr pressed disc and from a well-oriented sedimented deposit with the sedimentation plane at 0° incidence to the IR beam (Figure 4.3). Only the bands at 3676, 1044, 688 and 535 cm^{-1} show variation in intensity between the two orientations. They are clearly very much weaker with the film at normal incidence (Figure 4.3(b)) and, therefore, originate in vibrational modes which are perpendicular to the electric vector of the beam (i.e. parallel to the beam direction). The assignments of these bands have been discussed in detail by Russell *et al.* (1970), but the OH-stretching band occurring at 3676 cm^{-1} has the most diagnostic value, because this intensity behaviour, called pleochroism, is exhibited only by trioctahedral layer silicates. Dioctahedral minerals such as muscovite, pyrophyllite and montmorillonite, in which the O–H dipole lies at 75–80° to the *c** axis, show no change in OH intensity between random orientation and normal incidence. Bands such as those listed above for talc are designated as perpendicular bands, because their principal dipole moment change is perpendicular to the electric vector of the IR beam. Their intensity, in keeping with the relationship described above for random and oriented particles, increases when the angle of incidence between the film and the beam is increased by rotating the film. This is an important and rapid means of distinguishing di- and trioctahedral minerals.

4.5 Identification and characterization of clay minerals and associated minerals

It is proposed to present spectra of a selection of the layer silicates and associated minerals most likely to be encountered in IR analysis of clay materials. Where possible, the minerals have been selected for their high purity and authenticity. Characteristic absorption features in the spectra suitable for the identification and, if possible, quantification of particular mineral phases will be highlighted, along with any other relevant information. The minerals selected are ordered and disordered, crystalline and noncrystalline, the last-named being of considerable importance in some soils. A few examples of how to analyse and interpret soil-clay spectra for their mineral components will then be presented, and some guidance given on keeping misinterpretation to a minimum. Important compilations of spectra have been published by Van der Marel and Beutelspacher (1976), Moenke (1962, 1966), Ferraro (1982) and Kodama (1985), and the collection of spectra of inorganic salts by Nyquist and Kagel (1971) is extremely useful. It should be remembered however that none of these collections is perfect, incorrect identification or impurities being not uncommon.

4.5.1 *Clay minerals*

4.5.1.1 *Kaolins.* Minerals of this group are frequently encountered in clay materials. Kaolinite of variable order and crystallinity and halloysite are the most common, whereas dickite is rare. Nacrite, the other polymorph of kaolinite is even rarer. Spectra of all members show a strong family resemblance (Figure 4.4), particularly over the 1200–250 cm^{-1} range, with variations in absorption pattern in the 1100–1150 cm^{-1} range arising from effects of particle size and shape (Farmer and Russell 1966; Rendon and Serna, 1981). The three pure kaolinite polymorphs, kaolinite, dickite and nacrite, can be readily distinguished by differences in position and relative intensity of their OH-stretching bands (Figure 4.4(a), (b), (c)). The complete assignment of these bands is still in dispute, although it is generally accepted that the band near 3620 cm^{-1} arises from internal OH groups, and that near 3700 cm^{-1} arises from internal surface OH groups. Disorder in kaolinite is detectable mainly in the OH-stretching region, although some general broadening of all bands in the spectrum may also occur. While the 3620 and 3700 cm^{-1} bands are essentially unchanged, the 3669, 3652 cm^{-1} doublet is replaced by a single broad band at 3653 cm^{-1} (Figure 4.4(d)). This pattern is typical of disordered

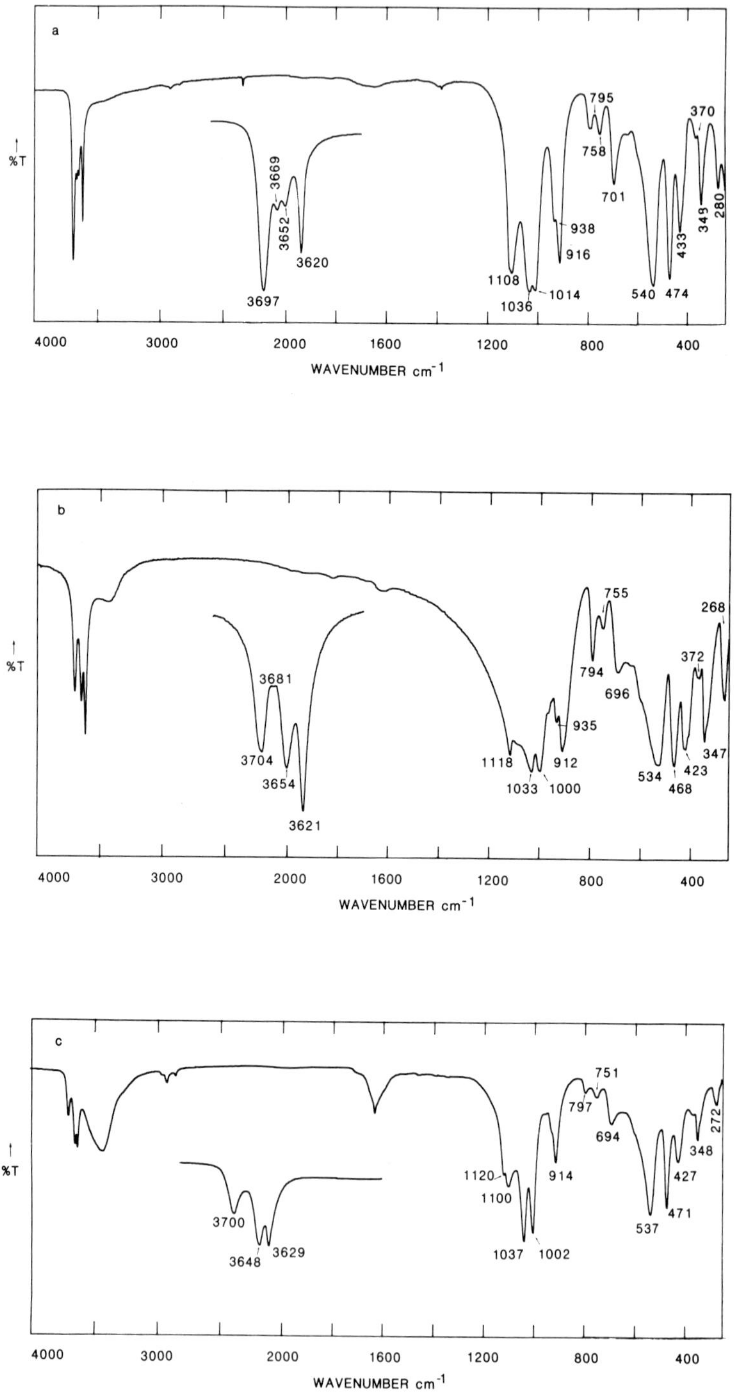
a
%T
795
370
758
3669
3652
3620
3697
701
938
916
349
280
433
1108
1036
1014
540
474
4000
3000
2000
1600
1200
800
400
WAVENUMBER cm⁻¹
b
%T
755
268
3681
3704
3654
3621
794
696
372
935
1118
912
1033
1000
534
468
423
347
4000
3000
2000
1600
1200
800
400
WAVENUMBER cm⁻¹
c
%T
751
797
694
272
348
1120
1100
914
427
471
537
3700
3648
3629
1037
1002
4000
3000
2000
1600
1200
800
400
WAVENUMBER cm⁻¹

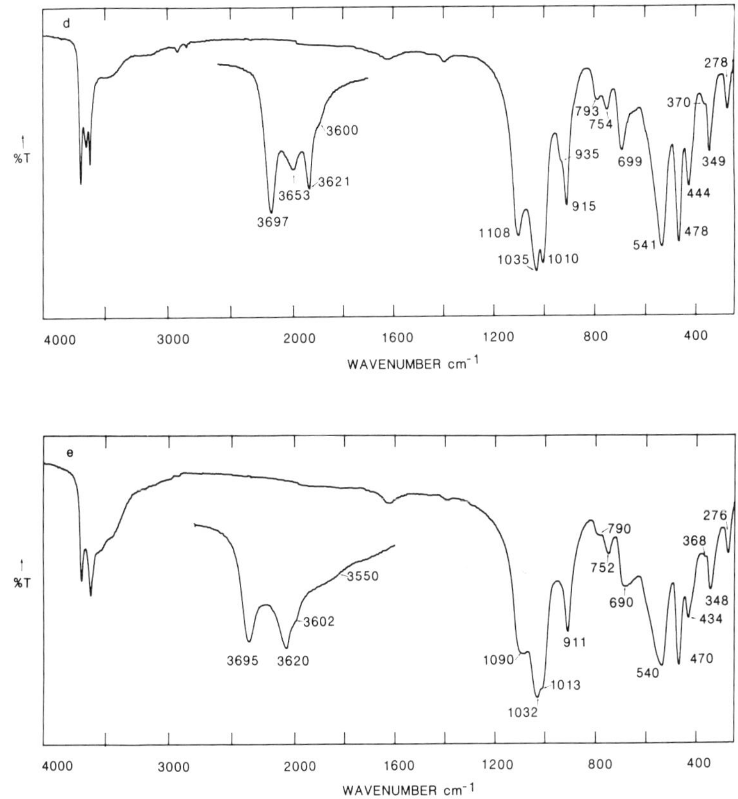

Figure 4.4 IR spectra of kaolin minerals: (a) kaolinite, well crystallized (International), (b) dickite (Anglesey), (c) nacrite (Germany), (d) disordered kaolinite (Frantex B), (e) halloysite (Eureka)

kaolinites and a comparison with dickite and nacrite spectra (Figure 4.4(b), (c)) suggests that such disorder could arise from a small amount of dickite- and/or nacrite-like stacking in the kaolinite structure. Halloysite shows significant broadening of bands throughout its spectrum (Figure 4.4(e)) particularly for the 3700, 3620 cm^{-1} OH-stretching bands because of structural distortion caused by variable hydration. The weak band near 3570 cm^{-1} is not found in all halloysites and may arise from H-bonding between surface OH groups and interlayer water (Kodama and Oinuma, 1963).

The 3700, 3620 cm^{-1} doublet is characteristic for the kaolin group in general, and a semiquantitative estimate of kaolin content can be made from the absorbance of the 3700 cm^{-1} band, which usually lies well separated from those of most other mineral bands. The OH deformation bands of kaolinite at 938, 916 cm^{-1} (Figure 4.4(a)) are also typical for the kaolin group, and arise from vibrations of the inner and inner surface OH groups respectively. These assignments have been made following deuterium-exchange studies (Rouxhet *et al.*, 1977; Ledoux and White, 1964). A useful spectral region to help to distinguish between well-crystallized kaolinite and halloysite is between 750 and 800 cm^{-1}. The two weak bands found here at 795 and 758 cm^{-1} are of about equal intensity for kaolinite (Figure 4.4(a)), whereas for halloysite, the 795 cm^{-1} band is reduced to a very weak inflection (Figure 4.4(e)). The relative intensity of this band in disordered kaolinite (Figure 4.4(d)) has an intermediate value.

4.5.1.2 *Serpentine minerals.* Minerals in this group include chrysotile and antigorite, trioctahedral analogues of kaolinite with very different morphologies. The spectra (Figure 4.5(a), (b)) show fundamental differences from those of the dioctahedral minerals, namely, a lower frequency for the principal Si–O stretching band at or below 1000 cm^{-1}, a single lower-frequency band for Si–

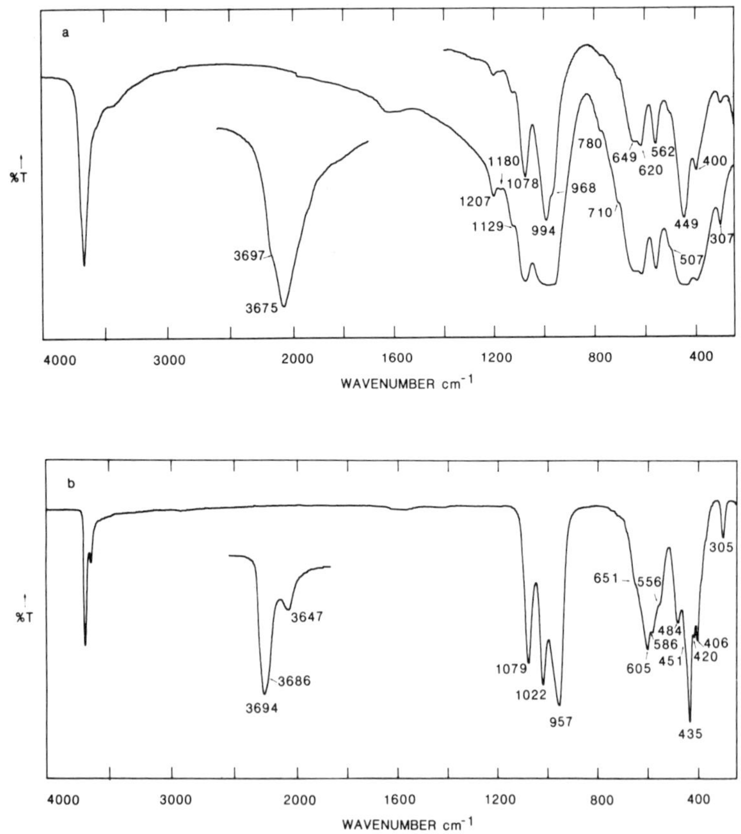

Figure 4.5 IR spectra of serpentine minerals: (a) antigorite (Aberdeenshire), (b) chrysotile (Canada)

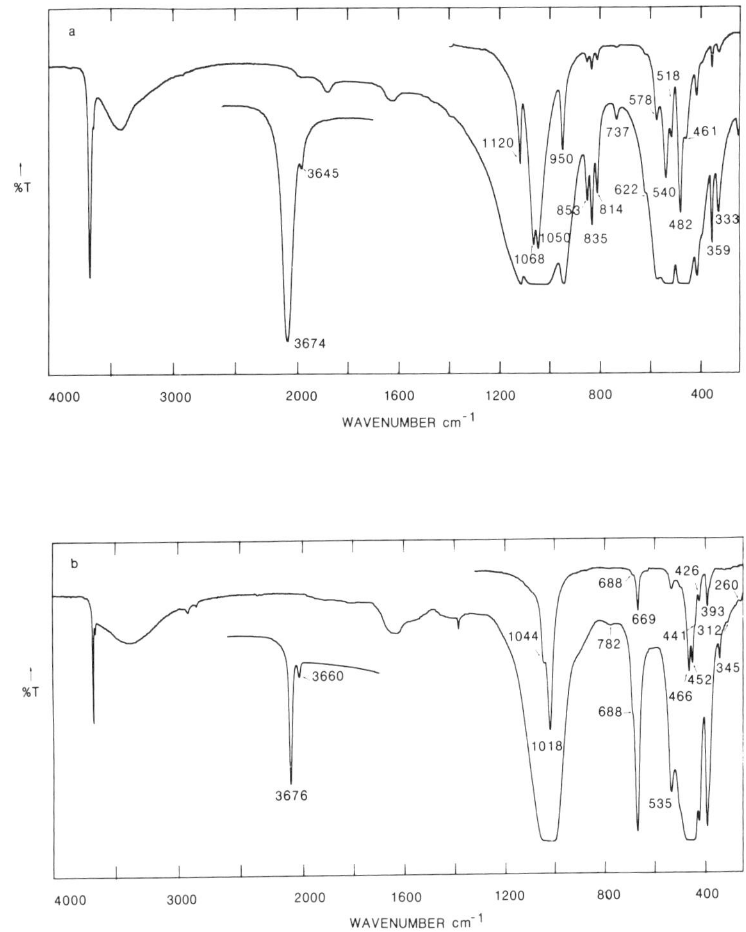

Figure 4.6 IR spectra of (a) pyrophyllite, (b) talc

O–Mg near 450–460 cm^{-1} compared with two or three strong bands in this area for the dioctahedral minerals, and an OH deformation band at a much lower frequency near 600–700 cm^{-1}, compared with 900–950 cm^{-1}. These di–tri differences are also well illustrated by pyrophyllite and talc (Figure 4.6). Chrysotile and antigorite spectra are significantly different from those of the kaolin minerals and are easily distinguished from each other, particularly from the OH-stretching region. Neither pattern, however, shows clear resolution of bands that might be assigned to inner and inner-surface OH groups, as in kaolinite. These serpentine minerals are readily identified in clays from their characteristic absorption bands, but as with all minerals, the whole spectrum must be considered for reliable identification.

4.5.1.3 *Pyrophyllite and talc.* Although these non-expanding layer silicates occur only rarely in clay materials, they are worthy of inclusion here because they can be regarded as the parent unsubstituted structures from which the smectites and mica minerals are derived by isomorphous substitution. Spectral shifts resulting from this substitution have been considered in detail by Farmer and Russell (1964) and Farmer (1974). Spectra of pyrophyllite and talc (Figure 4.6) show the differences in peak positions already reported for di- and trioctahedral minerals. The absorption bands are sharp, as anticipated for well-crystallized, ordered minerals, but it can be seen that the OH-stretching band for the dioctahedral pyrophyllite (Figure 4.6(a)) is considerably broader than that of talc. In talc, a much more symmetrical environment exists for OH groups because of full octahedral occupancy by Mg ions, higher symmetry and crystalline order. The OH groups in talc are pleochroic, as discussed in Section 4.4 (Figure 4.3), whereas in the dioctahedral pyrophyllite the OH groups show no such effect. The weak, low-frequency satellite bands at 3645 cm^{-1} for pyrophyllite, and at 3660 cm^{-1} for talc, arise from small amounts of iron in these minerals, producing the groupings $AlFe^{3+}OH$ and $Mg_2Fe^{2+}OH$ respectively. A very weak inflection, just discernible near 900 cm^{-1} (Figure 4.6(a)) may be the deformation vibration of the former grouping, although an alternative explanation is also possible (Russell *et al.*, 1970).

4.5.1.4 *Smectites.*

(i) Dioctahedral smectites. Montmorillonites contain both tetrahedral and octahedral isomorphous substitution, Al (and occasionally Fe^{3+}) for Si in the former case, and Fe^{3+} and Mg for Al in the latter. As a result of these substitutions, crystalline order is reduced and structural imperfections arise, these leading to considerable broadening of IR absorption bands compared with those of the essentially unsubstituted pyrophyllite. Low-iron montmorillonite, so-called Cheto-type, (Figure 4.7(a)) shows a typically broad OH

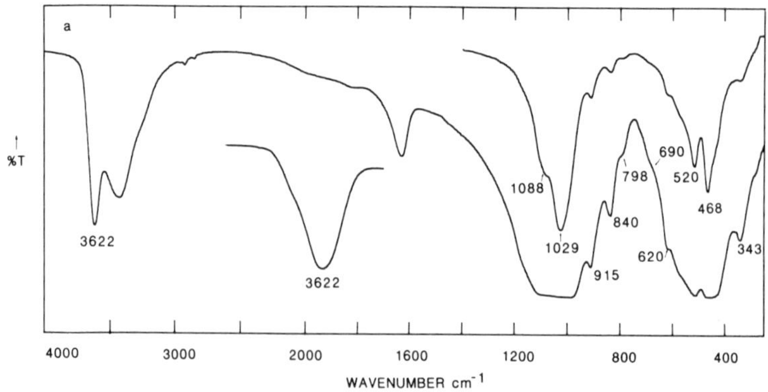

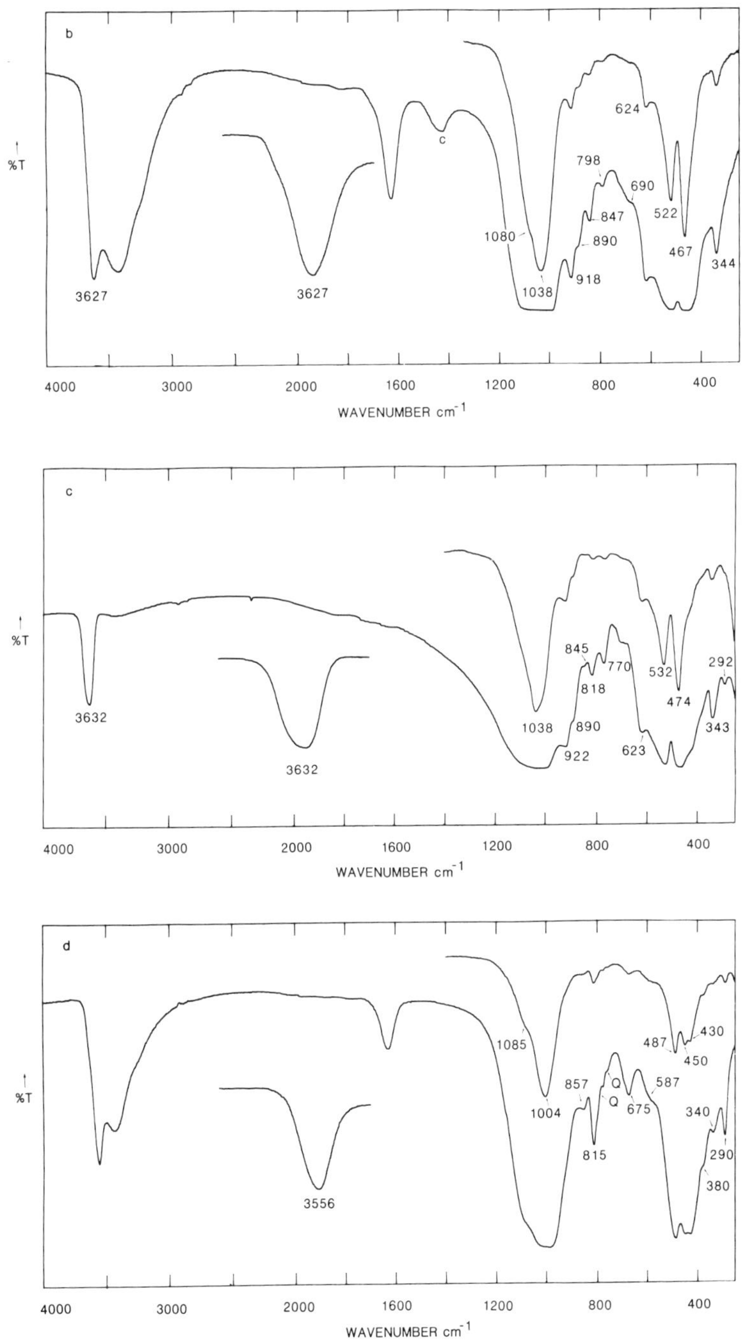

Figure 4.7 IR spectra of dioctahedral smectites: (a) montmorillonite (Otay), (b) montmorillonite, (Camp Berteaux[1]), (c) beidellite-rich clay, (Unterrupsroth[2]), (d) nontronite (California[3]).
[1] 1430 cm^{-1} band due to calcite. [2] Nadeau *et al.* (1985). [3] Goodman *et al.* (1976).

stretching band at 3622 cm^{-1}, which is the overall envelope for a wide range of AlAlOH and AlMgOH environments in the highly substituted and distorted structure. This band occurs close to that of the inner OH groups in kaolinite (Figure 4.4(a)) but can be distinguished by its much greater breadth. More characteristic, however, are the well-resolved OH deformation bands at 915 cm^{-1} (AlAlOH) and 840 cm^{-1} (AlMgOH), assignments that were confirmed by deuteration (Russell *et al.*, 1970). The spectrum of a so-called Wyoming-type montmorillonite (Figure 4.7(b)) in which there is some replacement of octahedral Al by Fe^{3+}, is very similar to that of the iron-free type (Figure 4.7(a)), apart from some relatively minor shifts in peak positions. The most significant difference between the two is the presence of the OH deformation band of $AlFe^{3+}OH$ groupings at 890 cm^{-1} in the Wyoming-type, this being diagnostic of iron substitutions in the smectite structure. The small amount of calcite (which absorbs near 880 cm^{-1}, Figure 4.21(a)) in the montmorillonite would make little or no contribution to the intensity of the 890 cm^{-1} band. It is uncertain whether the weak band at 798 cm^{-1} arises from montmorillonite, a band at this frequency having been ascribed to a platy form of tridymite in a montmorillonite (Wilson *et al.*, 1974).

In contrast to montmorillonite, substitution in beidellite is exclusively Al for Si in the tetrahedral sheet. This type of substitution gives rise to a quite different pattern of absorption bands. Although the beidellite spectrum generally has a strong montmorillonite character (Figure 4.7(c)), the two bands at 818 and 770 cm^{-1} are uniquely characteristic. Their origins are Al–O out-of-plane, and Al–O–Si in-plane, vibrations, and are analogous to similar tetrahedral sheet vibrations in muscovite (Farmer, 1974). The sample of beidellite illustrated contains some montmorillonite (Nadeau *et al.*, 1985) as shown by the OH deformation bands at 890 and 845 cm^{-1}, and the lower OH-stretching frequency at 3632 cm^{-1} compared with 3661 cm^{-1} in pure beidellite.

With increasing substitution of Al by Fe^{3+} in montmorillonite, the $AlFe^{3+}OH$ deformation band shifts from 890 to 870 cm^{-1} and, when sufficient iron is present to give $Fe^{3+}Fe^{3+}OH$ groupings as in nontronite, a band appears at 815 cm^{-1} (Figure 4.7(d)). This grouping is dominant in nontronite, and gives an OH-stretching frequency of 3556 cm^{-1}, some 100 cm^{-1} lower than the equivalent AlAlOH band in beidellite. This low-frequency shift of OH stretching and deformation bands with increasing Fe^{3+} content is a very important criterion for the determination of ferruginous smectites in clay materials. Other characteristic bands for nontronite are found at 675 cm^{-1}, which is an Fe^{3+}–O out-of-plane vibration, and 290 cm^{-1}. A series of iron-rich smectites and nontronites has been studied by Goodman *et al.* (1976).

(ii) Trioctahedral smectites. Hectorite is formed when about 1 in 9 of the octahedral Mg cations in talc is replaced by Li; it is the trioctahedral equivalent of montmorillonite. Its spectrum (Figure 4.8(a)) shows marked perturbation from that of talc (Figure 4.6(b)), i.e. the bands are broadened and

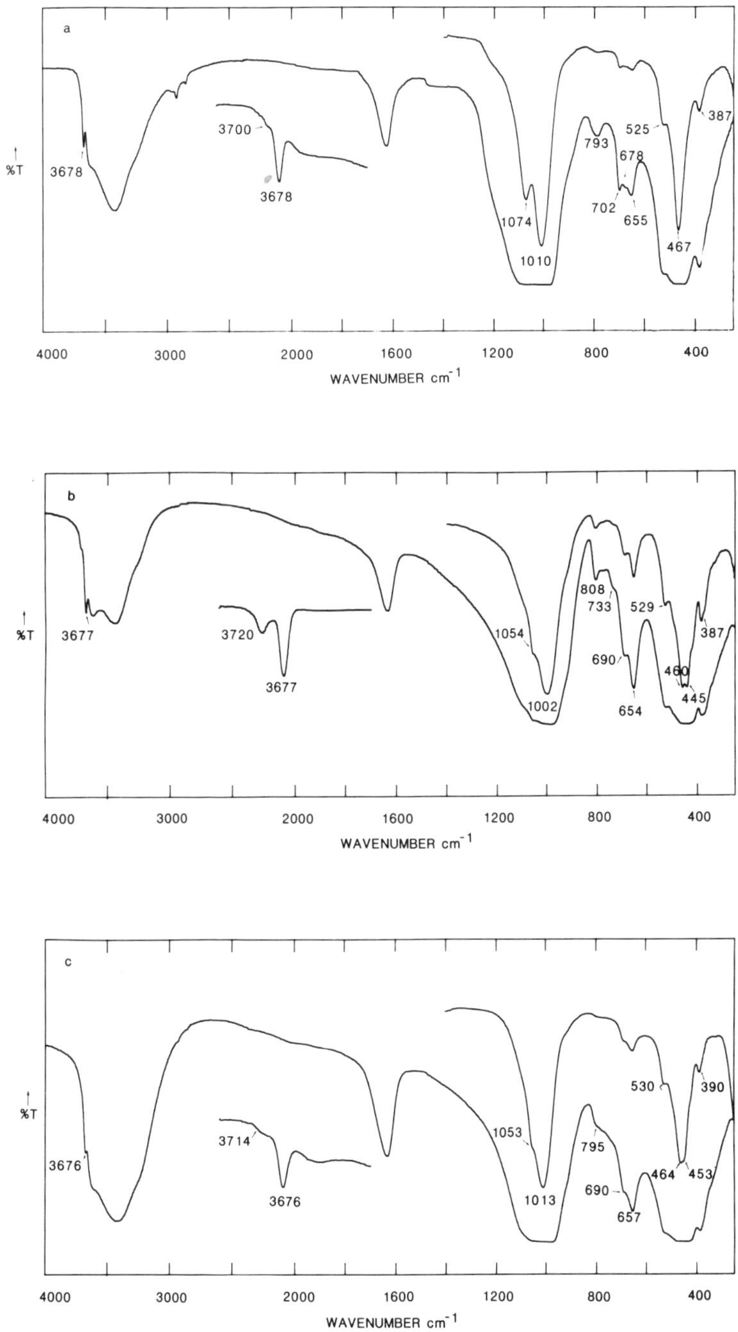

Figure 4.8 IR spectra of trioctahedral smectites: (a) hectorite (Hector, USA), (b) saponite (Allt Ribhean Scotland); (c) stevensite (Arizona)

shifted. Similarly, for saponite, where about 1 in 12 of the tetrahedral Si is replaced by Al, the spectrum (Figure 4.8(b)) shows significant displacement from that of talc but less than observed for hectorite. Both saponite and hectorite show the typical trioctahedral spectrum, but their OH-stretching bands are characteristic. These bands occur within a few wavenumbers of that of talc, but are markedly broadened, the result of lattice distortion and crystal imperfection caused by the isomorphous substitutions. Again, substitution of Mg by Fe^{2+} produces one or more of a group of three bands at lower frequency, depending on iron content. These bands are analogous to those in iron-substituted talc, and occur at 3663 (MgMgFe), 3646 (MgFeFe) and 3642 cm^{-1} (FeFeFe) (Wilkins and Ito, 1967). Saponite and hectorite are potentially difficult to distinguish from their IR spectra, but a reaction occurs in KBr pressed discs after heating to 150°C to remove adsorbed water, and this is diagnostic. Under these conditions, saponite produces a high-frequency satellite at 3720 cm^{-1} on the OH-stretching band (Figure 4.8(b)). It is an effect produced by interlayer K^+ ions (from the KBr), electrostatically repelling the proton of the OH group, thereby raising its frequency. The effect occurs in saponite because of the highly localized negative charge on the Si–O surface, resulting from Al-for-Si substitution. It is a much weaker effect in hectorite (Figure 4.8(a)) where the Li substitution for octahedral Mg results in a more delocalized surface charge. The effect is also very weak in stevensite (Figure 4.8(c)), where the layer charge is derived from a deficiency of octahedral Mg ions, again resulting in the same delocalization of charge at the Si–O interlayer surface. The spectrum of stevensite is otherwise very similar to those of saponite and hectorite, but the weakness of the K-effect band distinguishes it from saponite. Chemical analysis would also make this distinction because stevensite contains no Al, only Si and Mg, as for talc.

4.5.1.5 *Vermiculite, biotite and phlogopite.* Identification of trioctahedral vermiculite is difficult by IR spectroscopy, because there are no sharp, characteristic absorption bands that might be recognized in the complex pattern of a mineral mixture. This can be seen from its spectrum (Figure 4.9(a)) which, although obviously of the trioctahedral type, shows broad, rounded absorption bands that generally resemble saponite. The only recognizably different feature is the OH-stretching band, which is broad, strong and at low frequency. These properties indicate that, although vermiculite is nominally trioctahedral, some of the OH groups are associated with vacancies, and are in a locally dioctahedral environment. Such vacancies are produced by ejection of octahedral Fe ions during oxidation and vermiculitization of biotite (Farmer *et al.*, 1971). The resulting band at 3571 cm^{-1} (Figure 4.9(a)) is close to that for nontronite (Figure 4.7(d)).

Spectra of biotite, which vermiculite closely resembles, and phlogopite (Figure 4.9(b), (c)) show a pattern of OH-stretching bands that vary more or less continuously from the iron-free phlogopite to the iron-rich biotite shown

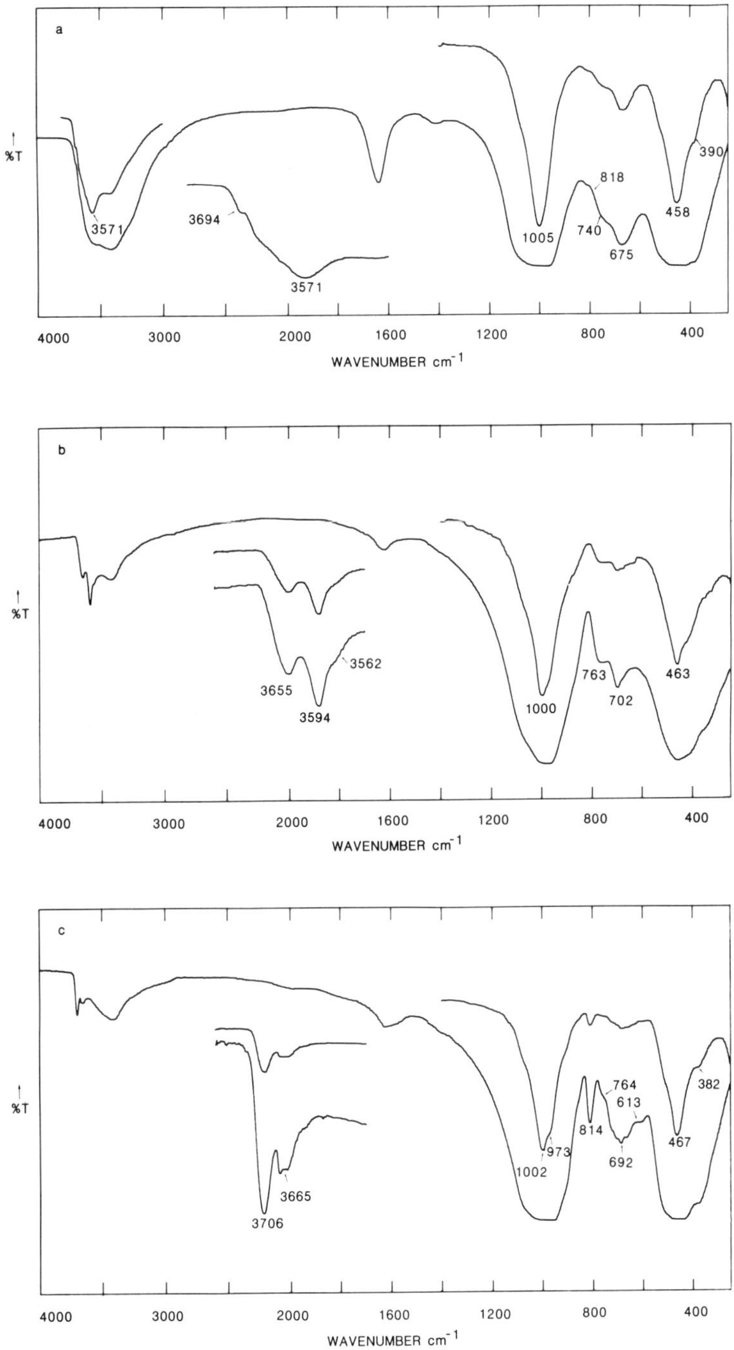

Figure 4.9 IR spectra of (a) vermiculite (G. Walker), (b) biotite, (c) phlogopite

here. Detailed treatments of these spectra can be found in Vedder (1964), Juo and White (1969), Farmer and Russell (1964).

4.5.1.6 *Illite.* The composition of illite is quite variable, as is its IR spectrum. Nevertheless, two main types of spectrum have been identified, being associated with muscovite-like (Figure 4.10(a)) and phengite-like illites (Figure 4.10(b)). An important chemical difference is that phengite has a much higher Mg content than muscovite. The illite spectra can range between these two types and may show even weaker absorption bands near 825 and 750 cm^{-1} than in the spectra illustrated. Consequently, it can be difficult to identify the type of illite, but determination of the presence of the mineral usually presents little difficulty. The broad OH-stretching band near 3625 cm^{-1}, ranging down below 3620 cm^{-1} in some instances, coupled with the 825, 750 cm^{-1} doublet is almost certainly diagnostic. The variety most frequently encountered in soil clays is the phengite type, with relatively weak

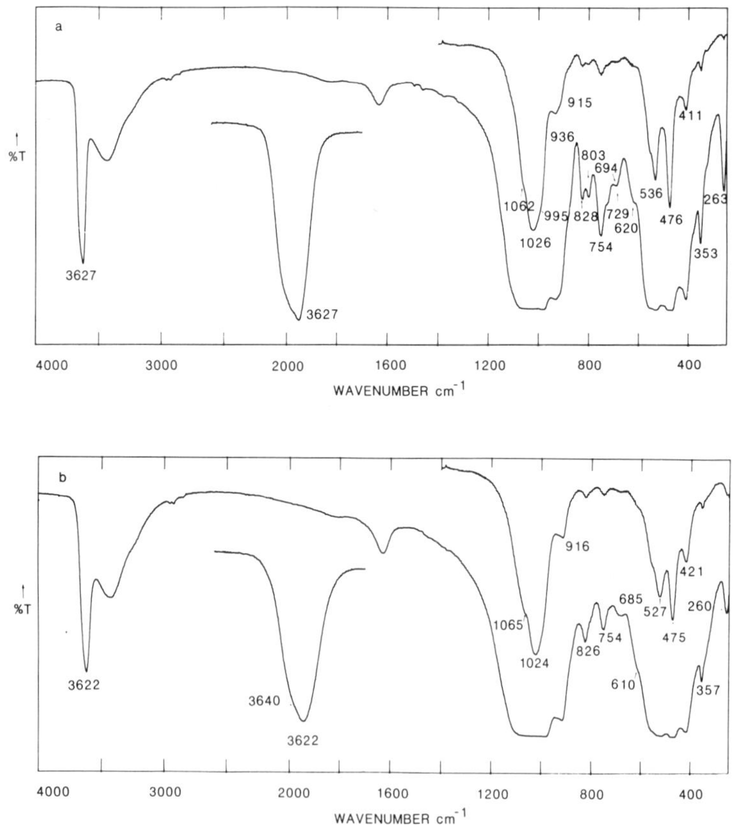

Figure 4.10 IR spectra of illite: (a) $2M_1$ illite (muscovite-like); (b) 1M illite (phengite-like)

bands at 825 and 750 cm^{-1}. These bands originate from Al–Mg–OH deformation and, by analogy with muscovite (Farmer, 1974), Al–O–Si in-plane vibration. Diagenetic illites sometimes contain interlayer NH_4^+ ions replacing K^+ (Nadeau and Bain, 1986), the NH_4^+ ions absorbing at 3260 and 1430 cm^{-1}.

4.5.1.7 *Glauconite and celadonite.* These iron-rich micas derive most of their layer charge from octahedral replacement of trivalent (Al, Fe^{3+}) by divalent cations (Mg, Fe^{2+}). Their structures are characterized by pairing of a trivalent and a divalent cation in the octahedral sheet. In glauconite there is also significant tetrahedral substitution and this can lead to disorder and broadening of bands, as can be seen in Figure 4.11(a). Nevertheless, the poorly resolved OH-stretching bands can be distinguished at 3601 (AlMgOH), 3558

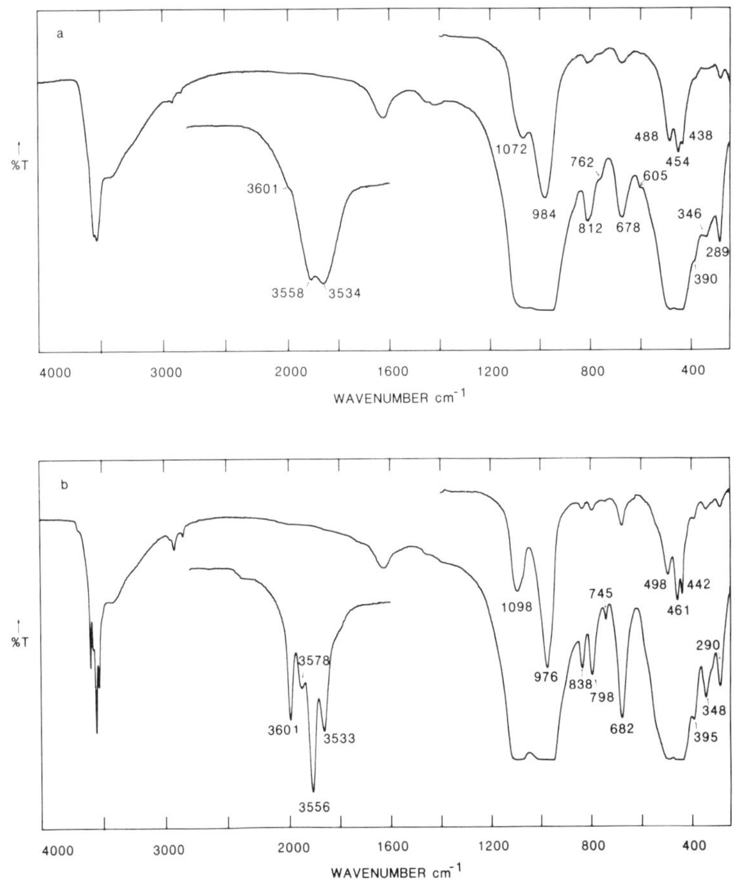

Figure 4.11 IR spectra of (a) glauconite; (b) celadonite, Buckley *et al.* (1978)

(Fe^{3+}MgOH) and 3534 cm^{-1} ($Fe^{3+}Fe^{2+}$OH) (Farmer *et al.*, 1967). The spectrum of the celadonite shown here is much sharper (Figure 4.11(b)) and testifies to a more highly ordered structure. The four clearly resolved OH-stretching bands correspond with all possible tri–divalent pairings from an octahedral population containing Al, Mg, Fe^{2+}, Fe^{3+}. The assignments are those detailed for glauconite, the band at 3578 cm^{-1} being attributed to (AlFe^{2+}OH). This is a very characteristic pattern, which varies with composition but which is easily recognized, as is the Si–O pattern. The OH deformation bands for the major octahedral pairings occur at 838 (AlMgOH and 798 cm^{-1} (Fe^{3+}MgOH) in celadonite (Figure 4.11(b)) whilst in glauconite the unresolved doublet at 812 cm^{-1} (Fe^{3+}MgOH and $Fe^{3+}Fe^{2+}$OH) is characteristic (Figure 4.11(a)). Buckley *et al.* (1978) proposed that celadonite was ordered (sharp spectrum), glauconite was disordered (broad spectrum), and that they were two distinct mineral types. Slonimskaya *et al.* (1986), however, have shown from IR spectroscopy that both minerals can be ordered or disordered, that the sharpness of the spectrum is not diagnostic, and that there is probably a continuous series between the two minerals.

4.5.1.8 *Chlorites.* The compositions and, therefore, the IR spectra of chlorites vary widely. Figure 4.12(a) shows the spectrum of a fairly iron-rich type, also containing aluminium (brunswigite). The strong broad doublet at 3565, 3434 cm^{-1} arises from OH-stretching vibrations of the interlayer hydroxide sheet. The exact position of this doublet varies with composition, shifting to lower frequency with increasing aluminium and iron content, but is distinctive for a fully trioctahedral chlorite. The weak feature at 3662 cm^{-1}, sometimes sharp and well-resolved at higher frequency, arises from OH groups in the trioctahedral 2:1 layer, similar to that observed in the spectra of biotite and phlogopite (Figure 4.9(b), (c)). The strong band at 667 cm^{-1} (Figure 4.12(a)) comprises OH deformation components from the 2:1 layer and the hydroxide sheet. The broad OH-stretching doublet of chlorite in clay materials is frequently masked by the stretching band of adsorbed water, removal of which is essential for unambiguous identification.

The IR spectrum of the magnesium-rich penninite (Figure 4.12(b)) is generally much sharper than that of brunswigite. This arises because of the lower levels of isomorphous substitution in penninite in both the 2:1 and the hydroxide sheet. The OH-stretching bands of the latter occur at 3625 and 3485 cm^{-1}, the higher-frequency band moving towards the unsubstituted brucite position 3698 cm^{-1}. The marked splitting of the principal Si–O band at 998, 960 cm^{-1} reflects the relatively low level of Al-for-Si substitution in the 2:1 layer (Tuddenham and Lyon, 1959), and this can be an important diagnostic feature.

With increasing substitution of Mg by Al, chlorites become more dioctahedral. The aluminous chlorite, sudoite, shows three strong bands in the OH-stretching region (Farmer 1974; Kodama 1985) as shown in Figure 4.12(c). The

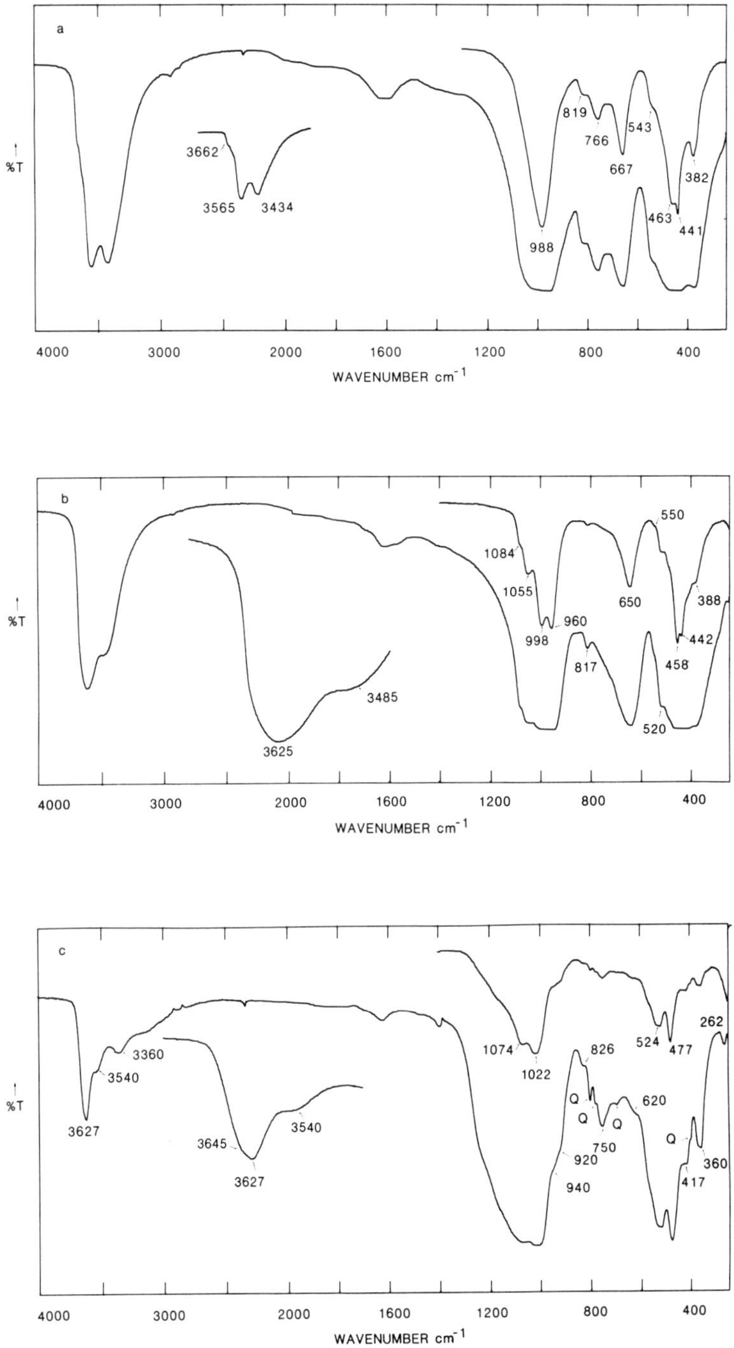

Figure 4.12 IR spectra of chlorites: (a) brunswigite, (b) penninite (Zermatt), (c) sudoite

band at 3627 cm^{-1} is much stronger than the other two bands, suggesting a stronger contribution from a more dioctahedral 2:1 layer. Another feature indicating that the 2:1 layer resembles that of muscovite is the band at 750 cm^{-1}. However, the weak 3540, 3360 cm^{-1} OH doublet (Figure 4.12(c)) suggests that the hydroxide sheet still retains some trioctahedral character.

4.5.1.9 *Interstratified minerals* Infrared spectroscopy responds to interstratified clay minerals as if they were a simple mixture, their IR components being additive. The only exception known to the author is rectorite with a spectrum showing unique bands not present in the spectra of its separate components.

The spectrum of an illite-smectite interstratification (Figure 4.13(a)) closely resembles that of a Cheto-type montmorillonite (Figure 4.7(a)), with a contribution from a phengite-like illite (Figure 4.10(b)). The weakness of the 756 cm^{-1} illite band should not necessarily be taken to indicate a minor amount of illite, because as indicated previously, this band may be inherently weak in some illites.

Similarly, the spectrum of aliettite (Figure 4.13(b)) shows features that can be ascribed to its separate components, talc and saponite (Figure 4.6(b), 4.8(b)). Its absorption bands have intermediate breadth and resolution. Specifically, saponite is indicated by the 3715 cm^{-1} K^+-effect band, and the relatively broad $MgFe^{2+}Fe^{2+}OH$ band at 3633 cm^{-1}. Talc is signalled by the overall sharpness of all bands, but particularly by the $MgMgFe^{2+}OH$ band at 3661 cm^{-1}, which has the same width at half height as the Mg_3OH 3767 cm^{-1} band (Figure 4.6(b)).

The spectrum of rectorite (Figure 4.13(c)), which is a regular alternation of the Na-mica paragonite and Na-saturated beidellite, obviously resembles those of muscovite (closely similar to paragonite) (Kodama, 1985) and beidellite (Figure 4.7(c)) (Farmer and Russell, 1964). However, there are many additional bands, and a spectral sharpness not shown by either component, for example, bands at 1120, 992, 920, 808, 730 and 337 cm^{-1}, suggesting structural integrity and order beyond that implied by simple interstratification.

In other interstratified minerals, for example kaolinite-smectite, both components are relatively easy to identify from their IR spectra. In chlorite-vermiculite, however, only one component can be recognized because the chlorite pattern dominates and completely overlies that of the vermiculite, to which it has many similarities. It should perhaps be stressed that IR cannot distinguish regular and random interstratifications.

Other examples of spectra of interstratified minerals can be found in Kodama (1985), Van der Marel and Beutelspacher (1976) and Oinuma and Hayashi (1968) which the interested reader should consult.

4.5.1.10 *Palygorskite and sepiolite.* Despite their unique channel structures, these minerals are similar to layer silicates intermediate between the di- and

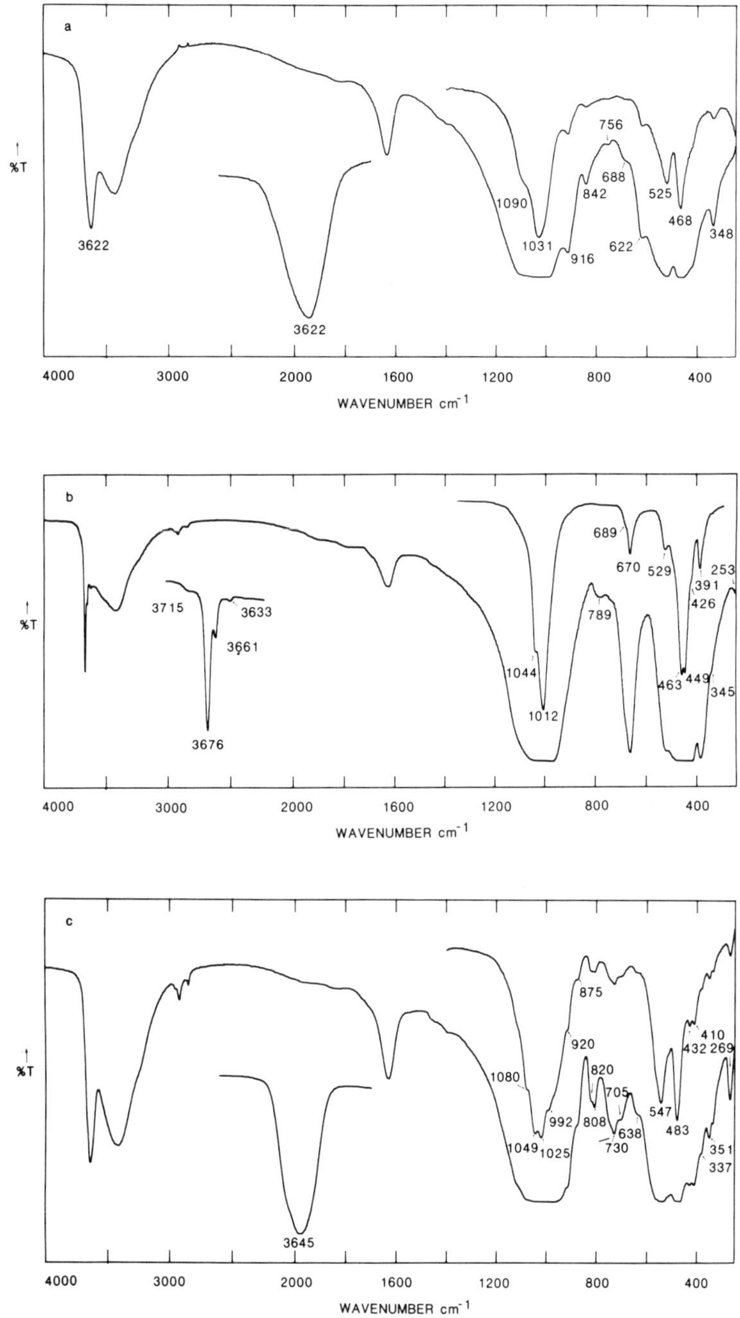

Figure 4.13 IR spectra of interstratified minerals: (a) illite-smectite (Nadeau, 1980), (b) aliettite, (c) rectorite (Baluchistan)

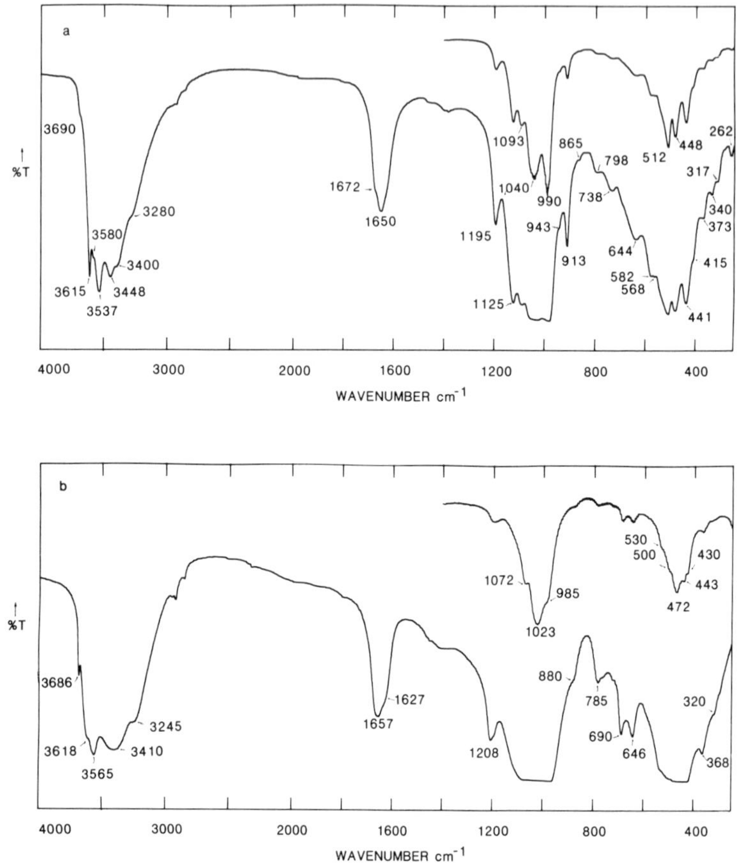

Figure 4.14 IR spectra of (a) palygorskite (Cabrach, Scotland), (b) sepiolite (Spain)

trioctahedral types respectively. In palygorskite, the structural OH-stretching bands at 3615 cm^{-1}, with a weak inflection at 3690 cm^{-1} (Figure 4.14(a)) originate in the aluminous octahedral layer, which must contain both vacancies and complete site filling. The dioctahedral character is dominant, as shown by the well-defined AlAlOH deformation band at 913 cm^{-1}. The unique nature of the 2:1 ribbon layers in the structure gives the complex Si–O stretching pattern, the most characteristic features of which are the high- and low-frequency components at 1195 and 990 cm^{-1}.

The more completely trioctahedral sepiolite shows a dominant OH-stretching band at 3686 cm^{-1} (Figure 4.14(b)), slightly higher than that of talc (Figure 4.6(b)) and antigorite (Figure 4.5(b)), with which it might be compared, and a typically trioctahedral OH deformation band within the 690, 646 cm^{-1} doublet. Again, the unique ribbon layer structure gives rise to the very characteristic Si–O vibration at 1208 cm^{-1}, one of the highest recorded for layer silicates. The bands near 1200 cm^{-1} for palygorskite and sepiolite may well have their origin in the virtually straight Si–O–Si bonds linking the 2:1 ribbons along their edges. The only other silicate mineral absorbing near 1200 cm^{-1} is sillimanite, the origin of the band being the straight Si–O–Al bond linking tetrahedral silicate and aluminate chains (Farmer, 1974).

A striking feature of palygorskite and sepiolite spectra is the complex nature of absorption bands in stretching and bending regions of water molecules. Both minerals contain water in channels in their structures, and even in KBr discs, as here, they display patterns of bands originating in bound molecular water directly coordinated to Al or Mg ions at the edges of the 2:1 ribbon layers (3580, 3537, 1650 cm^{-1} in palygorskite; 3618, 3565, 1627 cm^{-1} in sepiolite), and unbound zeolitic water molecules within the channels (3448, 3400, 3280, 1672 cm^{-1} for palygorskite; 3410, 3245, 1657 cm^{-1} for sepiolite). The characteristic IR pattern of the water molecules in these minerals has been studied by many workers, among them Hayashi *et al.* (1969), Tarasevich (1970) and, more recently, Serna *et al.* (1977).

4.5.2 *Other minerals*

4.5.2.1 *Feldspars.* These framework silicate minerals are frequent constituents of clay materials. Despite their extensive mutual solid solution and complexity of composition, they can be distinguished by their IR spectra, at least at principal species level. The spectra of microcline and albite (Figure 4.15(a), (c)), which are close to the K and Na end-member compositions respectively, are closely related, but the quartet of bands between 700 and 800 cm^{-1} and the sharp feature at 400 cm^{-1} are particularly characteristic of albite (Figure 4.15(c)). These bands can also be clearly seen in the spectrum of orthoclase (Figure 4.15(b)), identifying the presence of an albite component in this predominantly K-feldspar. Anorthite, the Ca-plagioclase end-member,

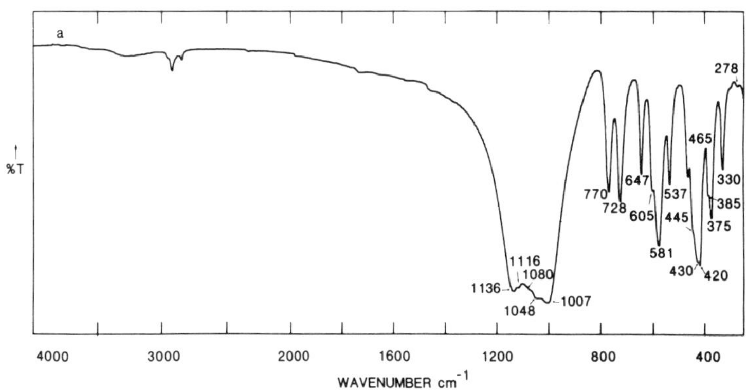

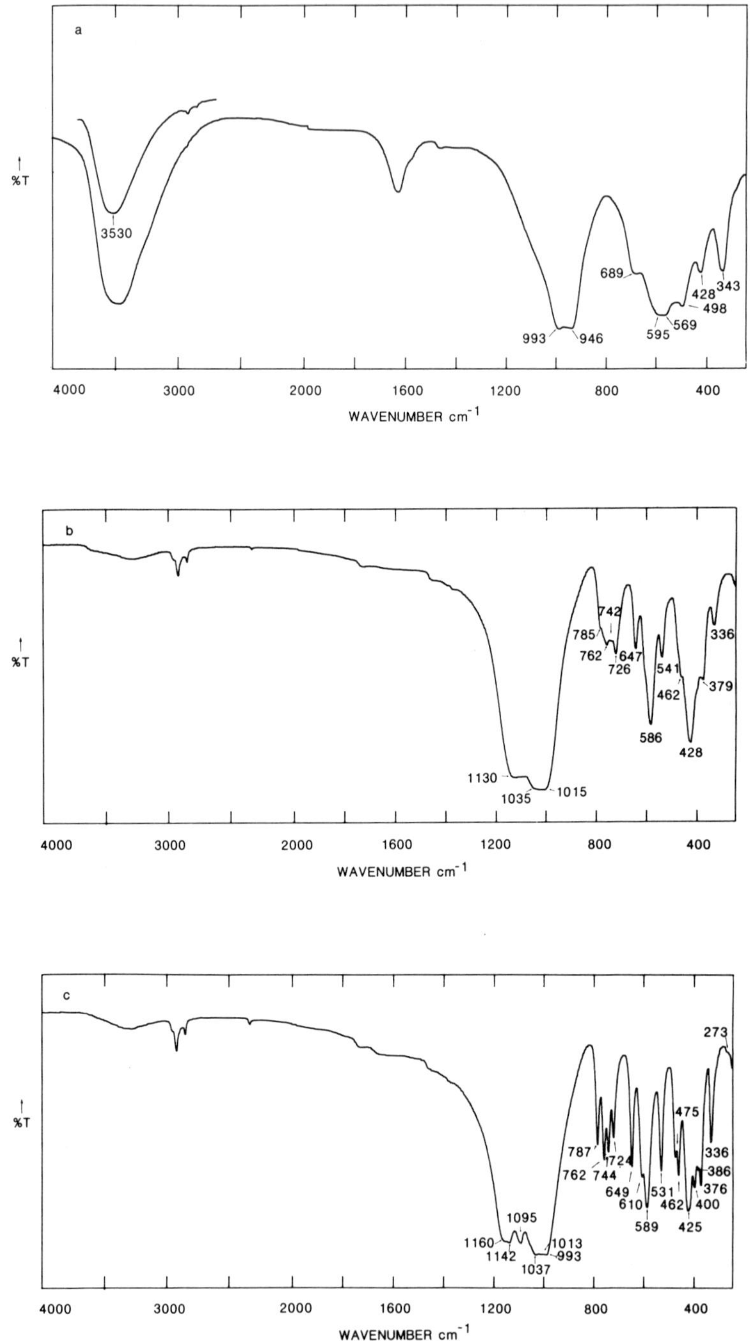

Figure 4.15 IR spectra of feldspars: (a) microcline, (b) orthoclase, (c) albite, (d) anorthite

has a related but uniquely different spectrum (Figure 4.15(d)). It too has a very complex pattern of absorption bands in the 800–200 cm^{-1} region, but at 929 cm^{-1} there is an intense Si–O vibration not seen in the other feldspars. Kodama (1985) has presented a short series of plagioclase-feldspar spectra, in which the steady enhancement of absorption at about 920 cm^{-1} correlates well with the increasing proportion of anorthite. Feldspars show wide variation in crystalline order, and this is reflected in the IR spectra, which can range from broad and diffuse to sharp and well defined.

4.5.2.2 *Imogolite and allophane.* Allophane is a general term referring to noncrystalline, hydrated aluminosilicates of variable composition. IR spectra of allophanes reflect this variation and indicate, from the position of the main Si–O absorption band, the degree of polymerization of the silicate anion and the likely coordination number of aluminium. Imogolite is a specific name for a cryptocrystalline aluminosilicate, with an Al:Si molar ratio of 2:1, and with a fibrous tubular morphology (Cradwick *et al.*, 1972). It contains orthosilicate anions coordinated to the inside surface of a tubular gibbsite sheet. Proto-imogolite allophane, a closely related noncrystalline aluminosilicate, has the same short-range structure as imogolite, but does not show its fibrous morphology. These minerals typically occur in acid soils and have very characteristic spectra.

The IR spectrum of imogolite (Figure 4.16(a)) shows well-resolved components at 993 and 946 cm^{-1} in the orthosilicate Si–O band, and clearly defined bands in the 800–200 cm^{-1} region, the one at 343 cm^{-1} (shifting to 348 cm^{-1} on heating to 150°C) being particularly useful for identification of imogolite structures (Farmer *et al.*, 1977). The spectrum of proto-imogolite allophane (Figure 4.16(b)) has substantially the same pattern as imogolite, but is more diffuse. The band at 348 cm^{-1} is nevertheless still well defined.

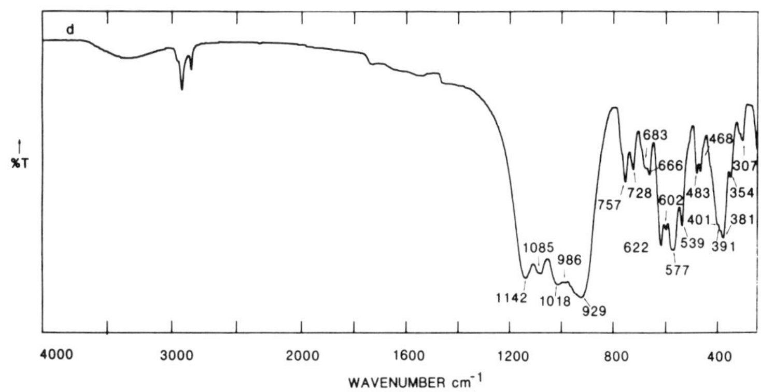

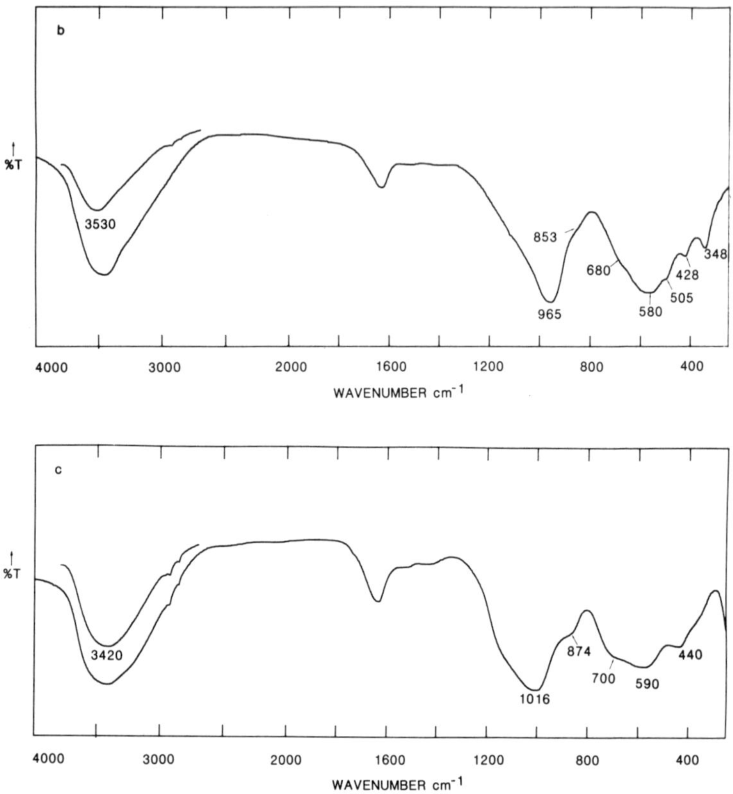

Figure 4.16 IR spectra of (a) imogolite (Kurayoshi), (b) proto-imogolite allophane (synthetic), (c) hydrous feldsphathoid allophane (Silica Springs, New Zealand)

Allophanes formed at high pH have a framework type of structure, with Al in 4- and sometimes in both 4- and 6-fold coordination. The spectrum of such an allophane, with the same 2:1 Al/Si ratio as proto-imogolite allophane (Figure 4.16(c)), shows the Si–O band at 1016 cm^{-1}, indicative of a polymerized silicate anion containing tetrahedral Al. The general pattern resembles that of a feldspathoid, and has been described as a hydrous feldspathoid (Farmer *et al.*, 1979). Other allophanes contain phyllosilicate anions which may be related to kaolins (Farmer *et al.*, 1979). Allophanes generally have broad, unresolved OH-stretching bands, and become impossible to distinguish in a mixture of minerals. However, the considerable adsorbed and hydration water associated with them can suggest their presence. A more positive indication of imogolite structures can be obtained from difference spectroscopy using computer subtraction, either for the clay before and after oxalate extraction of the allophane/imogolite, or before and after dehydroxylation of the allophane (Farmer *et al.*, 1984).

4.5.2.3 *Aluminium minerals.* The spectra of the trihydrate mineral gibbsite and the monohydrates diaspore and boehmite are shown in Figure 4.17. The monohydrates do not occur extensively, but are included for comparison with their iron analogues goethite and lepidocrocite, both of which have a widespread occurrence. The most characteristic feature of gibbsite is its

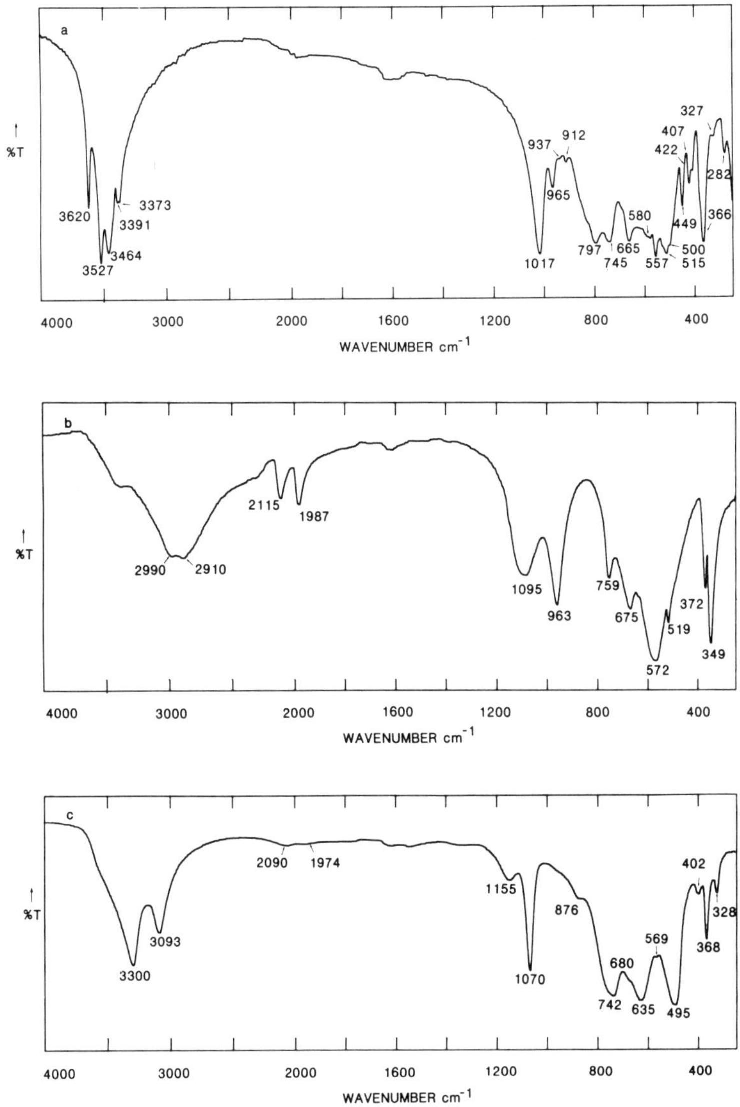

Figure 4.17 IR spectra of aluminium hydroxides: (a) gibbsite; (b) diaspore; (c) boehmite (synthetic)

pattern of OH-stretching bands. The specimen illustrated is well crystallized and shows splitting of the 3391, 3373 cm^{-1} doublet. This is not always observed, but the quartet of bands at 3620, 3527, 3464 and 3380 cm^{-1}, and the 1017 OH deformation and 366 cm^{-1} bands are quite diagnostic. Disappearance of the bands after heating to 350°C will provide confirmation of gibbsite.

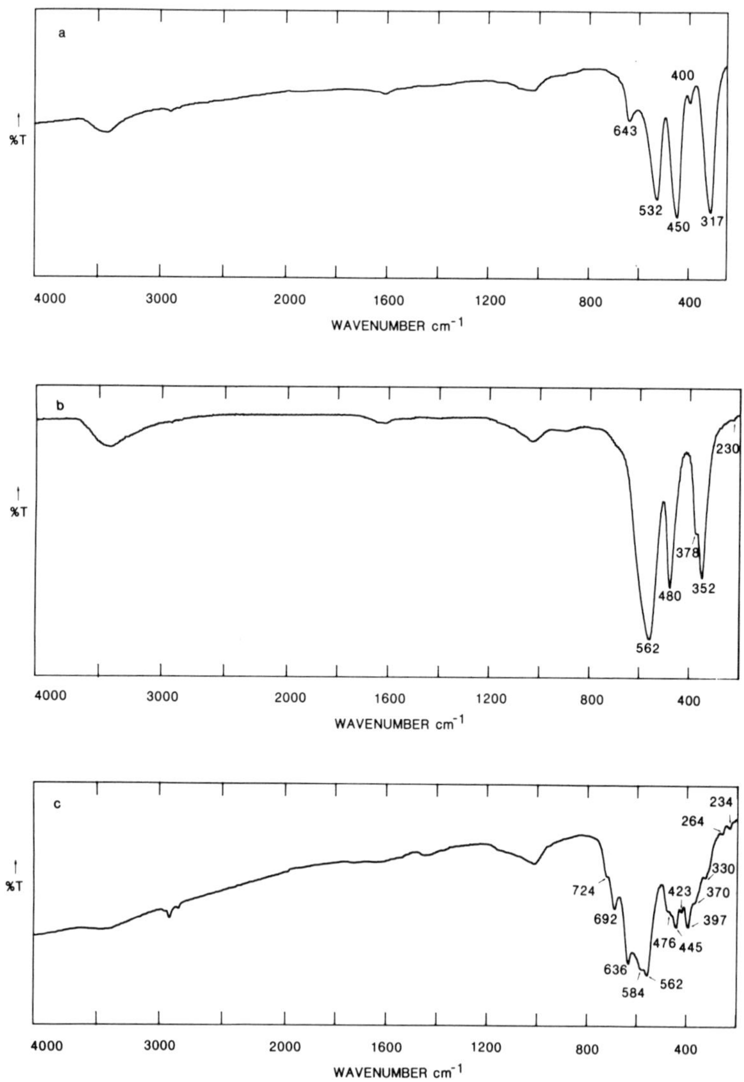

Figure 4.18 IR spectra of iron oxides: (a) hematite (kidney ore), (b) hematite (specularite, synthetic), (c) maghemite

4.5.2.4 *Iron minerals.* Hematite, an anhydrous oxide, occurs in two morphological forms, namely, a platy form (kidney ore) and a more cubic, equiaxial form (specularite). Their spectra (Figure 4.18(a), (b)) are generally similar, but show considerable differences in detail. Because of differences in crystal size, and particularly, shape, the bands at 532, 450 and 317 cm^{-1} in kidney ore shift

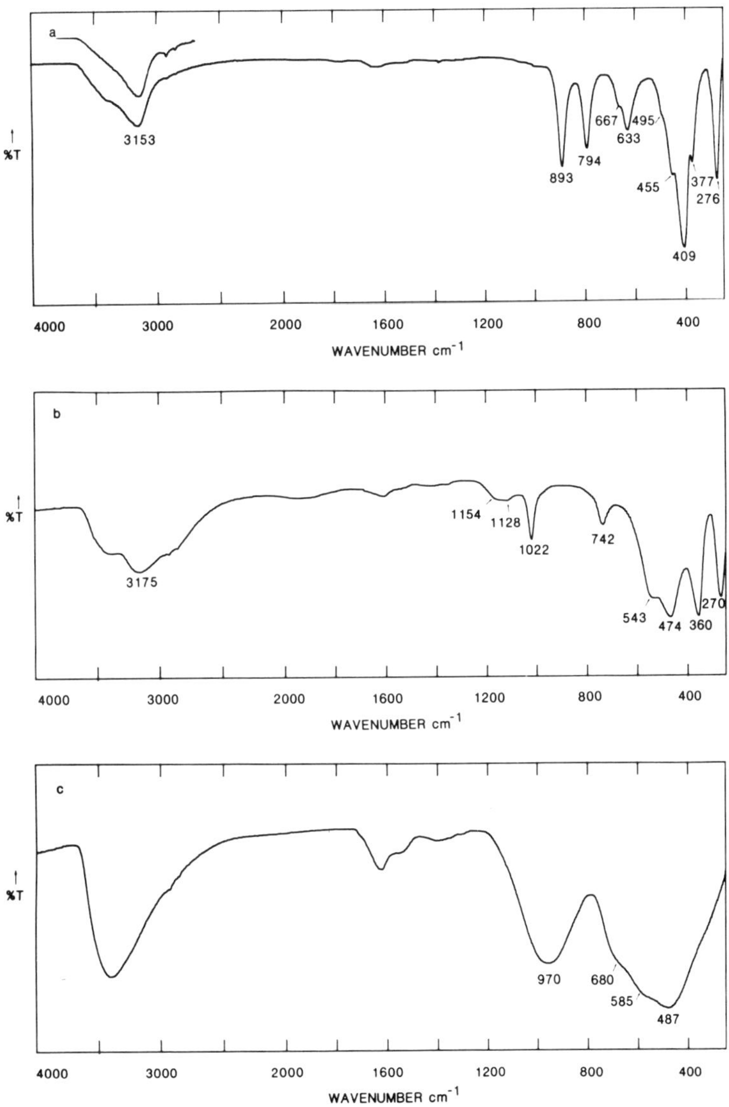

Figure 4.19 IR spectra of iron hydroxides: (a) goethite (synthetic), (b) lepidocrocite (synthetic), (c) ferrihydrite (Egmont, New Zealand)

to higher frequencies at 562, 480 and 352 cm^{-1} in specularite. Again, the weaker bands at 643 and 400 cm^{-1} in kidney ore shift to lower frequencies and merge with the stronger bands in specularite, the 378 cm^{-1} band being just resolved (Rendon and Serna, 1981; Serna *et al.*, 1982). The specularite form is the more common in clay materials. Solid-state dehydration of platy goethite also yields platy hematite, not a new form of hematite as was proposed by Yariv and Mendelovici (1979). This type of hematite morphology was also reported in the new mineral macaulayite (Wilson *et al.*, 1981, 1984). Maghemite, another anhydrous oxide, has a surprisingly sharp pattern of absorption bands (Figure 4.18(c)) with a general resemblance to magnetite, from which it is thought to be derived. The IR spectrum of magnetite is broad and diffuse, showing two ill-defined features near 570 and 347 cm^{-1} (Kodama, 1985). All the above oxides generally occur at low abundances in clay materials and may be difficult, if not impossible, to identify by their IR spectra because of overlap by silicates.

The IR spectra of the iron oxyhydroxides goethite and lepidocrocite (Figure 4.19(a), (b)) are easily distinguished. Thus, the strongly hydrogen-bonded OH absorbs at 3153 cm^{-1} in goethite, and at 3175 cm^{-1} in lepidocrocite, indicating a slightly weaker H-bond in the latter. Again, OH deformation bands occur at 893 (in-plane) and 794 cm^{-1} (out-of-plane) for goethite, and 1154, 1022, and 742 cm^{-1} respectively for lepidocrocite. The deformation bands and the lower-frequency lattice bands are difficult to identify in the spectra of clays rich in layer silicates, but the OH-stretching bands are usually diagnostic of an oxyhydroxide mineral, after removal of adsorbed water by heating. However, it would be difficult to distinguish goethite and lepidocrocite in these circumstances.

Ferrihydrite is a naturally occurring, poorly ordered, hydrated iron oxide, and its IR spectrum (Figure 4.19(c)) is a very simple, broad-banded pattern. Natural specimens are invariably associated with silicate, and from the position of the Si–O band at 970 cm^{-1}, it is present as orthosilicate. Hydroxyl groups are present in the structure (Russell, 1979), which from the band widths is very disordered. Recognition of this phase in a clay would be virtually impossible, except perhaps by difference spectroscopy as described for imogolite.

4.5.2.5 *Silica minerals.* Quartz is a common constituent of most clay materials, and its spectrum (Figure 4.20(a)) shows several sharp bands which have been used for identification, the 800, 781 cm^{-1} doublet being the most generally useful. In contrast, the spectrum of tridymite (Figure 4.20(b)) shows only broad bands, testifying to its relatively disordered structure. The single broad band at 792 cm^{-1}, with no accompanying absorption down to 600 cm^{-1} has been used for its positive identification in fine-grained siliceous materials (Wilson *et al.*, 1974). The tridymite spectrum closely resembles that of opaline silica. For comparison, the spectrum of cristobalite, the third

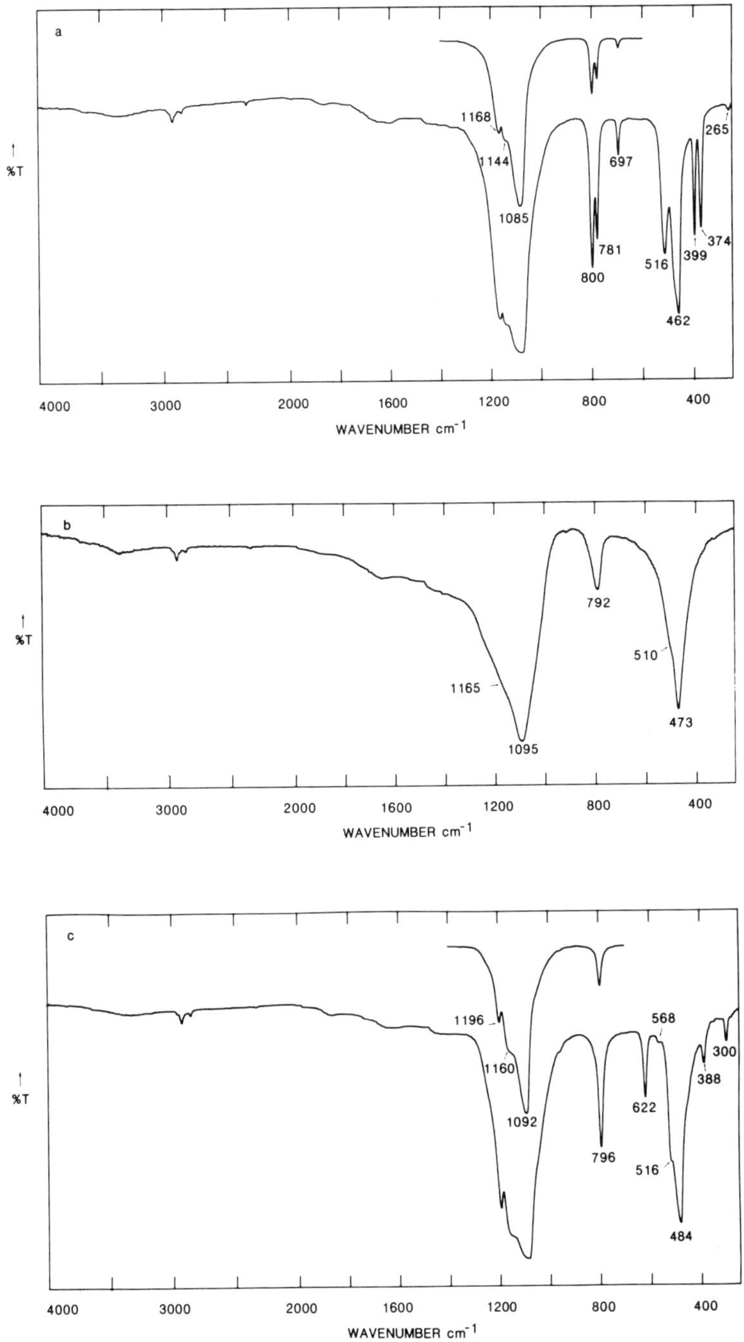

Figure 4.20 IR spectra of silicas: (a) quartz, (b) tridymite (Virginia), (c) cristobalite

polymorph of SiO_2 is included (Figure 4.20(c)), the diagnostic bands being at 1196 and 622 cm^{-1}.

4.5.2.6 *Carbonates.* Calcite is the most common carbonate mineral in clays. Its IR spectrum (Figure 4.21(a)) is easily recognizable, and the bands at 1428, 878 and 714 cm^{-1} are diagnostic. The spectrum of dolomite (Figure 4.21(c)),

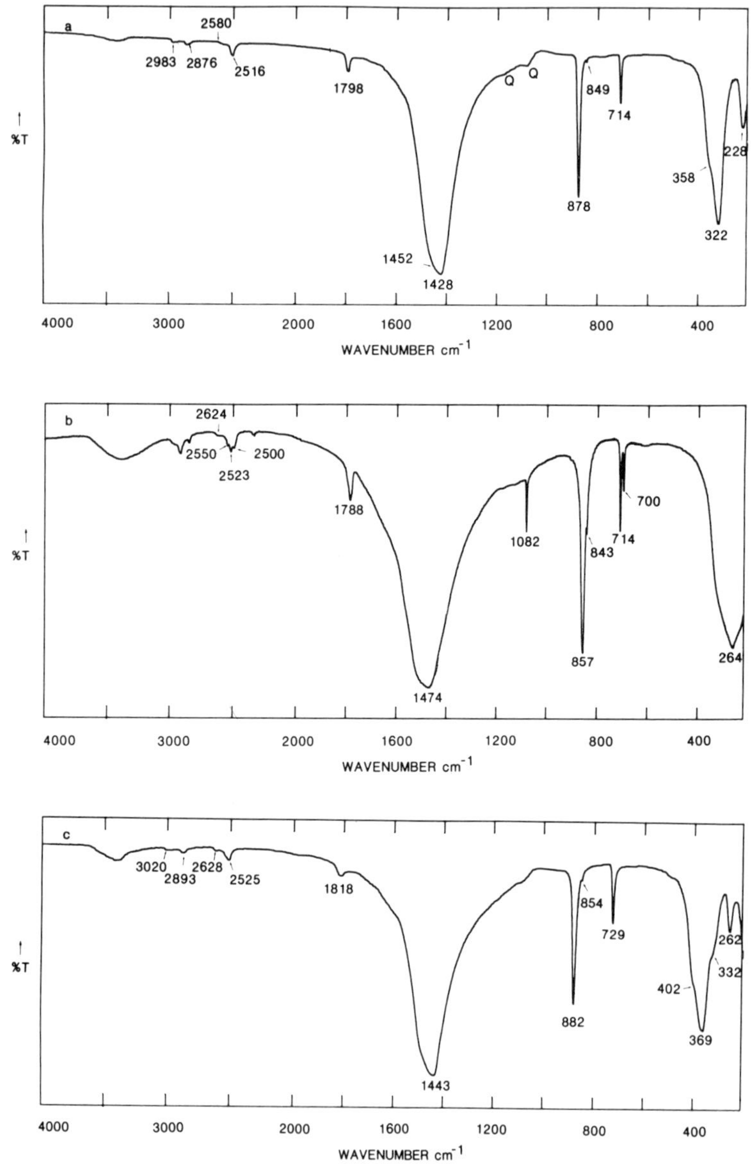

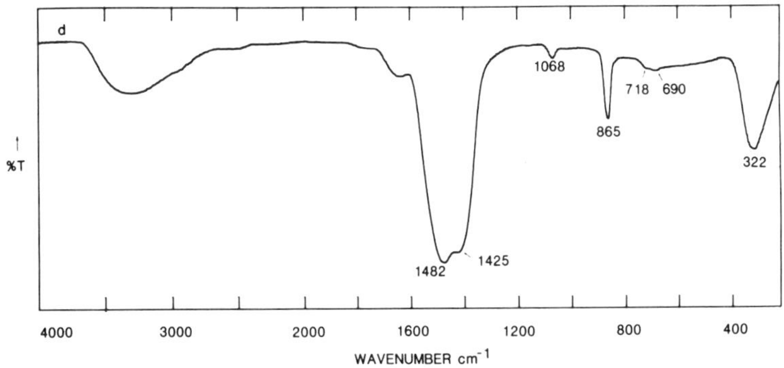

Figure 4.21 IR spectra of carbonates: (a) calcite, (b) aragonite (coral), (c) dolomite (Australia), (d) amorphous $CaCO_3$ (synthetic)

which has a calcite-type structure, shows a shift of all bands to higher frequency. The spectrum of the aragonite form of $CaCO_3$ (Figure 4.21(b)) is distinct from that of calcite, undoubtedly reflecting structural differences. For comparison with the crystalline carbonates, the spectrum of a non-crystalline $CaCO_3$ is included (Figure 4.21(d)), a form recently identified in thin, calcareous weathering crusts. (Dupuis *et al.*, 1984).

4.6 Interpreting the IR spectra of clay materials

This requires the positive identification of as many characteristic absorption bands of individual minerals as possible and will obviously depend on the complexity of the mineral suite present and the degree of overlap of the component spectra. The OH-stretching region is perhaps the most important region of the spectrum, but since some minerals have no hydroxyl groups in their structure, or show only weak absorption bands, the full spectrum is required for the best possible interpretation. High and low concentrations of sample must also be used, so that full weight can be given to weak and strong mineral bands. Finally, because of interference from the absorption bands of adsorbed water, KBr discs of clay samples should be dehydrated by heating at 150°C, in order that underlying absorption bands of mineral OH groups may be unambiguously identified.

To illustrate these points several spectra of soil clays have been selected for interpretation, such clays being somewhat more difficult to handle than clays from geological deposits.

(i) Soil clay 1: the spectrum (Figure 4.22(a)), even of the hydrated clay, clearly shows the sharp OH doublet of a kaolin mineral at 3696, 3620 cm^{-1}, with supporting bands at about 800, 700 and 280 cm^{-1} (Figure 4.4). Absorption bands at 828 and 750 cm^{-1} and slight broadening under the 3620 cm^{-1} band indicate the presence of illite (Figure 4.10(b)). After removal of adsorbed

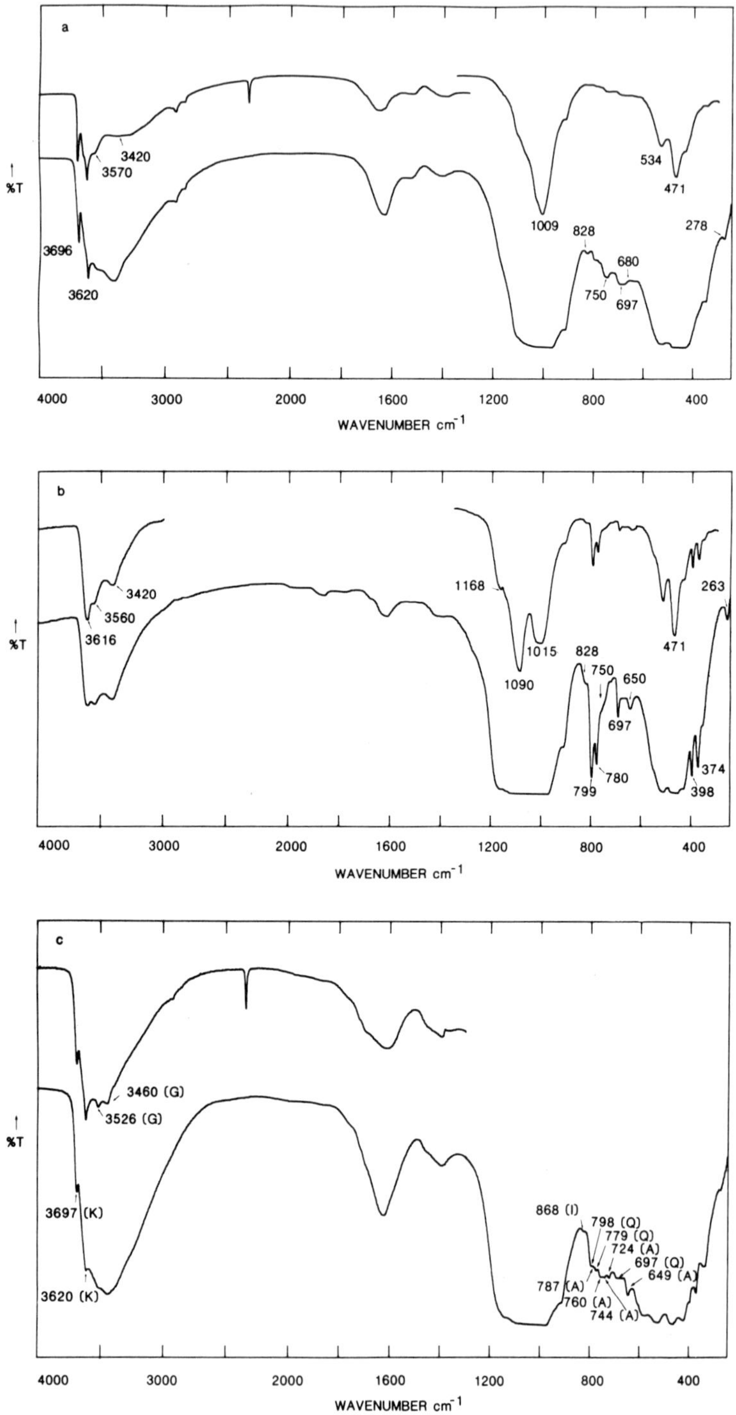
a
%T
3420
3570
3696
3620
534
471
1009
828
278
680
750
697
4000
3000
2000
1600
1200
800
400
WAVENUMBER cm⁻¹
b
%T
3420
3560
3616
1168
263
1015
828
471
1090
750
650
697
374
780
799
398
4000
3000
2000
1600
1200
800
400
WAVENUMBER cm⁻¹
c
%T
3460 (G)
3526 (G)
3697 (K)
3620 (K)
868 (I)
798 (Q)
779 (Q)
724 (A)
697 (Q)
649 (A)
787 (A)
760 (A)
744 (A)
4000
3000
2000
1600
1200
800
400
WAVENUMBER cm⁻¹

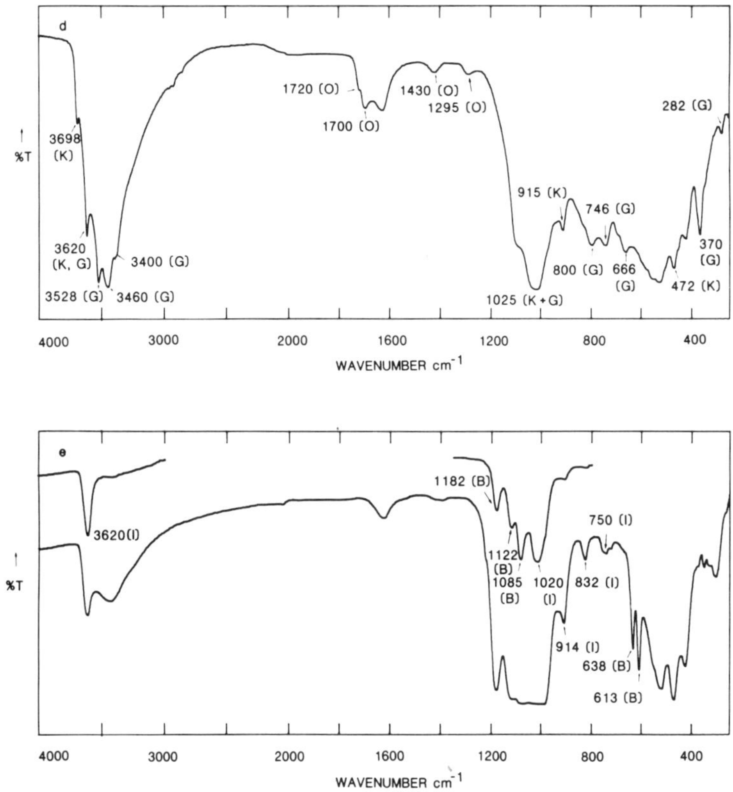

Figure 4.22 IR spectra of soil clays (a)—(e) Q—quartz, A—albite, G—gibbsite, K—kaolin, I—illite, O—complex oxalate, B—baryte. The sharp band near 2340 cm^{-1} arises from molecular CO_2 from thermally decomposed organics

water, the clay shows two broad bands at 3570 and 3420 cm^{-1}, which, together with enhancement at 750 cm^{-1} beyond the requirements for illite, a band at 680 cm^{-1} and marked enhancement of the Si–O bands at 1009 and 471 cm^{-1}, is consistent with the presence of an iron-containing chlorite (Figure 4.12(a)). The broad underlying absorption in the OH region, centred on about 3300 cm^{-1}, suggests that an amorphous aluminosilicate (allophane) might also be present. Soil organic matter is responsible for the absorption bands between 1300 and 1800 cm^{-1}.

(ii) Soil clay 2: the spectrum (Figure 4.22(b)) is dominated by the bands of quartz at 1168, 1090, 799, 780, 697, 398 and 263 cm^{-1}. The 3616 cm^{-1} band of a dioctahedral layer silicate is visible from the OH-stretching region after heating. Taken together with bands at 915, 828, 750 and 350 cm^{-1}, the presence of an illite is strongly indicated. The low OH-stretching frequency compared with those in Figure 4.10 suggests that the illite may be ferruginous.

The OH-stretching region also reveals the broad doublet of a chlorite at 3560, 3420 cm^{-1}, and this is to some extent supported by the enhanced intensity at 750 cm^{-1}, and the band at 650 cm^{-1}. The latter suggests a ferruginous chlorite, but there is no appreciable Si–O contribution present in the 950–1000 cm^{-1} range (Figure 4.12(a)). The main Si–O in Figure 4.22(b) is, however, quite broad and this may accommodate a low-frequency component. Kaolin minerals are absent, no absorption band being detectable at the characteristic 3700 cm^{-1} position.

(iii) Soil clay 3: again the sharp kaolin-mineral doublet at 3697, 3620 cm^{-1} is easily recognized from the spectrum (Figure 4.22(c)). In the spectrum of the dehydrated sample, broadening under the 3620 cm^{-1} band, and the weak feature at 828 cm^{-1} both suggest the presence of illite, but additional confirmation is difficult because of other overlapping absorption. The well-resolved OH bands at 3526 and 3460 cm^{-1} (a 3620 cm^{-1} component is overlain by the kaolin band) and absorption at $\sim$375 cm^{-1} strongly indicate gibbsite. In addition, sharp, weak features of quartz at 798 and 779 cm^{-1}, and albite at 787, 760, 744, 724 and 649 cm^{-1} can be easily identified. The clay contains significant soil organic matter, and possibly also noncrystalline allophanic material as judged by broad underlying absorption in the OH-stretching region, even after removal of molecular water.

(iv) Soil clay 4: the spectrum of this clay (Figure 4.22(d)) shows dominance of gibbsite (cf. Figure 4.17(a)), with minor kaolin mineral (3700, 915, 472 cm^{-1}). This clay has been selected because it has been peroxidized and its spectrum exhibits the pattern of absorption bands typical of the adsorbed complex oxalato-aluminate, the oxalate being complexed to structural Al ions (Mitchell and Farmer, 1964) and not removed by water washing.

(v) Soil clay 5: from the spectrum (Figure 4.22(e)), this soil clay is dominated by illite which appears to be of the phengite type (cf. Figure 4.10(b)). No other silicate mineral is detectable from the IR spectrum, but the clay is significantly contaminated by barium sulphate, whose absorption bands at 1182, 1122, 1085, 638 and 613 cm^{-1} are very readily identified by comparison with reference spectra in Moenke (1962, 1966), Nyquist and Kagel (1971), and Kodama (1985). This clay is included here because it illustrates one of the most important assets of the IR technique, that of revealing an unexpected component in a sample.

4.7 Quantitative analysis

Quantitative determination by IR spectroscopy of clay minerals is sometimes possible if the following conditions are met:

(i) Particle size of mineral components are all $< 2 \mu m$, to ensure maximum absorption and minimum scattering by the sample. This may be difficult if the sample contains minerals of widely differing hardness and grindability

(ii) Good control over weighing and sample transfer during preparation of KBr discs

(iii) Uniform dispersion and mixing with KBr, and uniform quality of pressed discs

(iv) One (or more) absorption bands in the spectrum sufficiently isolated from bands of other components to allow easy measurement of absorbance or the peak area

(v) The availability of *appropriate* standard minerals of the same particle size. That is, the standard mineral should have the same crystallinity, state of order, and similar composition to the component to be determined

(vi) Reproducibility of technique.

From this, it is obvious that truly quantitative determination of minerals is difficult. Among the more demanding of the conditions listed are (iv), (v) and (vi). For example, when measuring the kaolinite content of a soil clay from the absorbance of the 3700 cm^{-1} band of structural OH groups, estimates can vary by factors of 2 or 3, depending on the crystallinity of the selected standard. Assessment of the crystallinity of the kaolin mineral in a soil clay from the IR spectrum is not easy, because of overlapping absorption bands of other minerals.

Nevertheless, certain minerals can be determined quantitatively, among them quartz (Chester and Green, 1968), quartz and cristobalite (Adamek *et al.*, 1969), quartz and asbestos (Coates, 1977) and biogenic opal (Chester and Elderfield, 1968). Semiquantitative estimates of kaolin minerals (3700 cm^{-1}), carbonates (1400, 870 cm^{-1}), imogolite (348 cm^{-1}), illite (3620, 830 cm^{-1}) and others, are possible with care. Detection levels may be as low as 1–2% for quartz and kaolinite, but for many other minerals they are 5–10%.

References

Adamek, P., Linhova, R., and Ruzek, J. (1969) Infrared spectroscopy in the study of silica phases. II *Sborník Vysoké Skoly chemicko-technologické v. Praze*, **13**, 259–283.

Adams, M.J. and Black, I. (1983) Application of a microcomputer system in an infrared laboratory. *J. Automat. Chem.*, **5**, 9–13.

Buckley, H.A., Bevan, J.C., Brown, K.M., Johnson, L.R. and Farmer, V.C. (1978) Glauconite and celadonite: two separate mineral species. *Miner. Mag.*, **42**, 373–382.

Chester, R. and Elderfield, H. (1968) The infrared determination of opal in siliceous deep-sea sediments. *Geochim. Cosmochim. Acta*, **32**, 1128–1141.

Chester, R. and Green, R.N. (1968) The infrared determination of quartz in sediments and sedimentary rocks. *Chem. Geol.*, **3**, 199–213.

Coates, J.P. (1977) IR Analysis of Toxic Dusts. Analysis of Collected Samples of Quartz and Asbestos. Part 1. *Am. Lab.* (Fairfield, Connecticut, USA), **9**, 105–108, 110–111.

Cradwick, P.D.G., Farmer, V.C., Russell, J.D., Masson, C.R., Wada, K. and Yoshinaga, N. (1972) Imogolite, a hydrated aluminium silicate of tubular structure. *Nature*, **240**, 187–189.

Dupuis, T., Ducloux, J., Butel, P. and Nahon, D. (1984) Etude par spectrographie infrarouge d'un encroûtement calcaire sous galet. Mise en evidence et modelisation expérimental d'une suite minèrale évolutive à partir de carbonate de calcium amorphe. *Clay Miner.*, **19**, 605–614.

Durig, J.R. (1980) Analytical applications of FTIR to molecular and biological systems. *NATO Adv. Study Inst. Ser.*, D. Reidel, Dordrecht.
Farmer, V.C. (1964) Infrared absorption of hydroxyl groups in kaolinite. *Science*, **145**, 1189–1190.
Farmer, V.C. (1974) *The Infrared Spectra of Minerals*. Mineralogical Society, London.
Farmer, V.C. and Mitchell, B.D. (1963) Occurrence of oxalates in soil clays following hydrogen peroxide treatment. *Soil. Sci.*, **96**, 221–229.
Farmer, V.C. and Russell, J.D. (1964) The infrared spectra of layer silicates. *Spectrochim. Acta*, **20**, 1149–1173.
Farmer, V.C. and Russell, J.D. (1966) Effects of particle size and structure on the vibrational frequencies of layer silicates. *Spectrochim. Acta*, **22**, 389–398.
Farmer, V.C., Fraser, A.R., Robertson, L. and Sleeman, J.R. (1984) Proto-imogolite allophane in podzol concretions in Australia: possible relationship to aluminous ferralitic (lateritic) cementation. *J. Soil Sci.*, **35**, 333–340.
Farmer, V.C., Fraser, A.R., Russell, J.D. and Yoshinaga, N. (1977) Recognition of imogolite structures in allophanic clays by infrared spectroscopy. *Clay Miner.*, **12**, 55–57.
Farmer, V.C., Fraser, A.R. and Tait, J.M. (1979) Characterization of the chemical structures of natural and synthetic aluminosilicate gels by infrared spectroscopy. *Geochim. Cosmochim. Acta*, **43**, 1417–1420.
Farmer, V.C., Russell, J.D., Ahlrichs, J.L. and Velde, B. (1967) Vibrations of the hydroxyl group in layer silicates. *Bulletin de la Groupe Française des Argiles*, **19**, 5–10.
Farmer, V.C., Russell, J.D., McHardy, W.J., Newman, A.C.D., Ahlrichs, J.L. and Rimsaite, J.Y.H. (1971) Evidence of loss of protons and octahedral iron from oxidized biotites and vermiculites. *Miner. Mag.*, **38**, 121–137.
Ferraro, J.R. (1982) *The Sadtler Infrared Spectra Handbook of Minerals and Clays*. Sadtler/Heyden, London.
Goodman, B.A., Russell, J.D., Fraser, A.R. and Woodhams, F.W.D. (1976). A Mössbauer and IR spectroscopic study of the structure of nontronite. *Clays & Clay Miner.*, **24**, 53–59.
Hadni, A. (1967) *Essentials of Modern Physics Applied to the Study of the Infrared*. Pergamon, Oxford.
Hayashi, H., Otsuka, R. and Imai, N. (1969) Infrared study of sepiolite and palygorskite on heating. *Am. Miner.*, **54**, 1613–1624.
Juo, A.S.R. and White, J.L. (1969) Orientation of the dipole moments of hydroxyl groups in oxidized and unoxidized biotite. *Science*, **165**, 804–805.
Kodama, H. (1985) *Infrared Spectra of Minerals. Reference Guide to Identification and Characterization of Minerals for the Study of Soils*. Techn. Bull. **1985-1E**, Research Branch, Agric. Canada.
Kodama, H. and Oinuma, K. (1963) Identification of kaolin minerals in the presence of chlorite by X-ray diffraction and infrared absorption spectra. *Clays & Clay Miner.*, **11**, 236–249.
Lazarev, A.N. (1974) 'The dynamics of crystal lattices,' in *The Infrared Spectra of Minerals*, (V.C. Farmer ed.), Mineralogical Society, London, 69–86.
Ledoux, R.L. and White J.L. (1964) Infrared study of selective deuteration of kaolinite and halloysite at room temperature. *Science*, **145**, 47–49.
Martin, A.E. (1980) In *Vibrational Spectra and Structures*. Vol. 8 *Infrared Interferometric Spectrometers*, (J.R. Durig ed.), *Elsevier, Amsterdam*.
Moenke, H. (1962) *Mineralspektren I*. Akademie-Verlag, Berlin.
Moenke, H. (1966) *Mineralspektren II*. Akademie-Verlag, Berlin.
Nadeau, P.H. (1980) *Burial and Contact Metamorphism in the Mancos Shale*. Ph.D. Thesis, Dartmouth College, Hanover, New Hampshire, 200 pp.
Nadeau, P.H., Farmer, V.C., McHardy, W.J. and Bain, D.C. (1985) Compositional variations of the Unterrupsroth beidellite. *Am. Mineral.*, **70**, 1004–1010.
Nyquist, R.A. and Kagel, O. (1971) *Infrared Spectra of Inorganic Compounds*. Academic Press, London.
Oinuma, K. and Hayashi, H. (1968) Infrared spectra of clay minerals. *J. Toyo Univ. Gen. Educ. (Nat. Sci.)*, **9**, 57–98.
Rendon, J.L. and Serna, C.J. (1981) IR spectra of powder hematite: effects of particle size and shape. *Clay Miner.*, **16**, 375–382.
Rouxhet, P.G., Samudacheata, N., Jacobs, H. and Anton, O. (1977) Attribution of the OH stretching bands of kaolinite. *Clay Miner.*, **12**, 171–180.

Russell, J.D. (1974) 'Instrumentation and techniques,' in *The Infrared Spectra of Minerals*, (V.C. Farmer ed.), Mineralogical Society, London.

Russell, J.D. (1979) Infrared spectroscopy of ferrihydrite: evidence for the presence of structural hydroxyl groups. *Clay Miner.*, **14**, 109–114.

Russell, J.D. (1987) 'Infrared spectroscopy of inorganic compounds, in *Laboratory Methods in Infrared Spectroscopy*, (H. Willis ed.), Wiley, New York. In press.

Russell, J.D., Farmer, V.C. and Velde, B. (1970) Replacement of OH by OD in layer silicates and identification of the vibrations of these groups in infrared spectra. *Miner. Mag.*, **37**, 869–879.

Serna, C.J., Rendon, J.L. and Iglesias, J.E. (1982) Infrared surface modes in corundum-type microcrystalline oxides. *Spectrochim. Acta*, **38A**, 797–802.

Serna, C.J., Vanscoyoc, G.E. and Ahlrichs, J.L. (1977) Hydroxyl groups and water in palygorskite. *Am. Miner.*, **62**, 784–792.

Slonimskaya, M.V., Besson, G., Dainyak, L.G., Tchoubar, C. and Drits, V.A. (1978) Interpretation of the IR spectra of celadonites and glauconites in the region of OH-stretching frequencies. *Clay Miner.*, **21**, 377–388.

Tarasevich, Yu.I. (1970) Spectral study of the thermal dehydration of palygorskite. *Dopov. Akad. Nauk Ukr. Rsr Ser. B*, **32**, 938–942.

Theng, B.K.G., Russell, M., Churchman, G.J. and Parfitt, R.L. (1982) Surface properties of allophane, halloysite and imogolite. *Clays & Clay Miner.*, **30**, 143–149.

Tuddenham, W.M. and Lyon, R.J.P. (1959) Relation of infrared spectra and chemical analysis for some chlorites and related minerals. *Anal. Chem.*, **31**, 377–380.

Turrell, G. (1972) *Infrared and Raman Spectra of Crystals*. Academic Press, London.

van der Marel, H.W. and Beutelspacher, H. (1976) *Atlas of Infrared Spectroscopy of Clay Minerals and their Admixtures*. Elsevier, Amsterdam.

Vedder, W. (1964) Correlations between infrared spectrum and chemical composition of mica. *Am. Mineral.*, **49**, 736–768.

Wilkins, R.W.T. and Ito, J. (1967) Infrared spectra of some synthetic talcs. *Am. Mineral.*, **52**, 1649–1661.

Wilson, M.J., Russell, J.D. and Tait, J.M. (1974) A new interpretation of the structure of disordered α-cristobalite. *Contrib. Mineral. Petrol.*, **47**, 1–6.

Wilson, M.J., Russell, J.D., Tait, J.M., Clark, D.R., and Fraser, A.R. (1981) A swelling hematite/layer silicate complex in weathered granite. *Clay Miner.*, **16**, 261–278.

Wilson, M.J., Russell, J.D., Tait, J.M., Clark, D.R., Fraser, A.R. and Stephen, I. (1984) Macaulayite, a new mineral from North-East Scotland. *Miner. Mag.*, **48**, 127–129.

Yariv. S.H. and Mendelovici, E. (1979) The effect of degree of crystallinity on the infrared spectrum of hematite. *Appl. Spectroscopy*, **33**, 410–411.

5 Scanning electron microscopy

W.J. McHARDY AND A.C. BIRNIE

5.1 Introduction

Among the first published scanning electron microscope (SEM) micrographs of clay minerals were those of Borst and Keller (1969) who studied many of the API Project 49 reference clays, and Gillot (1969) who explored the potential of the SEM for petrographic investigation of fine-grained rocks. These early papers showed the instrument's potential in the study of clays, stressing its large depth of field at high magnification, its ease of operation and simplicity of sample preparation. However, these papers were restricted to descriptions of the various clay mineral groups. Bohor and Hughes (1971) concluded that SEM was indispensable for the study of clay-mineral configuration, fabric, texture and growth mechanics and speculated that increased resolution and built-in microanalysis facilities would improve the usefulness of the SEM in clay studies. Recently, the attainable resolution has improved, and add-on energy-dispersive X-ray analysers are regarded as almost essential. It is also now appreciated that the specimen preparation method can have an important influence on the information obtained and that the SEM should be used to try and answer specific predetermined questions.

This chapter aims firstly to describe the instrument and how it operates in a way that is helpful in achieving optimum performance. The various methods of specimen preparation are described in detail, and a separate section is devoted to microanalysis by SEM. Finally, a section illustrates SEM applications to clay minerals and the more important accessory minerals. For a more detailed account of the preparation and examination of soils and sediments by electron microscopy, readers are referred to an excellent book by Smart and Tovey (1982). The same authors (Smart and Tovey, 1981) have produced a separate volume, with examples of how electron microscopy can be applied to study of soils and sediments. The examples consist of both transmission and scanning electron micrographs.

5.2 Instrumentation

5.2.1 *Principles*

The first commercial scanning electron microscope, based on the principles illustrated in Figure 5.1, was made available by the Cambridge Scientific

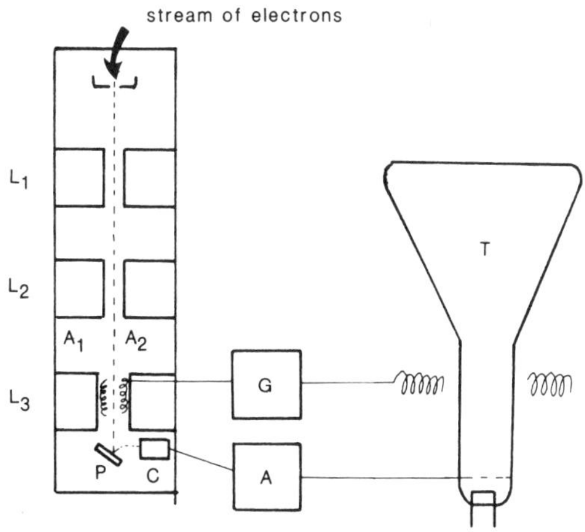

Figure 5.1 Schematic diagram of the scanning electron microscope

Instrument Company in 1965. A stream of electrons generated from a heated tungsten filament, and accelerated by a potential of up to 50 kV, are focused by a series of electromagnetic lenses, L_1, L_2 and L_3, into a fine probe, so that when this strikes the object, P, it may have a diameter of $\sim$ 10 nm. The probe is caused to move over the surface of P in a zig-zag raster by two pairs of deflecting coils, A_1 and A_2, which carry current from a scan generator, G. The same current traverses the coils of a cathode-ray tube, T, to produce on the screen an identical, but very much larger, raster. The electron probe striking the specimen creates secondary electrons that may be accelerated to a collector, C, and the resulting signal amplified at A, and used to modulate the brightness of the cathode-ray tube display, T. Thus, there is a point-by-point correspondence between the raster on P and the display raster, with the brightness of each point of the latter governed by the number of secondary electrons leaving the corresponding point on the surface of P. The picture created on the cathode-ray tube is thus a secondary electron image of the area of the specimen surface scanned by the primary beam. Magnification is determined by the relative sizes of the two rasters, and this can be varied by simply controlling the magnitudes of the currents flowing in two sets of deflecting coils. Since the electron beam would be scattered by residual air molecules, the column is held at high vacuum and specimens are changed by an air-lock system.

The topographical properties of the specimen surface control the secondary electron emission, and so the arrangement just described gives a picture of the specimen surface that is similar to that observed by reflected light microscopy, but with potential for much greater resolution and depth of field. However, other events take place where the primary electron beam is stopped, and these include the generation of X-rays, light, back-scattered electrons and Auger electrons, as well as the dissipation of energy as electrical current in the sample. Variations in magnitude of any of these excitations may be converted to corresponding variations in an electrical signal, which in turn can be used to control the display. Just as secondary electron emission gives information on topography, back-scattered electron emission can be related to topography or atomic number, depending on the circumstances; also, the X-ray emission is characteristic of chemical composition. The processing of the X-ray emission is dealt with in detail in the microanalysis section, later in this chapter.

The essential features of the scanning electron microscope are:

(i) The electron gun (equivalent to the light source in optical microscopy)

(ii) Two or three electromagnetic lenses to focus the electron beam into as fine a probe as possible at the point where it strikes the specimen

(iii) A scan generator/scanning coil system which deflects the focused beam across the specimen

(iv) A specimen stage, allowing movement, tilt and rotation of the specimen

(v) A method of collecting and amplifying the emitted electrons

(vi) A high-capacity vacuum system to maintain a pressure below 1.3×10^{-3} Pa and permit rapid evacuation after a change of specimen.

In addition, there are cathode-ray tubes or monitors to display the image for visual observation or photographic recording, and a variety of electronic circuitry to supply, stabilize and control the beam-accelerating voltage, lens currents, scan generator and signal amplification.

5.2.2 *The electron gun*

The various types of electron guns, with some typical performance characteristics are shown in Table 5.1.

It is probable that most working scanning electron microscopes still use the simple hairpin filament, in spite of the brighter electron sources that are now available. Twenty years ago, Oatley (1966) expressed some doubt concerning

Table 5.1 Various types of electron guns, with performance characteristics.

	Field emission	LaB_6	Tungsten point thermionic	Tungsten hairpin
Brightness ($A/cm^2.Sr$)	10^9	10^7	10^6	10^6
Source diameter (μm)	0.01–1	1–10	1–10	50
Service life (hours)	> 1000	1000	160*	40†
Operating vacuum (Pa)	10^{-7}	10^{-5}	3×10^{-7}	10^{-3}

* from Swift and Brown (1979), assuming their run-in schedule has been followed.
† the life of a tungsten hairpin can vary considerably depending on the state of the vacuum and the design of the gun.

the development of high-brightness sources such as the field emission gun, since resolution was limited by penetration of electrons into the object. At an accelerating voltage of 20 kV, the primary electron beam will penetrate an average specimen to a depth of $\sim 1\ \mu m$, producing secondaries along its path, which move off in random directions, few actually reaching the surface. Of the secondary electrons which do escape, most will have originated within $\sim$ 10 nm of the surface. Calculations suggest that more than half will emerge within a distance of 5 nm from the point where the primary electron entered. It is this spreading of the secondaries that limits the achievable resolution, at least in SEM of bulk specimens. Using calculations based on such arguments, Oatley (1966) showed that at an accelerating voltage of 20 kV, the resolution limit imposed by electron penetration considerations was $\sim$ 10 nm.

5.2.3 *Accelerating voltage*

All but the cheapest, most basic, commercial scanning electron microscopes have a range of accelerating voltages. A typical top-of-the-range model offers 0.2–40 kV, with 0.2–1 in 0.2 kV steps and 1–40 in 1 kV steps. Very low accelerating voltages are desirable where it is necessary to study uncoated, nonconducting specimens. Higher voltages are needed if X-ray analysis is required.

The choice of accelerating voltage for secondary electron imaging should be as low as possible, and compatible with adequate resolution. According to Chapman (1986), using an excessively large accelerating voltage results in loss of fine surface detail because of excessive beam penetration. The effect of an increasing accelerating voltage on the image of a gold-coated mineralogical specimen is illustrated in Figure 5.2. There is a slight but definite improvement in sharpness (or resolution) in moving from 10 kV through 20 kV to 30 kV, and the 10 kV images are marginally less contrasty than the higher voltage images. However, it is unlikely that a scientific description of the features observed in the 10 kV images would differ from a similar description of the 30 kV images. In retrospect, the results of studying mineralogical and soil-derived samples suggest that three accelerating voltages, 5, 15 and 25 kV, would have satisfied the majority of our requirements.

5.2.4 *The electron lenses*

The purpose of the lens system is to produce a greatly demagnified image of the cathode at the specimen. The first commercial SEM used three pre-aligned electromagnetic condenser lenses, and the same system is adopted by most manufacturers today, although there have been considerable design improvements to reduce the various aberrations. The probe size is decreased by increasing the current through the first two lenses. Reducing the probe size increases image sharpness or resolution. For some purposes, such as X-ray

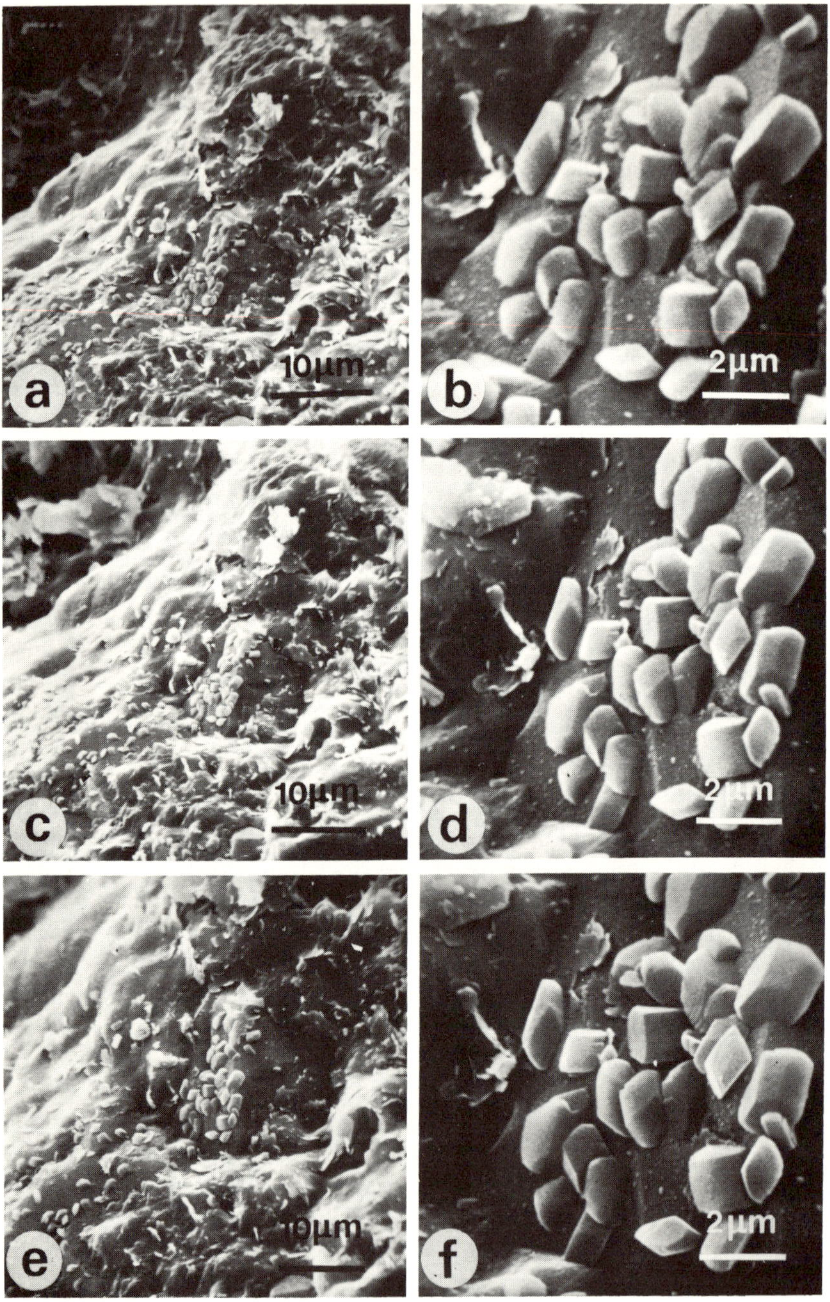

Figure 5.2 The effect that increasing accelerating voltage has on the secondary electron image: (a) and (b)—10 kV; (b) and (c)—20 kV; (c) and (d)—30 kV. The micrographs feature tiny crystals of glauberite—a sodium-calcium sulphate—growing on pyrites in a sandstone.

microanalysis, a large spot size or probe current is required, and this is achieved by decreasing the current through the first two lenses. The function of the final lens, sometimes referred to as the objective, is to focus the image. Another essential component, closely associated with the final lens, is the final or objective aperture. There are usually three or four such apertures, which are, mounted on a rod mechanism so that they can be readily interchanged without switching off the beam. The choice of the aperture size will depend on what is required. X-ray wavelength spectrometry requires maximum probe current, which needs a relatively large aperture, e.g. 400 μm. For high resolution, a 100 μm aperture would be normal, although in the case of a poor signal a 200 μm aperture could be used, with some loss of resolution due to a corresponding increase in aberrations. For maximum depth of field, the smallest aperture compatible with an adequate signal level is necessary. Working distance, i.e. the distance between the plane of the aperture and the specimen also needs to be considered. For high resolution, the shortest possible working distance is required. Conversely, increasing the working distance improves the depth of field.

5.2.5 *The electron collection system*

For practical purposes, the electron emission from a specimen can be divided into two energy regions, known as secondary and back-scattered electrons. The former originate from atoms in the specimen, by collision with high-energy electrons. Because of their low energies, only those secondary electrons generated within a few nm of the specimen surface will escape and contribute to the signal; their yield is little affected by the chemistry of the specimen material and is almost entirely related to topography. Reflected or back-scattered electrons have energies approaching that of the incident electron beam, and so are able to escape from a somewhat greater depth of the excitation volume. The yield of such electrons is strongly dependent on atomic number.

5.2.5.1 *Secondary electrons.* The design of the secondary electron detector has remained virtually unchanged since its invention by Everhart and Thornley (1960). The transducer is usually a plastic scintillator, coated with a thin layer of aluminium, which is maintained at a potential of 12 kV and mounted within a Faraday cage (Figure 5.3). The grid of this cage can be biased either to pull as many secondary electrons as possible into the cage (positive bias) or to repel them completely (negative bias), so allowing only the high-energy electrons to participate in image formation. Each electron produces a photon of light on striking the scintillator, and this is conducted along a light-pipe to a photomultiplier, where it is converted into an electrical signal and amplified. This system produces a noise-free amplification of about $10^6 \times$, which is virtually 100% efficient.

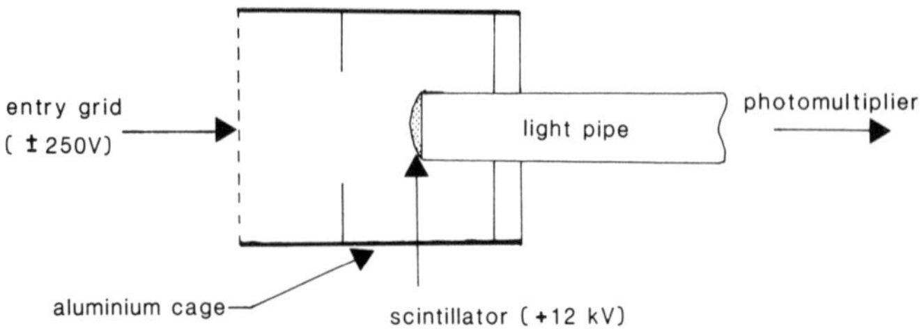

Figure 5.3 Schematic diagram of the Everhart-Thornley secondary electron detector

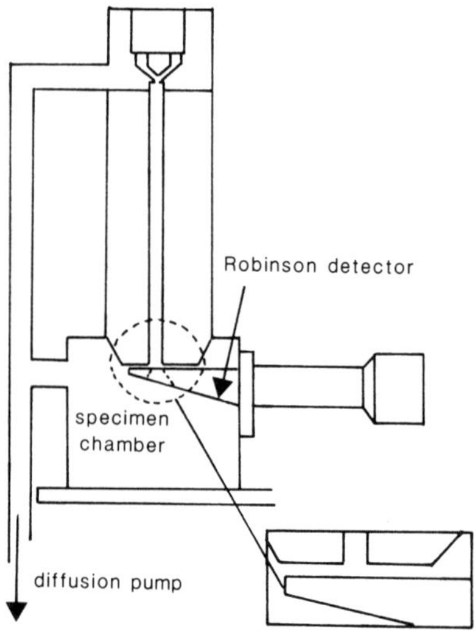

Figure 5.4 Diagram of the Robinson-style scintillator back-scattered electron detector, in relation to the SEM column and specimen chamber

5.2.5.2 *Back-scattered electrons.* Two types of back-scattered electron (BSE) detectors are now available. Both can produce images of a quality approaching secondary electron (SE) images. The scintillator, or Robinson (1974) back-scattered electron detector works on the same principle as the Everhart/Thornley SE detector, but the scintillator material is on the flat, wedge-shaped end of the light-guide, positioned directly above the specimen (Figure 5.4). The other type uses a semiconductor device for the electron detector, and this can be fixed directly to the bottom of the objective lens or mounted on a movable arm. There is considerable argument as to which type of back-scattered detector is the best. For mineralogical applications there would appear to be little to choose between them. One advantage of the more popular semiconductor type is its small size; it is a disc-shaped device 3–4 cm in diameter and only 3 mm thick. The scintillator type, although it can be mounted on a retractable arm, needs more space between the objective lens and the specimen, as can be seen in Figure 5.4, and may be quite incompatible with some of the older scanning electron microscopes. The application of BSE imagery to the study of clay materials is illustrated by the work of Pye and Krinsley (1983) and Pye (1984). Here, the relationship between the BSE signal and specimen atomic number was determined experimentally, these data being used as a basis for quantitative atomic number determination, light-element detection at high resolution, and elemental mapping at normal SEM scan rates.

The use of a scintillator-type BSE detector, in conjunction with a low-vacuum mode of SEM operation was developed by Robinson (1974) and could be a useful technique for clay materials (Robinson and Nickel, 1979). The relatively high gas pressure in the specimen chamber allows the dissipation of charge from the specimen surface, and consequently, no conductive coatings are required for insulating the specimens, even at high accelerating voltages. It is claimed that drill-core chips, hand specimens, polished sections or porous wet samples can all be imaged directly as received—with the possibility of displaying an image and X-ray spectrum within a minute of receipt of the specimen.

5.2.6 *The specimen stage*

The requirements for the specimen stage are that it should be mechanically stable, accept reasonably large specimens, and provide movements in three perpendicular directions, together with rotation and tilt. In this way, different areas of the specimen can be observed, and the working distance between the specimen and the final lens can be adjusted. The design of stages has improved greatly since the first scanning electron microscopes, where the specimen tended to be 'buried' in the mechanism. Modern stages present the specimen to the objective lens well-clear of any mechanical parts, which greatly facilitates the introduction of X-ray detectors and other specimen-chamber accessories. Stages employing eucentric geometry are offered as an optional extra for most microscopes, as are motorized, computer-controlled stage movements. The eucentric stage is of particular advantage for the recording of stereopairs, since it can be set up so that tilt and rotation are about the axis of focus, and so that the field of interest remains static while these manoeuvres are taking place.

5.2.7 *The pumping system*

The original commercial scanning electron microscope had a vacuum system using two oil-diffusion pumps. The larger of these was mounted directly under the specimen chamber, and the smaller one was connected to the column and electron gun through a manifold in the column plinth. Both diffusion pumps were backed by a single rotary pump via a vacuum reservoir. The pump-down sequences were controlled automatically by a complex system, consisting of vacuum relays, electromagnetic valves and interlock circuits. Modern pumping systems, although based on the same principle of a diffusion pump backed by a rotary pump, are more simple, and only one diffusion pump may be used, with no isolating valve between column and specimen chamber. Specimen exchange is effected either through an air-lock mechanism or by venting the whole column. Turbomolecular pumping offers an alternative to the diffusion pump system, but it is only recently that a leading manufacturer has been sufficiently confident about the reliability of such pumps to offer turbomolecular pumping as the only option.

5.2.8 *Computer control*

Probably the most significant recent innovation in the area of instrumentation has been computer control. A computerized system can be used to control the currents in each of the three lenses independently, thus optimizing resolution and signal-to-noise performance under selected operating conditions. Computer technology can also be used to automate the focusing operation and to set up the brightness and contrast levels for photo recording. In general, good results are now attainable by relatively inexperienced operators.

5.2.9 *Digital image processing*

Another important recent innovation, now rapidly gaining ground, is digital image processing and image store. SEMs may be run at a range of scan rates; in low-contrast or high-noise conditions, or for photographic recording, a slow scan is used to produce an integrated signal. Unfortunately, for uncoated, nonconducting specimens, this may build up the charge, which can eventually deflect the electron beam itself. For such materials it is often possible to obtain an acceptable image at fast scan rates, but quite impossible to make a photographic record of the same image. With an image-store system, however, the 'raw' (charge-free but noisy) TV image can be integrated over several hundred frames and stored on disc, from which it can be photographed or further refined by image processing. Martin *et al.* (1985) have described an instrument where an SEM has been equipped with a dedicated image store and an on-line image analyser.

5.3 Specimen preparation

One of the many merits claimed for the SEM is the ease with which specimens can be prepared and examined. However, since the microscope operates under high vacuum, the specimen must be dry and must make good electrical contact with the support (usually an aluminium stub). This is so that the surface becomes electrically conducting with no build-up of electrical charge. Clay specimens require special care in preparation for SEM examination, because of their hydrous nature; some useful techniques will now be discussed.

5.3.1 *Methods of water removal*

5.3.1.1 *Air-drying.* For many specimens, e.g. rock fragments or sand grains, simple air- or oven-drying is adequate, but clay materials will shrink with such treatment. This will inevitably result in some alterations in the fabric of the specimen, and here, air-drying cannot really be recommended.

5.3.1.2 *Freeze-drying.* This involves plunging a specimen into a cryogenic liquid such as liquid nitrogen (–196°C), liquid-nitrogen slush (–210°C), Freon 22 at its melting point (–146°C) or liquid propane at its melting point (–190°C), and then vacuum-drying from the frozen state. Theoretically, if it is possible to cool a specimen quickly enough, water should be vitrified in place, in an amorphous state, and the potentially structure-damaging growth of ice crystals should not occur. Of the liquid cryogens just mentioned, liquid nitrogen itself, although the most convenient, is the least effective. Since it is a liquid at its boiling point, a specimen plunged into it will be surrounded by a layer of gas immediately. This acts as an insulator for the transfer of heat from the specimen to the cryogen. Liquid propane achieves the highest rates of cooling, but it can readily form explosive mixtures with air and needs to be handled with caution. In general, to obtain sufficiently high cooling rates, a specimen must be as small as possible ($< 0.1\ mm^3$), or only its surface regions should be examined (Robards, 1984). For freeze-drying clays and soils, Greene-Kelly (1973), cites 0.1 mm as being the maximum permitted thickness for the very rapid cooling rate required for vitrification. Thus, the size of the sample, rather than the efficiency of the liquid cryogen, is the limiting factor. Even if satisfactory freezing is accomplished, sublimation of the ice in a reasonable time is possible only if the specimen is warmed to at least –40°C. Freeze-drying of clay soils may cause only small amounts of shrinkage, but can produce a distorted pore-size distribution, the extent of distortion depending on the rate of freezing (Lawrence *et al.*, 1979).

5.3.1.3 *Critical-point drying.* This technique, used for many years for drying biological specimens, involves exchanging the water in the sample for liquid carbon dioxide, via an intermediate miscible fluid such as methanol, acetone or 1,4 dioxan. Sometimes two intermediate fluids are used, e.g. ethanol followed by amyl acetate. The sample is placed in a pressure-vessel, immersed in liquid CO_2 and heated to a temperature a few degrees above the critical point (31°C and 7.6 MPa). The supercritical carbon dioxide is allowed to vent slowly to the atmosphere, so as to avoid adiabatic cooling. There is no liquid–vapour interface in supercritical liquids, and the surface-tension forces at these interfaces, as they move through a porous system, actually cause the structural damage. Several technical variations on the same theme have been described in the literature. Newman and Thomasson (1979) suspended soil aggregates on a gauze rack above a methanol–water mixture, in the bottom of a sealed plastic box. This ensured that there was no contact between the soil and the liquid, thus lessening the risk of sample disintegration. Exchange occurred through the vapour phase, and the progress of exchange was monitored by determining the density of the aqueous methanol.

The detailed procedure used by McHardy *et al.* (1982) for critical-point drying of preserved reservoir-rock samples containing sensitive clays, is as follows. The formation fluid (salt solution) in subcentimetre rock pieces was gradually replaced with methanol, by passing it successively through 1:3, 1:1 and 3:1 methanol–water mixtures, and finally into 100% methanol. The methanol was then changed daily until free from chloride (the silver nitrate test), at which point it was assumed that all water in the pores had been replaced by methanol. The critical-point drying apparatus used was a *Polaron model E 3000*, equipped with a liquid-transfer boat with integral drain valve. The boat was filled with methanol, the rock samples were added—6 to 8 pieces could be accommodated—and the assembly was loaded into the pressure chamber of the apparatus, which had been precooled to ~ 10°C by mains cold water. (The exact temperature of the apparatus is not important at this stage, as long as it is a few degrees colder than the CO_2 cylinder, which should be of the siphon type). The chamber was flushed with liquid CO_2 for about 10 min and then allowed to stand for several hours, ensuring that the CO_2 level was always above the samples. The chamber was then flushed with CO_2 for ~ 5 min periods, twice a day for about a week; the objective was to ensure that the methanol was completely replaced by liquid CO_2. Subsequently, the pressure apparatus was heated at a slow, controlled rate to 37°C by circulating water from a 40°C thermostat. (This temperature is adequately above the critical point of CO_2). After maintaining the temperature at 37°C for at least half an hour, thus ensuring equilibrium throughout the samples, the CO_2 was slowly vented from the chamber at a rate of 100–200 cm^3/min, i.e. sufficiently slowly to prevent recondensation of carbon dioxide in the chamber.

Alternative critical-point drying procedures have been described by Koller and Bernhard (1964), Parsons *et al.* (1974), Boyde and Tamarin (1984) and Diamond (1970).

No complete agreement exists on how to solve the problems involved in removing water from clays and soils before study of their fabric. Thus, Erol *et al.* (1976), in preparing Na-montmorillonite samples for SEM, rejected critical-point drying in favour of freeze-drying, because the former required replacement of the original porewater with an organic liquid, which could result in swelling. Again, clays with high surface activity have ions adsorbed on their surface, in equilibrium with the ion concentration in the interstitial fluid. This means that if the original fluid is replaced, ion exchange may occur on the surface of the particles, leading to changes in the arrangement of the fabric of

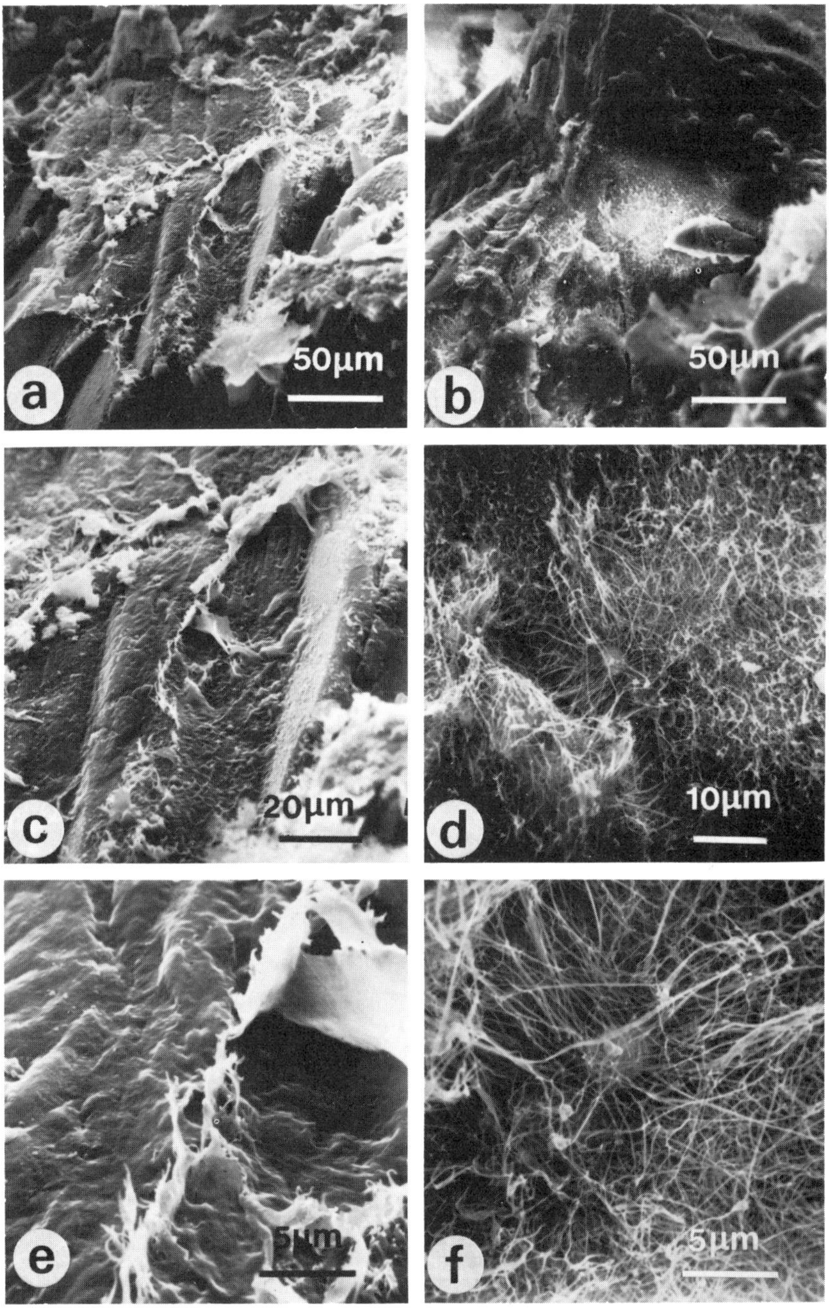

Figure 5.5 Fibrous illite in sandstone from the Magnus Field. (a), (c) and (e) are increasing magnifications of an air-dried sample. (b), (d) and (f) are increasing magnifications of a critical-point dried sample

the solid phase. However, Greene-Kelly (1973), Lawrence (1977) and Murray and Quirk (1980) all considered critical-point drying was the most promising technique for study of pore-size distribution in clay soils. Lawrence *et al.* (1979) compared pore-size distribution in critical-point- and freeze-dried clay aggregates. They found that freeze-drying gave little shrinkage, but produced pore distributions distorted to larger proportions of coarse pores, whereas critical-point drying produced some shrinkage, depending on the initial wetness of the samples.

McHardy *et al.* (1982) showed that the SEM morphology of diagenetic, filamentous illite, occurring in the pores of a reservoir sandstone, is very dependent on the specimen-drying method. After air-drying, the lath-like illite mineral occurred as mats, densely packed against the pore walls, whereas after critical-point drying an open, tangled web of very long, thin ribbons, virtually filling the pore space was observed (Figure 5.5). They suggested that these observations could explain the discrepancies that often occur between permeabilities of the wet and air-dried cores of such rocks.

Thus, there is no single answer to the problem of removing water from specimens prior to SEM study. The method chosen will depend on the type of information sought, practical convenience, time and equipment available. However, the effect of the drying method chosen on the basis of the observations, should always be taken into account.

5.3.2 *Resin embedding and polishing*

Many materials containing clays become incoherent when dried. Therefore, for some purposes, such as quantitative X-ray microanalysis or back-scattered electron imaging, resin impregnation and preparation of a polished surface or thin-section are required. Various epoxy and polyester resin kits suitable for the purpose are available, and a comprehensive account of impregnation methods and materials is given by Smart and Tovey (1982). A resin used successfully in the authors' laboratory is 'L.R. White', from the London Resin Company. This is a blend of acrylic monomers, formulated specifically for its low toxicity and ultralow viscosity. It can be cold-cured by the addition of a catalyst, but the reaction is rapid, exothermic and difficult to control. However, it may be thermally cured without a catalyst: recommended conditions are 60°C for between 18 and 24 hours. The procedure is to immerse the specimen in the resin, in a suitable polythene container in a vacuum desiccator, which is then evacuated with a water pump. When bubbling (due to escape of air from pores within the sample) ceases, the vacuum is released. After a few minutes—to allow the resin to penetrate into the sample—the desiccator is re-evacuated and put into an oven for the curing period. Polymerization is inhibited by the presence of oxygen, so that curing must be done under vacuum, or in a nitrogen (or similarly inert) atmosphere.

After curing, the resin-impregnated specimen is cut through with a diamond saw. The face is then ground and polished on standard lapping machines, using a range of diamond-impregnated bronze lapping plates, and polished with diamond pastes. A slab 2–3 mm thick may then be cut off and mounted on an SEM stub, polished side up. For some microanalytical applications this may be adequate, but the process may be taken further, with a thin-section suitable for examination by polarizing microscope being prepared in the usual way. Such thin-sections can be utilized for complementary optical microscopy and SEM studies. Where the SEM is equipped with X-ray microanalysis, an obvious possibility is the acquisition of analytical information from optically-observed features, provided these features can be identified in the secondary or back-scattered electron image of the SEM.

The most useful information from back-scattered electron imaging is derived from atomic number contrast, and for this to be optimized topographic contrast must be reduced as much as possible by polishing. Sections or polished surfaces of resin-impregnated material can be used successfully; however, Pye and Krinsley (1984) point out that conventional polishing techniques cause smearing and formation of a distorted layer several tens of nanometres thick, which can be evident at magnifications $> 3000\times$. With some specimens, the condition of the surface can be improved by gentle etching with hydrofluoric acid, but in other cases the etching may be mineral specific and introduce too much relief. Smart and Tovey (1982) recommend ion-beam etching for removing layers of smeared debris from polished surfaces, although care is needed in setting up the conditions distinguishing whether features 'revealed' by ion-beam etching are real, or artefacts. The whole subject of ion-beam etching, as applied to soil materials, is dealt with fully by Smart and Tovey (1982).

5.3.3 *Low-temperature ashing*

Price and Jenkins (1980) and Jenkins (1981) have also addressed the problem of correlating optical observations on standard soil thin-sections with SEM, their approach being to selectively remove resin from the thin-section by low-temperature ashing. This involves partial ionization of oxygen at low pressures by high-frequency (radio or microwave) excitation, to produce a nonequilibrium plasma. This partial ionization can achieve controlled oxidation and volatilization of organic materials at relatively low temperatures. The actual sample-surface temperature depends on oxygen flow rate and rf (radio frequency) power input, but appears to be ~ 150–200°C.

The procedure and operating conditions recommended by Price and Jenkins (1980) are as follows. Having selected features for further study in thin-section, optical micrographs are taken at a range of magnifications, depending on the feature. It is important to include micrographs at low magnification, encompassing the whole area to be examined by SEM, in order to locate easily the feature(s) of interest. To remove the resin from the thin-section, Price and

Jenkins (1980) used a low-temperature ashing instrument, operating at 14 MHz and an oxygen flow rate of 200 ml/minute, at a pressure of 130 Pa. The power input was kept at the lowest value (1950 W), to minimize temperature elevation and possible distortion of the sample. Under these instrumental conditions, complete removal of resin was achieved in periods ranging from 5 to 15 min, but this time will vary according to the particular equipment, the type of resin used, the nature of the soil sample, and whether partial or complete removal of resin is desired.

Jenkins (1981) points out that specimens could be damaged or distorted, and artefacts introduced, by the partial-ionization technique. Although the low temperatures ($<200°C$) involved safeguard most minerals, gypsum suffers badly, presumably because of dehydration to hemihydrate, and Na-smectites are vulnerable to some degree (Frazer and Belcher, 1973). Another problem is the residual ash from oxidation of the resin, which is assumed to account for the stringy or frothy features sometimes observed at high magnification. Generally, organic features of the samples will also be removed with the resin.

5.3.4 *Specimen mounting*

Having prepared a satisfactory surface, the specimen is glued onto the aluminium support stub with electrically conducting glue, of which several types are available. DAG is a fast-drying, fine-flake, silver suspension in methyl isobutylketone. LEIT-C is a conducting carbon cement, consisting of a suspension of very fine, conducting, carbon particles in a volatile carrier. Larger specimens may be mounted in two stages. First, the specimen is pressed gently into a thin layer of cement spread on the stub. When this has hardened, more cement, thinned to a paint consistency by a solvent (supplied with the glue kit) is smeared round the edges. Smaller sand-size particles can be attached to the stubs with double-sided tape. Fine silt and clay may be dried from suspension onto thin, circular glass cover-slips or—if microanalysis is envisaged—polished carbon discs, which are fixed to the stub with conductive cement. Carbon supports can easily be made by cutting sections 2–3-mm thick from a carbon rod, the surface then being smoothed with a file and, if necessary, polished by rubbing on a piece of velin tissue stretched over a glass slide.

5.3.5 *Coating*

For nonconducting specimens—and most clay materials come into this category—the surface to be examined by SEM must be coated with a layer of conducting material, such as C, Au, Au–Pd alloy or Pt. The conducting film must be continuous, and thick enough to conduct away the excess charge, but not so thick as to modify the underlying surface morphology. The choice of coating material depends on the information required. For microscopy, the efficiency of secondary electron production should be high, so that any chemical contrast is suppressed and the signal will depend only on the topography of the specimen. For X-ray microanalysis and back-scattered electron imaging, the opposite is required. Thus, for topographical information, gold has been used most extensively, although gold–palladium alloy is preferred by some workers since it supposedly gives a film of slightly better integrity and finer grain size than pure gold. It also wets tungsten filaments where an evaporative method of deposition is to be used, whereas gold does not. Platinum is also being used increasingly; it has the advantage of fine grain size if high-resolution studies are to be attempted.

Coating of specimens for SEM is now done largely by sputter coating, a relatively simple procedure which occurs when a low-pressure ionized gas plasma (argon) of sufficient energy is allowed to etch a suitable target of low binding energy, such as gold. Atoms of the target are eroded away, and because of their short mean-free-path in the argon atmosphere of ~ 10 Pa, they suffer multiple collisions before being deposited on the specimen. This scattering of the target atoms results in omnidirectional coating, so that no specimen-rotation rig is necessary. Many commercial sputter coaters are now available. They are simple, easy-to-use items of equipment, requiring only a double-stage rotary pump. Once the operating conditions (i.e. argon pressure, high voltage and operating current) required for satisfactory coating have been established, these same conditions can reliably produce an identical thickness of film for the life of the target, which with normal usage ought to be several years. Many manufacturers supply a formula with the instruction booklet, so that a rough estimate can be made of the film thickness from the time of sputtering, the voltage and the current. A film-thickness monitor can be included, but is unnecessary unless an exact coating thickness is important for a particular study. Modern sputter coaters now incorporate magnetic deflection and focusing systems which serve to deflect high-energy electrons away from the specimen, and focus the metal plasma onto the specimen. This gives a fast, cool coating process, with little specimen damage. Gold and gold–palladium alloy have been most widely used in sputter coating, although platinum, with its finer grain size, is preferred by some workers.

Unfortunately, carbon cannot be sputtered and, until recently, was almost always evaporated by resistive heating, short bursts of current from a 30 V, 30 A source being passed through a pair of pointed carbon rods which are pressed together (Bradley, 1965). The trend now is to use carbon fibre instead of carbon rods, and several carbon evaporation devices, either self-contained or designed as accessories to sputter coaters, are now available. A typical system, using only carbon fibre and a rotary pump, claims to be a rapid and reproducible method of carbon coating. The fibre is clamped between two fixed-distance clamps in a single length, and it is necessary to preheat the filament for 1–2 minutes at red heat to outgas the fibre. A shutter is usually provided to protect the specimen during this period. There is then a single 'one shot' button for the final evaporation, which is performed at a pressure of 10^{-4} Pa, and which seems to be controlled by the fibre burning through.

5.4 Microanalysis

5.4.1 *Introduction*

The generation of X-rays is one of the events that takes place when the electron beam strikes the specimen. Although the relationship between X-ray line energy and atomic number had been known for some time, it was Castaing and Guinier (1950) who first described the instrument which they called the electron microprobe. This led to the realization that high-resolution imaging, with simultaneous elemental analysis was potentially, a very powerful research technique.

5.4.2 *Generation of X-rays*

In the electron microscope, the primary electron beam ejects electrons from the inner shells of atoms from the specimen under examination. The subsequent electron transitions result in a series of X-ray lines, with energies characteristic of the elements present and with intensities characteristic of the amounts of these elements. The maximum efficiency of X-ray production will be obtained when $E_0 > 2.7\ E_c$, where E_0 is the accelerating voltage of the primary beam, and E_c is the critical excitation voltage of the element. In practice, however, the usable X-ray intensity from a specimen will depend on the amount of the specimen being irradiated, the atomic number and the electron-beam current.

The electrons from a beam striking a solid target will penetrate several micrometres into the sample, and before coming to rest they will influence a volume, as shown in Figure 5.6. It is apparent that the volume from which X-rays are produced is much greater than would be expected from the diameter of the electron beam striking the surface. The absorption and scattering of the primary electrons within this irradiated volume will determine the analytical resolution achieved. The relative importance of the various processes taking place has been discussed by Beaman and Isasi (1972). The *diameter* of the volume of interaction is a function of the primary beam accelerating voltage, whereas the *shape* is a function of the atomic number of the specimen material. Thus, the easiest way of improving analytical resolution is to decrease the accelerating voltage as much as possible, but unfortunately, this will be accompanied by a corresponding reduction in peak intensity. A useful guide to calculating analytical resolution is found in Beaman and Isasi (1972). For example, at an accelerating voltage of 25 kV, the penetration and X-ray resolution for Fe in a typical clay-mineral matrix would be 2–3 μm.

The effect of lateral beam spreading is quite small in a thin specimen

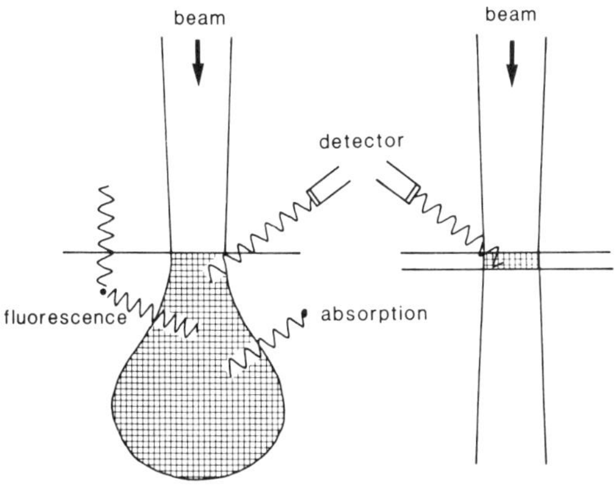

Figure 5.6 Diagrammatic representation of the processes occurring in the volume of interaction between primary electron beam and specimen: (left) bulk specimen; (right) thin specimen

(Figure 5.6). A single flake of clay mineral may be only $\leqslant 10$ nm thick and consequently, the analytical resolution can be calculated from the size of the area illuminated by the beam. However, longer analysis times are required compared with bulk specimens, because of the reduced volume of irradiated material (Chandler, 1977).

5.4.3 *Electron beam/specimen interactions*

Within the volume of interaction between primary electron beam and specimen (Figure 5.6) a number of processes occur which affect the production and collection of X-rays. These effects will only be described briefly here. For a more detailed mathematical treatment the reader is referred to Beaman and Isasi (1972) or Long (1977).

5.4.3.1 *X-ray absorption.* In a bulk specimen, the emerging X-rays will be partially absorbed by the specimen before reaching the detector. High-energy X-rays will be absorbed less than low-energy X-rays, and the further the X-rays have to travel in the specimen the more they will be absorbed. This length of travel will depend on the angular relationships between the primary beam, specimen and detector, and a high take-off angle will therefore result in less absorption than a low take-off angle. There will, of course, be an added absorption factor if there is a window between the emergent X-rays and the detector.

5.4.3.2 *X-ray fluorescence.* Provided that their energy is high enough, X-rays passing through a specimen will excite X-rays characteristic of the other elements present. This secondary fluorescence will increase the total X-ray output.

5.4.3.3 *Atomic number and back-scatter.* Low atomic number elements have a greater number of orbital elements per unit mass, and as a result, absorb a disproportionately large amount of the incident electron energy. The generated intensity of an element in a light matrix is, therefore, less than the same amount of the element in a heavy matrix. On the other hand, electrons are back-scattered more in a heavy matrix, and consequently, less of their energy is available for producing characteristic radiation. These two effects produce a nonlinear relationship between concentration and X-ray intensity (Sweatman and Long, 1969).

The four interactions mentioned in the last three sections form the correction procedure for the X-ray emission from a thick specimen, and collectively are known as the *Z* (atomic number), *A* (X-ray absorption) and *F* (X-ray fluorescence) correction.

One other effect must be considered, namely the 'white radiation' or Bremsstrahlung produced by the deceleration of the primary electron beam in the specimen. Eventually, the electrons will have no energy, and so there will be a band of background radiation showing all energies from zero to the energy of the primary beam. The characteristic lines will be superimposed on this continuum.

5.4.4 *Wavelength-dispersive spectrometers*

In the wavelength-dispersive (WD) system, a crystal of known interplanar spacing, *d*, is used. A divergent beam of X-rays emanating from the specimen strikes the curved crystal and is diffracted according to the Bragg Law,

$$\lambda = 2d \sin \Theta,$$

where λ is the wavelength of the X-rays and Θ the angle of incidence of the X-rays to the crystal. The curvature of the crystal converges the diffracted beam onto a detector, usually a gas-flow proportional counter. For any element, there is a unique set of Bragg angles at which there will be X-rays detected by the counter. To detect a second wavelength, the angle of the crystal to the incident beam (i.e. the Bragg angle) must be changed. When the instrument is in use, the crystal is rotated on a motorized goniometer, while the counter sweeps through an arc, collecting X-rays and registering a signal as the Bragg condition is satisfied.

The great advantage that WD spectrometers have over energy-dispersive spectrometers is their excellent resolving power. However, they are large and mechanically complex: their size means that the detector will be at some distance from the specimen, so necessitating lengthy analysis times. Each crystal in the spectrometer is only useful for a small range of wavelengths. This means that there must be an arrangement which allows selection from several crystals if the whole X-ray spectrum is to be investigated. WD spectrometers are still the definitive form of microprobe analysis and are used to calibrate standards, although their use has been supplanted by the introduction of energy-dispersive detectors. However, computer control, which relieves the operator of most of the spectrometer manipulation, has revived interest in wavelength spectrometry.

5.4.5 *Energy-dispersive spectrometers*

The spread of interest in microprobe analysis, particularly as an accessory to scanning electron microscopy, can be attributed to the introduction of energy-dispersive spectrometry (EDS). The detector is a small, solid-state device cooled by liquid nitrogen. It produces a voltage pulse which is proportional in energy to the energy of the incident photon. The pulses are sorted, measured and counted by a multichannel analyser system whose output is displayed as a spectrum on a cathode-ray tube. EDS systems are mechanically very simple and can be readily interfaced with most electron microscopes, both the scanning and the transmission types. They display simultaneously most of the X-ray energies emitted from the specimen, which makes them ideal for a fast, preliminary analysis. The lower limit of the detectable energy level is set by the thin beryllium window, through which X-rays must pass to reach the detector. Normally, elements of atomic number 11 (Na), and greater, are readily detectable when using one of the K, L, or M lines. Recently, detectors with a removable window have been developed, where, it is claimed, the range of detection has been extended down to and including beryllium.

The EDS detector can be positioned very close to the specimen, collecting X-rays over a large solid angle, which allows short analysis times (of ~ 100 seconds) and small beam currents, both of which reduce the chance of specimen damage or volatilization of labile elements. In addition, the use of small beam currents means that X-ray microanalysis is compatible with the electron beam conditions that are required for reasonably high image resolution on an SEM. The large collection angle means that the semiconductor detector is relatively insensitive to specimen height, and is therefore more appropriate than a WD system for the study of rough surfaces.

One serious disadvantage of EDS systems is their inferior spectral resolution, leading to severe peak overlap. Failure to account for overlaps can result in grossly inaccurate determinations. Fortunately, the output from an EDS detector readily lends itself to computer manipulation, and software is available to overcome this problem. The typical resolution of a modern semiconductor detector is ~ 155 eV, but as the theoretical limit of detection is now being approached, future improvements will mean a greater sophistication of the data reduction techniques.

A useful summary of the relative advantages and disadvantages of solid-state detectors and crystal spectrometers is given by Chandler (1977). Velde (1984) has studied the reliability of electron-probe analysis for clay-mineral investigations. He showed that beam currents should be kept to a minimum and the probe size to a maximum, in order to minimize loss of the labile elements—such as Na and K. For many minerals, a beam diameter of 2 μm at 1–2 nA for 200–400 seconds is acceptable, but even this may result in some loss of, for example, the sodium signal. In general, Velde concluded that an EDS system was better than a WD system because it used shorter counting times, and with a spot size of 5 μm, most alkali losses could be avoided.

5.4.6 *Qualitative analysis*

Even when a full quantitative analysis is not required, all the peaks present in a spectrum should be identified, not just the ones of immediate interest. The automatic peak identification routine included in most systems may not be complete and reference should be made to tables of X-ray energies (Bearden, 1964) or plots of the energies normally found in practice (Myklebust *et al.*, 1979). The count rate should be kept below 5000 counts/sec while accumulating spectra, to reduce the possibility of spectral artefacts. Identification should start with the largest peaks, and all possible KLM combinations should be checked. This is particularly important with elements such as Ba, Ti and Ce, where the sequence of minor peaks is diagnostic.

5.4.7 *X-ray distribution images*

If the X-ray spectrometer is tuned to the wavelength or energy of a chosen element while the electron beam is scanning a specimen, and the resulting signal is used to modulate the microscope cathode-ray-tube display, then a map of that element's distribution can be obtained. This technique is best illustrated with an example. The scanning electron micrograph in Figure 5.7(a) is of a freshly cleaved biotite flake isolated from the C horizon of a soil profile. A distinctive feature of this flake when viewed optically was a white core which can be distinguished in the electron micrograph by its slightly rougher texture compared with the edge. An image of the same field, produced by using X-rays characteristic of potassium to modulate the microscope display, shows clearly that the central part of the flake is markedly potassium depleted (Figure 5.7(b)).

As a general guide, the exposure time required for recording such images can be based on the time taken for accumulation of 20 000 X-ray counts from the area being scanned. The X-ray distribution display can also be used for locating small regions within a specimen where the element of interest is concentrated.

5.4.8 *Line profiles*

Another way in which chemical information can be displayed is illustrated in Figure 5.7(c) and (d). The scanning electron micrograph shows a soil section

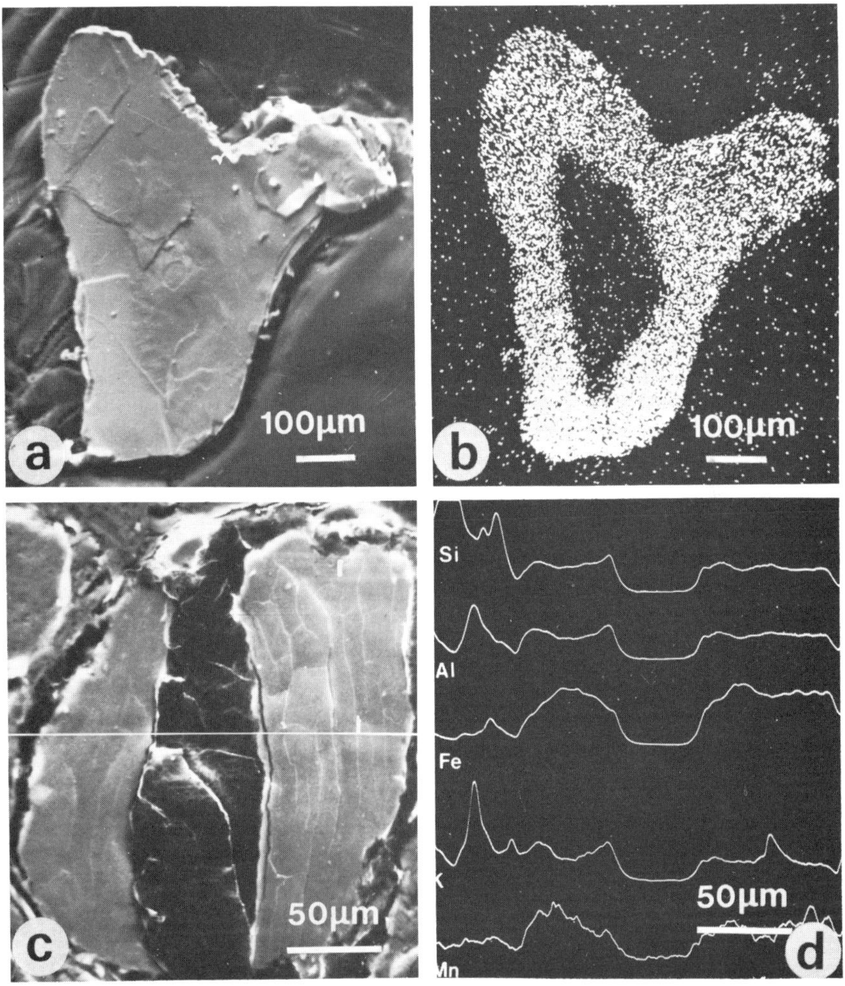

Figure 5.7 (a) Scanning electron micrograph of a partially weathered biotite particle, which has been freshly cleaved, mounted on a stub with double-sided tape and carbon coated; (b) K X-ray distribution image of the same field of view as (a) ; (c) Resin-impregnated soil section, polished and carbon coated, featuring a partially infilled pore; (d) Si, Al, Fe, K and Mn concentration profiles pertaining to the line of analysis along the middle of (c)

where a pore has been partially filled with fine clay. The concentration profiles for Si, Al, Fe and Mn (Figure 5.7(d)) refer to the line across the centre of Figure 5.7(c). The profiles were obtained by driving the beam slowly along this line and modulating it (in the y-direction) by the output from a ratemeter tuned to the appropriate X-ray signal for each of the elements of interest.

A line profile is superior to a distribution image, because it can be superimposed on a secondary electron image, which allows an easy correlation between morphology and chemical analysis. It is more sensitive to small changes in concentration and it is also readily quantified.

5.4.9 *Quantitative analysis*

(i) Bulk specimens To a first approximation, the concentration of an element A (C_A) is given by:

$$C_A = \frac{I_A}{I_S} C_S$$

where I_A is the X-ray intensity or peak integral of the element A in the sample, I_S is the X-ray intensity of element A in the standard and C_S is the concentration of element A in the standard. If the standard is a pure element, then $C_S = 1$ and:

$$C_A = \text{constant} \times I_A$$

When several elements are present, the *ZAF* corrections already mentioned must be taken into account. Thus, the corrected concentration of element A in a compound specimen is given by:

$$C_{A(\text{corr.})} = \frac{\text{constant} \times I_A}{Z \times A \times F}$$

where *Z*, *A* and *F* are the atomic number, absorption and fluorescence correction factors. These correction factors are applied to an initial concentration, given by the ratio of X-ray intensity of the element in the specimen to that in the standard, and a new corrected concentration results from this. The correction process is repeated using these new values, until the results of successive calculations converge. Where the standard is not a single element, matrix effects within the standard must also be taken into account. This kind of complex interactive calculation is obviously a task for a computer, either on-line, or using one of several readily available commercial routines. Myklebust *et al.* (1979) have described a data-evaluation procedure for carrying out quantitative electron-probe microanalysis with a lithium-drifted silicon detector.

An important parameter in the foregoing mathematical treatment of X-ray spectra is the take-off angle, that is, the angle between the incident electron beam and the emergent X-rays. Best results are obtained from highly polished surfaces; rough fracture surfaces with poorly defined specimen geometry present special problems. Statham (1979) has used peak-to-background ratios in a correction procedure that minimizes the effect of specimen geometry, which could be a considerable improvement on conventional '*ZAF*' procedures. The peak-to-background approach is also insensitive to spectrometer parameters and beam current fluctuations.

5.4.10 *Single-particle analysis*

Microprobe analysis of clay minerals and related materials will often involve single particles on an inert substrate. Methods of data reduction for analysis of particles of variable geometry have been devised by Armstrong and Buseck (1975) and Aden (1981). If the clay mineral being analysed is a thin plate lying normal to the electron beam, the X-ray absorption path is readily calculated, provided that the particle thickness can be estimated by SEM. For more complicated particle shapes, thickness and density must be taken into account, and an estimate must be made of the idealized geometric model (Armstrong, 1978). Where two or more structurally similar but chemically distinct species are being studied, provided that the particle diameters are $< 5\,\mu$m, the composition ratios may be calculated without the use of matrix corrections (Pyman *et al.*, 1978).

5.4.11 *Thin specimens*

For thin specimens, i.e. particles < 200 nm thick, lateral diffusion of the X-ray beam is minimal, and absorption and fluorescence effects can be neglected. The characteristic X-ray intensity is therefore proportional to the mass per unit area of the element, and the background radiation is proportional to the mass per unit area of the whole irradiated region (Hall, 1971). The relative concentration of an element will be:

$$\mathrm{C_x} = K\frac{I_\mathrm{x}}{\mathrm{W}}$$

where I_x is the characteristic X-ray intensity and W is the white radiation. K can be calculated from a standard. The characteristic X-ray intensity ratio for any two elements can be related to the corresponding weight-fraction ratio $\mathrm{C_1/C_2}$ by the equation:

$$\frac{\mathrm{C_1}}{\mathrm{C_2}} = K_\mathrm{s}\frac{I_1}{I_2}$$

where K_s is a constant at a given accelerating voltage and is independent of specimen thickness and composition (Champness *et al.*, 1982). There are many commercial software packages available for thin-sample analysis. Quantitative measurements can also be made using suitable standards, and a calibration curve can be constructed by analysing the relevant elements and normalizing the signals against one element. An application of a standardless, thin-film technique has been tested by White *et al.*, (1982) on three clay minerals of well-known composition (kaolinite, talc and pyrophyllite) and has been found to give reasonable results.

5.4.12 *Analysis using back-scattered electrons*

As previously stated, elements of atomic number < 11 cannot be detected by an ordinary X-ray energy-dispersive detector with a beryllium window. This restriction can be partly overcome by exploiting the fact that the BSE signal is proportional to the average atomic number of the specimen. By standardizing the important operating conditions, and calibrating the back-scattered signal variation with atomic number, a quantitative measure of the apparent atomic number of any region of a specimen can be obtained (Hall and Lloyd, 1981; Ball and McArtney, 1981; Pye and Krinsley, 1984). The average atomic number will of course be influenced by the light elements, and if the heavier elements can be identified then a library of compounds can be matched with the BSE signal to find the closest fit to the average atomic number. A system based on this type of backscattered detection, called Composition Analysis using a Robinson Detector (CARD) is available from Princeton Gamma-Tech. A similar Image and Micro Analysis System (IMAS) is available from Cambridge Technology.

5.4.13 *Specimen preparation*

Specimen preparation methods for microanalysis are similar to those described for SEM imaging, although carbon-based conducting pastes should be used. Small samples of clay can be analysed after pressing a few milligrams of the clay into a plug and fixing it to a specimen stub with double-sided tape. The sample can then be recovered for further use. As a general rule, specimens should be flat, but this may not be possible and rough surfaces may have to be reoriented to leave an unobscured path to the detector. Morphological examination and microprobe analysis of individual clay flakes are greatly simplified if the clay can be deposited onto a carbon planchet; Berkheiser and Monsees (1982) have described a surfactant technique enabling clay particles to be so dispersed for X-ray microanalysis. If a specimen must be recovered uncontaminated, or if a quick qualitative analysis is all that is required, microanalysis may be carried out on an uncoated specimen. Generally, however, clay materials require a conductive coating, and carbon is commonly used because no interfering X-ray lines arise. Coatings of ~20 nm are adequate. For quantitative analysis, reproducible film thickness is desirable, although serious errors attributable to film-thickness differences will arise only for light elements and low excitation voltages (Kerrick *et al.*, 1973; Springer, 1974). An initial carbon coating can be useful for SEM microanalysis, followed by 20 nm of gold if good secondary electron images are required. Gold-coated specimens can be used quite successfully for qualitative microanalysis, provided that there are no peaks of interest likely to be masked by the characteristic gold peaks.

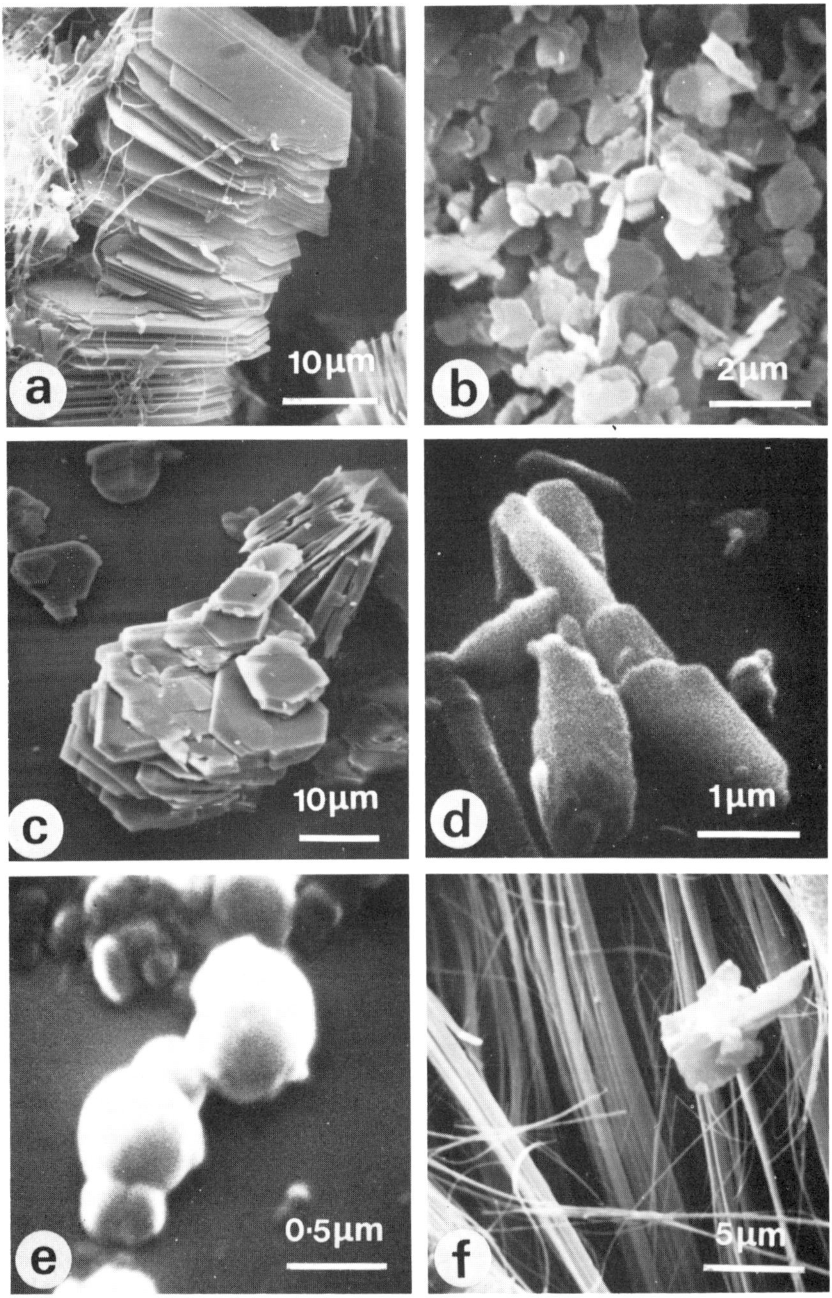

Figure 5.8 All the specimens have been gold coated. (a) A well-crystallized book of kaolinite crystals in a reservoir sandstone pore. The fine strands are of filamentous illite. (b) Fine-grained kaolinite occurring as grain coatings in a white sandstone of Carboniferous age from a borehole in the Midland Valley of Scotland. (c) Dickite (source unknown). Powdered material dried from suspension onto a glass slide. (d) Halloysite from Wagon Wheel, Colorado, USA. Material has been dried from suspension onto a glass slide. (e) Spherical or ellipsoid halloysite from New Zealand. (f) Chrysotile from a serpentine rock, Leslie, Aberdeenshire.

5.5 Examples

5.5.1 *The kaolinite group*

Kaolinite is by far the most common mineral of this group. Kaolinite can occur as well-formed, large crystals of clear hexagonal outline, often in stacks or vermicules such as those illustrated in Figure 5.8(a). However, it can also occur as much smaller crystals of less clearly defined outline (Figure 5.8(b)). The SEM study of kaolinite and its related minerals has been the subject of numerous publications by W.D. Keller (1978, 1982, etc.).

Dickite is much rarer than kaolinite itself, its morphology being typically that of well-formed crystals (Figure 5.8(c)), which can be up to 0.5 mm wide.

A tubular habit is generally distinctive of halloysite, although considerable variations in morphology have been reported (Tazaki, 1979). Halloysite from Wagon Wheel, Colorado is exceptionally distinctive and shows a morphology consisting of stepped tapering tubes (Figure 5.8(d)). Beyond the resolution of the SEM, but clearly seen in TEM replicas, is the observation that these 'tubes' are often polygonal in cross-section, rather than spherical. An example of a scanning electron micrograph of an ellipsoid halloysite, as described by Kirkman (1981), is given in Figure 5.8(e). Again the particles are too small for their true multifaceted nature to be resolved, although this is readily observed in TEM replicas (Figure 6.15c).

5.5.2 *The serpentine group*

The serpentine minerals are structurally and chemically analogous to the minerals of the kaolinite group. Variation from equidimensional plates to laths is also similar, although the platy varieties commonly lack the crystallographic features that characterize kaolinite. Chrysotile and halloysite tubes have many of the same shape characteristics, but the former has a much greater length/width aspect (Figure 5.8(f)).

5.5.3 *Illite/smectite*

Illite, smectite and various interstratified forms of illite/smectite are probably the most commonly encountered clay minerals, both in soils and in reservoir sandstones. Soil illites tend to be platy, for example, the irregular but sharply defined flakes surrounding the central quartz grain in Figure 5.9(a). Very thin illites, such as that illustrated in Figure 5.9(b), often show signs of curling at the irregular edges of individual plates. This particular mineral, which occurs in a reservoir sandstone, is classified (according to X-ray diffraction) as an interstratified illite-smectite with $\sim 10\%$ smectite. Figure 5.9(c) illustrates the morphology of a diagenetic, filamentous, illitic clay often associated with the pores of hydrocarbon reservoir sandstones. The filaments in this example are really ribbons $< 0.5\ \mu m$ wide, and (as measured by TEM) average 3 nm thick.

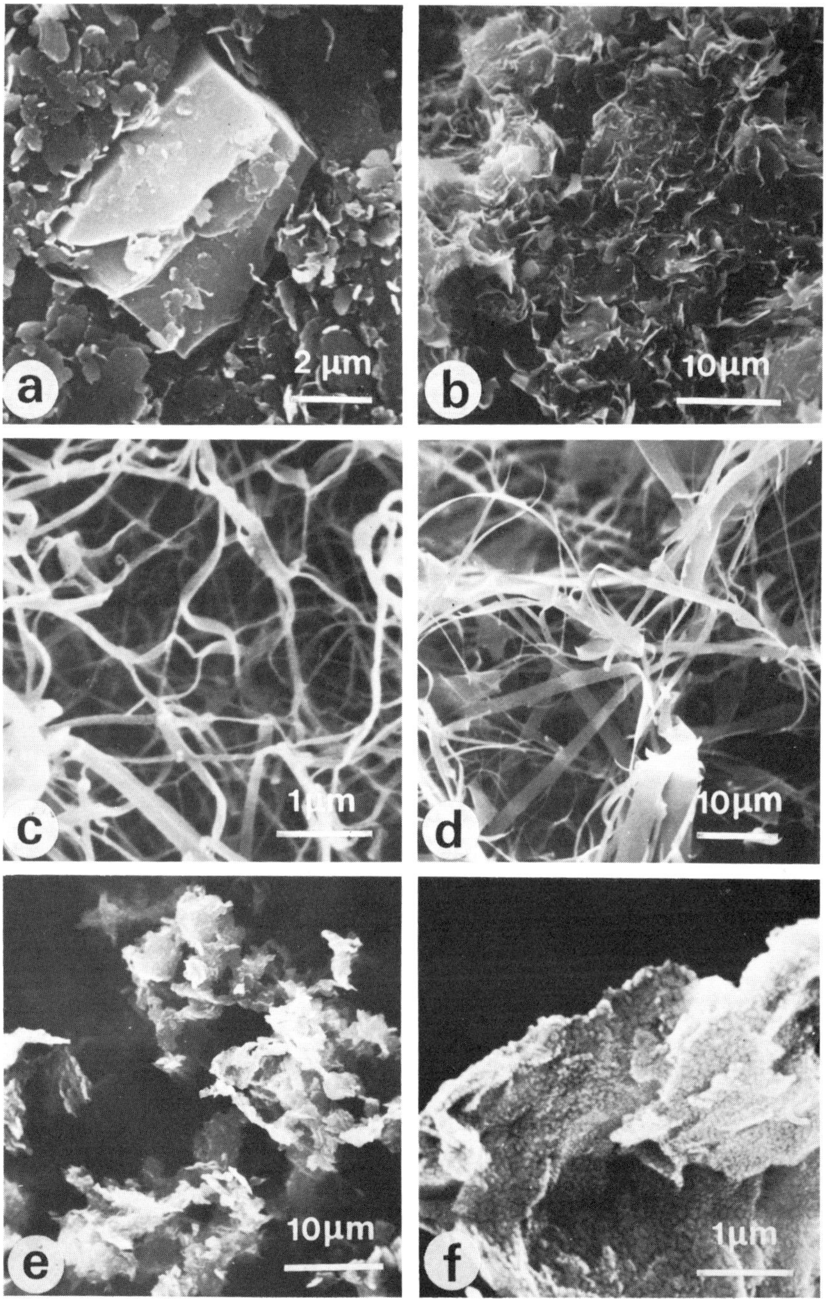

Figure 5.9 All the specimens have been gold coated. (a) Illite in a soil clay (from Tipperty, Aberdeenshire). The large grain in the middle of the micrograph is quartz. The material was critical-point dried prior to preparation of a fracture surface (b) Interstratified illite/smectite (with 10% smectite layers) in a reservoir sandstone (c) Filamentous illite (illite/smectite with 10% smectite layers) in reservoir sandstone from the Magnus Field (d) Filamentous illite from the Rotliegendes sandstone. (e) Smectite (from a Greek soil) freeze-dried from a sediment (f) Higher magnification micrograph of the central area of (e)

The X-ray diffraction pattern is that of an ordered interstratification of illite and smectite with ~20% smectite. Figure 5.9(d) illustrates an illite from a similar environment, which possesses a similar morphology of interwoven ribbons. However, the ribbons in Figure 5.9(d) are an order of magnitude wider and thicker than those in Figure 5.9(c), and this material gives an X-ray pattern of a pure dioctahedral illite.

As was pointed out previously, the SEM morphology of diagenetic illite in the pores of sandstones associated with hydrocarbon reservoirs is very dependent on the drying method of the specimens. According to McHardy *et al.* (1982), the critical-point drying technique appears to give the most realistic picture of the natural arrangement of illite particles.

The morphology of smectite itself is also strongly dependent on the method of drying. Figure 5.9(e) is a micrograph of a smectite that has been freeze-dried at −80°C from a thick suspension. The morphology is that of an extremely delicate aggregate of very thin, crumpled flakes. At very high magnification, the structure of the gold coating film is resolved, but not individual crystallites of smectite (Figure 5.9(f)).

5.5.4 *Chlorite*

Like minerals of the illite/smectite group, authigenic chlorite is a common constituent of hydrocarbon reservoir sandstone rocks, and the SEM morphology in such an environment is usually quite diagnostic. This is illustrated in Figure 5.10(a), where the primary grains of a sandstone are coated with a film of chlorite several μm thick. The slightly curved, bladed crystals are arranged in a honeycomb network. The individual crystals are thicker and more rigid than the illite laths found in similar environments so that chlorite morphology is less influenced by the method of drying.

5.5.5 *Sepiolite and palygorskite*

Both these minerals are characterized by a fibrous habit. An example of the SEM morphology of sepiolite is shown in Figure 5.10(b). This is the fracture surface of a Meerschaum-type rock. The individual particles of sepiolite are just resolved as a randomly oriented network of densely packed fibres. The study of such fine fibres is a task more appropriate for the transmission electron microscope.

Palygorskite has an SEM morphology similar to that of sepiolite and it would be difficult to distinguish the two on SEM observation alone. The SEM micrograph of palygorskite given in Figure 5.10(c) was obtained from a Turkish soil, and shows the delicate meshwork of fibres coating and binding carbonate grains.

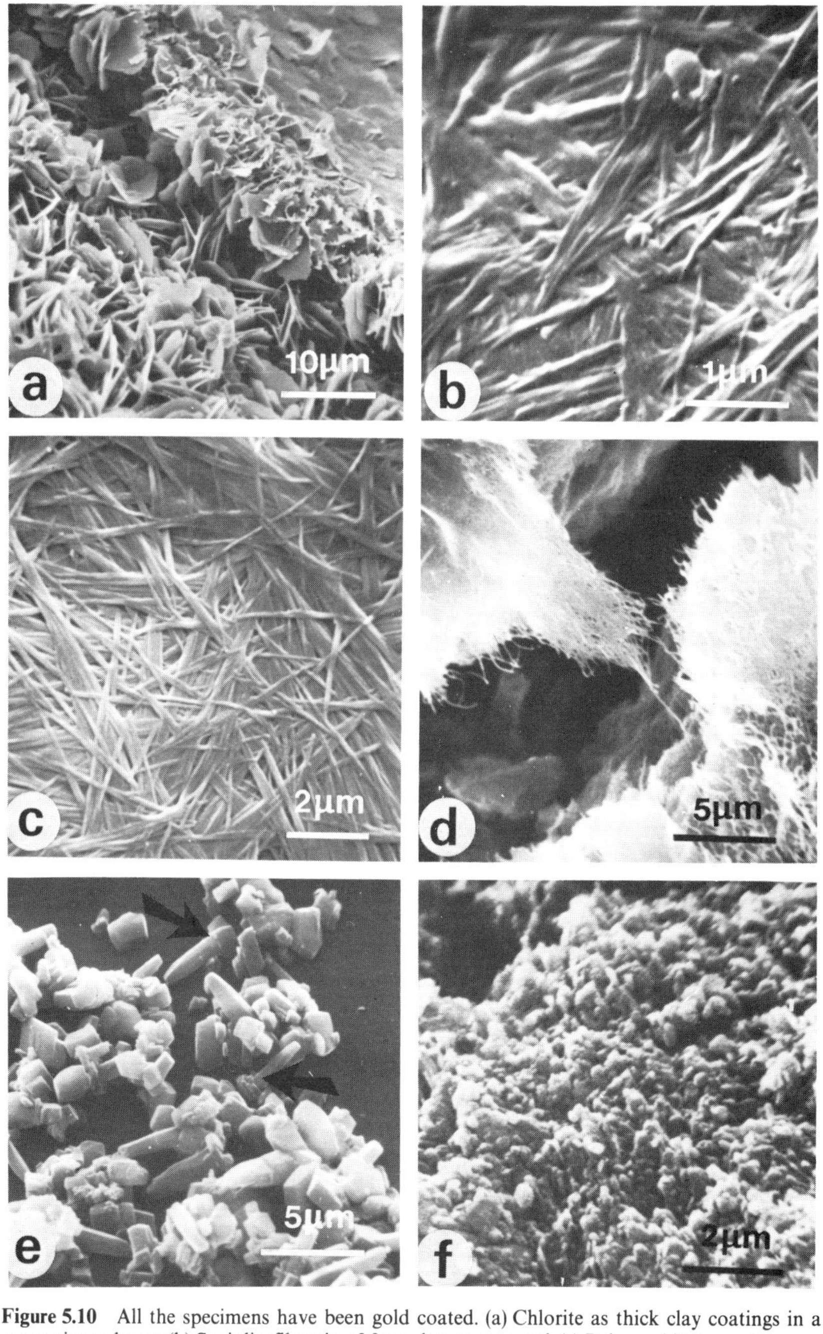

Figure 5.10 All the specimens have been gold coated. (a) Chlorite as thick clay coatings in a reservoir sandstone (b) Sepiolite fibres in a Meerschaum-type rock (c) Palygorskite in a calcareous Turkish soil (d) Imogolite in the B_s cemented horizon of a podzol (e) Synthetic nordstrandite and gibbsite particles sedimented onto a glass slide (f) Fine-grained gibbsite particles in a bauxite

5.5.6 *Allophane and imogolite*

The extremely fine-grained and structureless nature of allophane renders its study by SEM somewhat meaningless. On the other hand, the fibrous nature of imogolite can sometimes be recognized in favourable circumstances. The example of imogolite illustrated in Figure 5.10(d) an occurrence in the B_s cemented horizon of a podzol. This fibrous morphology, indicative of imogolite, was only observed in specimens which had been critical-point dried and then only in places where the cementing film had cracked and pulled apart.

5.5.7 *Oxides of aluminium*

Most SEM observations of the various polymorphs of aluminium hydroxides have been made on synthetic materials. The occurrence of bayerite and nordstrandite in nature has been only rarely reported. The micrograph Figure 5.10(e) is of a synthetic preparation described by Tait *et al.* (1983), featuring a mixture of tabular nordstrandite crystals and rod-shaped gibbsite. The hammer-shaped particle top centre in Figure 5.10(e) is composed of a tapered prism of gibbsite which has cocrystallized with tabular nordstrandite.

Gibbsite and the oxyhydroxides boehmite and diaspore are found in varying proportions and morphologies in bauxitic deposits. A scanning electron micrograph of densely packed, very fine-grained gibbsite crystals, which occur in the Ayrshire (Scotland) bauxite deposit, is given in Figure 5.10(f). Bardsossy *et al.* (1978) have published a comprehensive scanning electron microscope study of bauxites of different ages and origins, in which 42 micrographs illustrate the size and form of individual crystals and grain aggregates, as well as the different types of microtextures.

5.5.8 *Oxides of iron and manganese*

The variable redox conditions in soils result in many different forms of iron and manganese oxides. For example, poorly drained soils may be characterized by ochreous mottles where the predominant iron oxide is lepidocrocite. An example (Figure 5.11(a)) shows the lepidocrocite to consist of ragged platelets orientated face to face. Again, soil goethite may have a granular morphology consisting of rather ill-defined flakes (Figure 5.11(b)). Field drains can become blocked by accumulations of deep-brown gelatinous iron oxides, commonly referred to as 'ochre', (Birnie, 1984). Sometimes, this deposit consists of a mass of hollow tubes (Figure 5.11(c)) where the mineral (ferrihydrite) has precipitated as sheaths around iron-oxidizing bacteria. Ferrihydrite occurring in voids in poorly drained soils occasionally exhibits a globular morphology. Such globules can be up to 2 mm in diameter, and when fractured, show internal radiating spines with a curved outer surface (Figure 5.11(d)). In contrast, iron oxides associated with weathering over prolonged periods of time may occur in distinctive forms. Figure 5.11(e) shows

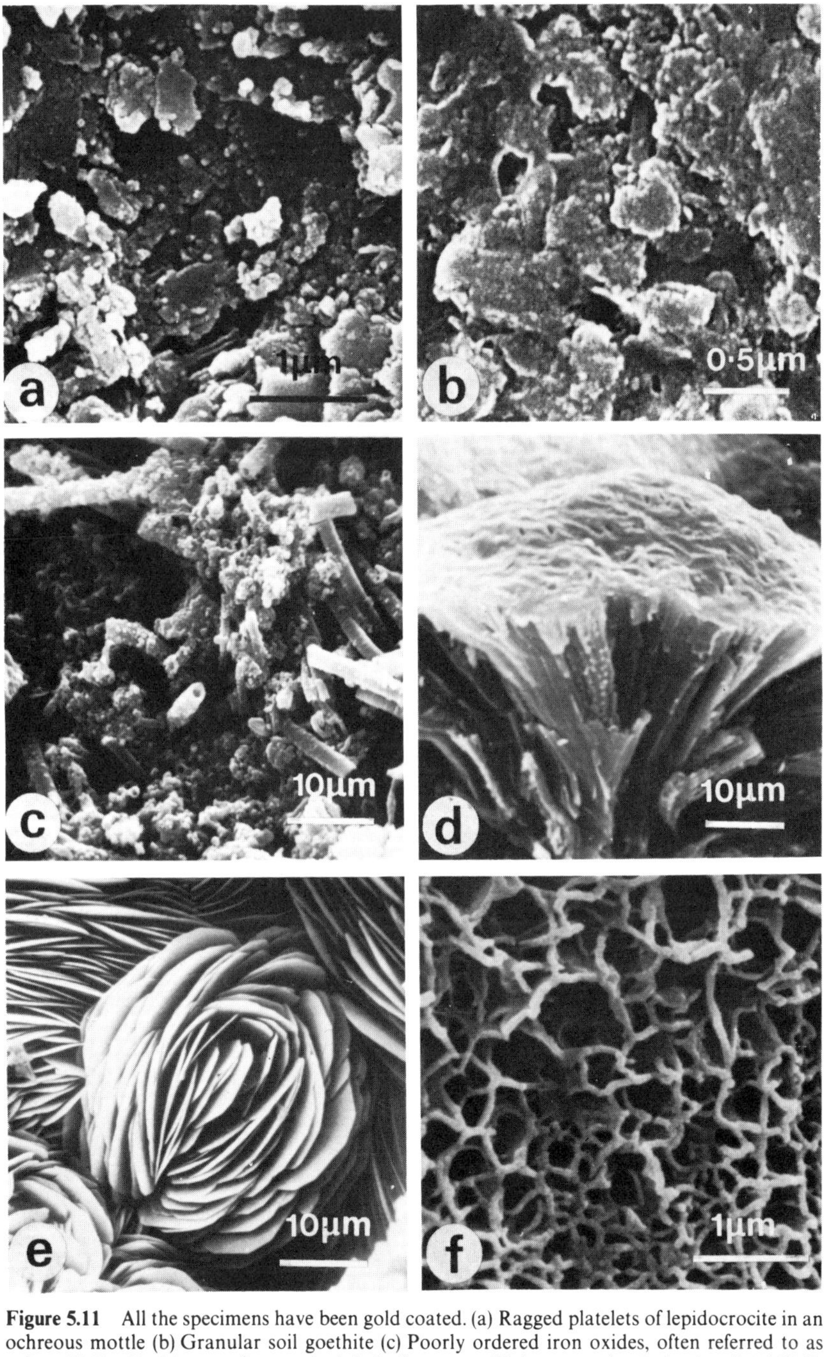

Figure 5.11 All the specimens have been gold coated. (a) Ragged platelets of lepidocrocite in an ochreous mottle (b) Granular soil goethite (c) Poorly ordered iron oxides, often referred to as ochre (d) Ferrihydrite—a fractured globule from a void in a poorly drained soil (e) Well-crystallized hematite in the cracks of a quartzite rock (f) Typical morphology of black manganese deposits and stains

highly crystalline hematite in a rosette form, this example being from deeply weathered schists in north-east Scotland.

Manganese oxides in soils commonly exhibit the wrinkled morphology shown in Figure 5.11(f). Individual platelets cannot be resolved, and deposits of this type are usually amorphous to X-rays. This morphology is typical of the black manganese deposits which occur as concretions and coatings in poorly drained soils in Scotland.

References

Aden, G.D. (1981) *Quantitative Energy Dispersive Analysis of Small Particles.* Ph.D. Thesis, Arizona State Univ., 459pp.

Armstrong, J.T. (1978) 'Methods of quantitative analysis of individual microparticles with electron beam instruments,' in *Scanning Electron Microscopy*, **1** (O. Johari, ed.) AMF O'Hare, SEM Inc., 455–468.

Armstrong, J.T. and Buseck, P.R. (1975) Quantitative chemical analysis of individual microparticles using the electron microprobe: theoretical. *Anal. Chem.*, **47**, 2178–2192.

Ball, M.D. and McArtney, D.G. (1981) The measurement of atomic number and composition in an SEM using backscattered detectors. *J. Microscopy*, **124**, 57–68.

Bardossy, Gy., Csanady, A. and Csordas, A. (1978) Scanning electron microscope study of bauxites of different ages and origins. *Clays & Clay Miner.*, **26(4)**, 245–262.

Beaman, D.R. and Isasi, J.A. (1972) Electron beam microanalysis. *STM Spec. Publ.*, **506**, ASTM, Philadelphia Pa.

Bearden, J.A. (1964) *X-ray Wavelengths*, **NYO-10586**, US Atomic Energy Comm., Oak Ridge, Tennessee, USA.

Berkheiser, V.E. and Monsees, M.B. (1982) Dispersion of clays on graphite support for X-ray microprobe analysis. *Soil Sci. Soc. Am. J.*, **46(3)**, 663–666.

Birnie, A.C. (1984) *Morphological, Physical and Chemical Changes occurring within a Hydrosequence of Soils of the Strichen Association.* M.Sc. Thesis, Univ. of Aberdeen, 152pp.

Bohor, B.F. and Hughes, R.E. (1971) Scanning electron microscopy of clays and clay minerals. *Clays & Clay Miner.*, **19**, 49–54.

Borst, R.L. and Keller, W.D. (1969) Scanning electron micrographs of API reference clay minerals and other selected samples. *Proc. Intern. Clay Conf. Tokyo*, Vol. I, Israel Univ. Press, Jerusalem, 871–901.

Boyde, A. and Tamarin, A. (1984) Improvement to critical point drying technique for SEM. *Scanning*, **6**, 30–35.

Bradley, D.E. (1965) 'Replica and shadowing techniques,' in *Techniques for Electron Microscopy*, (D. Kay, ed.) Blackwell Scientific Publications, Oxford, 96–152.

Castaing, R. and Guinier, A. (1950) Sur l'exploration et l'analyse elementaire d'un échantillon par une sonde electronique. *Proc. First Intern. Conf. Electron Microscopy*, Martinus Nijhoff, Delft, 60–63.

Champness, P.E., Cliff, G. and Lorimer, G.W. (1982) Quantitative analytical electron microscopy of metals and minerals. *Ultramicroscopy*, **8**, 121–132.

Chandler, J.A. (1977) 'X-ray microanalysis in the electron microscope,' in *Practical Methods in Electron Microscopy*, **5(2)**, (A.M. Glauert ed.) North-Holland, Amsterdam, 230 pp.

Chapman, S.K. (1986) *Working with a Scanning Electron Microscope.* Lodgemark Press, Chislehurst, Kent.

Diamond, S. (1970) Pore size distribution in clays. *Clays & Clay Miner.*, **18**, 7–23.

Erol, O., Lohnes, R.A. and Demirel, T. (1976) 'Preparation of clay-type, moisture containing samples for scanning electron microscopy,' in *Proc., Workshop on Techniques for Particulate Matter Studies.* Illinois Inst. of Technol. Press, Chicago, 769–776.

Everhart, T.E. and Thornley, R.F.M. (1960) Wide-band detector for micro-microampere low-energy electron currents. *J. Sci. Instruments*, **37**, 246–248.

Frazer, F.W. and Belcher, C.B. (1973) Quantitative determination of the mineral matter content of coal by a radiofrequency oxidation technique. *Fuel*, **52**, 41.

Gillot, J.E. (1969) Study of the fabric of fine-grained sediments with the scanning electron microscope. *J. Sedim. Petrol.*, **39**, 90–105.

Greene-Kelly, R. (1973) The preparation of clay soils for determination of structure. *J. Soil Sci.*, **24**, 277–283.

Hall, T.A. (1971) 'The microprobe assay of chemical elements,' in *Physical Techniques in Biological Research*, **1A**, (G. Ostern, ed.) Academic Press, New York, 393pp.

Hall, M.G. and Lloyd, G.E. (1981) The SEM examination of geological samples with a semi-conductor back-scattered electron detector. *Am. Mineral.*, **66**, 362–368.

Jenkins, D.A. (1981) 'The transition from optical microscopic to scanning electron submicroscopic studies on soils,' in *Submicroscopy of Soils and Weathered Rocks: 1st Workshop of the International Working Group on Submicroscopy of Undisturbed Soil Materials (IWGSUSM), 1980, Wageningen, the Netherlands.* (E.B.A. Bisdom ed.) Wageningen Pudoc, 163–172.

Keller, W.D. (1978) Classification of kaolins exemplified by their textures in scan electron micrographs. *Clays & Clay Miner.*, **26(1)**, 1–20.

Keller, W.D. (1982) Anauxite viewed by scanning electron microscopy. *Clays & Clay Miner.*, **30(5)**, 391–393.

Kerrick, D.M., Eminhizer, L.B. and Villaume, J.F. (1973) The role of carbon film thickness in electron microprobe analysis. *Am. Mineral.*, **58**, 920–925.

Kirkman, J.H. (1981) Morphology and structure of halloysite in New Zealand tephras. *Clays & Clay Miner.*, **29(1)**, 1–9.

Koller, T. and Bernhard, W. (1964) Sechage de tissus au protoxyde d'azote (N_2O) et coupe ultrafine sans matière d'inclusion. *J. Microscopie*, **3**, 589–606.

Lawrence, G.P. (1977) Measurement of pore sizes in fine-textured soils: a review of existing techniques. *J. Soil Sci.*, **28**, 527–540.

Lawrence, G.P., Payne, D. and Greenland, D.J. (1979) Pore size distribution in critical point and freeze dried aggregates from clay subsoils. *J. Soil Sci.*, **30**, 499–516.

Long, J.V.P. (1977) 'Electron probe microanalysis,' in *Physical Methods in Determinative Mineralogy*, J. Zussman ed.) Academic Press, London, 273–341.

Martin, J.P., Jenkinson, G. and Bulgin, D. (1985) Quantitative scanning electron microscopy using integrated digital image store for on-line image analysis. *Scanning*, **7**, 239–242.

McHardy, W.J., Wilson, M.J. and Tait, J.M. (1982) Electron microscope and X-ray diffraction studies of filamentous illitic clay from sandstones of the Magnus field. *Clay Miner.*, **17**, 23–29.

Murray, R.S. and Quirk, J.P. (1980) Clay–water interactions and the mechanism of soil swelling. *Colloids & Surfaces*, **1**, 17–32.

Myklebust, R.L., Fiori, C.E. and Heinrich, K.F.J. (1979) Frame C: a compact procedure for quantitative energy-dispersive electron probe X-ray analysis. *National Bureau of Standards Techn. Note*, **1106**, Washington, U.S. Government Printing Office, 106pp.

Newman, A.C.D. and Thomasson, A.J. (1979) Rothamsted studies of soil structure. III: pore size distributions and shrinkage processes. *J. Soil Sci.*, **30**, 415–439.

Oatley, C.W. (1966) The scanning electron microscope. *Science Prog.*, **54**, 483–495.

Parsons, E., Bole, B., Hall, D.J. and Thomas, W.D.E. (1974) A comparative survey of techniques for preparing plant surfaces for the scanning electron microscope. *J. Microscopy*, **101**, 59–75.

Price, F.R. and Jenkins, D.A. (1980) Removal of resin from standard soil thin-sections by low temperature ashing as a means of following transmitted optical by scanning electron microscopy. *Clay Miner.*, **15**, 309–315.

Pye, K. (1984) Rapid estimation of porosity and mineral abundance in backscattered electron images using a simple SEM image analyser. *Geol. Mag.*, **121**, 81–136.

Pye, K. and Krinsley, D.H. (1983) Mudrocks examined by backscattered electron microscopy. *Nature*, **301(5899)**, 412–413.

Pye, K. and Krinsley, D.H. (1984) Petrographic examination of sedimentary rocks in the SEM using backscattered electron detectors. *J. Sedim. Petrol.*, **54(3)**, 877–888.

Pyman, M.A.F., Hillyer, J.W. and Posner, A.M. (1978) The conversion of X-ray intensity ratios to composition ratios in the electron probe analysis of small particles using mineral standards.

Robards, A.W. (1984) Fact or artefact—a cool look at biological electron microscopy. *Proc. Roy. Microscopical Soc.*, **19**, 195–208.

Robinson, B.W. and Nickel, E.H. (1979) A useful new technique for mineralogy: the backscattered-electron/low vacuum mode of SEM operation. *Am. Miner.*, **64**, 1322–1328.

Robinson, V.N.E. (1974) The construction and uses of an efficient backscattered electron detector for SEM. *J. Phys. E: Sci. Inst.*, **2**, 650–652.

Smart, P. and Tovey, N.K. (1981) *Electron Microscopy of Soils and Sediments: Examples.* Clarendon Press, Oxford, 178pp.

Smart, P. and Tovey, N.K. (1982) *Electron Microscopy of Soils and Sediments: Techniques.* Clarendon Press, Oxford, 400pp.

Springer, G. (1974) The role of carbon film thickness in electron microprobe analysis: a comment. *Am. Miner.*, **59**, 1121–1122.

Statham, P.J. (1979) Measurement and use of peak-to-background ratios in X-ray analysis. *Mikrochimica Acta, Suppl.*, **8**, 229–242.

Sweatman and Long, (1969) Quantitative electron-probe microanalysis of rock-forming minerals. *J. Petrol.*, **10(2)**, 332–379.

Swift, J.A. and Brown, A.C. (1979) SEM electron source: pointed tungsten filaments with long life and high brightness.

Tait, J.M., Violante, P. and Violante, A. (1983) Co-crystallization of gibbsite and bayerite with norstrandite. *Clay Miner.*, **18**, 95–99.

Tazaki, K. (1979) 'Micromorphology of halloysite produced by weathering plagioclase in volcanic ash,' in *Proc. VI Intern. Clay Conf. 1978.* (M.M. Mortland and V.C. Farmer eds.) Elsevier, Amsterdam, 415–422.

Velde, B. (1984) Electron microprobe analysis of clay minerals. *Clay Miner.*, **19**, 243–247.

White, G.N., Berkheiser, V.E., Blanchard, F.N. and Hallmark, C.T. (1982) Thin-film analysis of clay particles using energy dispersive X-ray analysis. *Clays & Clay Miner.*, **30(5)**, 373–382.

6 Transmission electron microscopy

P.H. NADEAU AND J.M. TAIT

6.1 Introduction

6.1.1 *General statement*

The transmission electron microscope (TEM) is a high-resolution instrument which provides information concerning the sizes and shapes of clay particles and aggregates. Clay materials are well suited for examination by TEM, because most specimens are composed of particles with planar dimensions 10 to 10^3 nm within the optimum operating magnification range, (10^3 to 10^5), of these instruments. The construction of the microscope is analogous to the conventional optical microscope (Figure 6.1), but the electrons provide the illumination and electromagnetic 'lenses' condense the electron beam and magnify the specimen image. The strong interactions between electrons and matter require a high-vacuum environment (10^{-2}–10^{-4} Torr) in the microscope column, ultrathin specimen support films (1–10 nm) and thin specimens (< 100 nm). The instrument can also be operated in a diffraction mode, producing electron-diffraction (ED) patterns which are useful in determining the crystalline nature of clay particles and aggregates. Some instruments are also fitted with analytical systems which can determine the chemical composition of the specimen. Thus, the characterization of materials by TEM provides valuable information concerning the size, shape, structure and composition of clay particles on a scale which is not approachable by other analytical techniques.

6.1.2 *Historical development*

The first electron micrographs of clay materials were published by W. Eitel and his associates in 1940, these early studies being primarily concerned with the size and morphology of kaolinite and montmorillonite. The fibrous morphology of palygorskite (attapulgite) was observed by other workers in 1942, and the tubular nature of chrysotile and halloysite in 1943. In addition, the problem of determining particle thickness using stereoscopic techniques, and the combination of ED with microscopy for the study of thermal modifications of kaolinite were also demonstrated. Thus, many of the capabilities of TEM for clay-mineral studies were manifest during this early period.

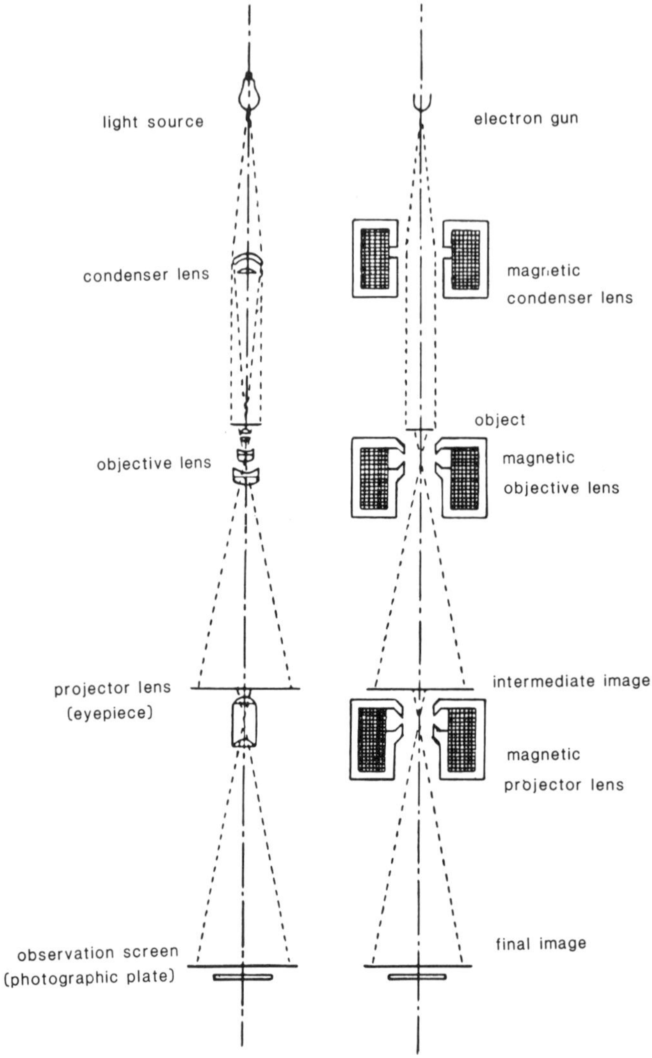

Figure 6.1 Comparison of the light and electron microscopes

Important contributions in later years included extensive ED studies of clay minerals (Zvyagin, 1967) using electron-diffraction camera and oblique texture techniques. More recent improvements in resolution have allowed examination of lattice-fringe images, which provide one-dimensional, and in certain cases, two-dimensional structure images (Yada, 1967; Iijima and Buseck, 1978), with a lattice resolution of 2 Å. A notable atlas of clay mineral micrographs was provided by Beutelspacher and van der Marel (1968). More detailed compilations of electron-optical data of clays, including specimen preparation techniques, were provided by Gard (1971) and more recently by Smart and Tovey (1981, 1982) and Sudo *et al.* (1981).

6.2 Principles

6.2.1 *Image formation*

The goal of electron microscopy is to obtain an exact image of a specimen. In practice, however, such an absolute image is unattainable, due to the scattering between the incident radiation and the specimen. The magnitude of the scattering effect determines the limit of resolution and is directly proportional to λ, the wavelength of radiation. For visible-light microscopy, where $\lambda = 5200–5800$ Å, the theoretical maximum resolution is 0.2 μm (2000 Å). The TEM employs an electron beam which can be regarded as having $\lambda = 0.02–0.05$ Å, and can obtain images of specimens with a resolution up to 2 Å. This increase in resolution is of great value for the study of clay minerals and other microcrystalline phases.

The contrast of micrographs recorded in transmission geometry is produced by variation in the electron density in the specimen. High electron density areas of the specimen scatter and absorb more of the electron beam and thus appear dark, whereas low electron density areas appear light. For clay particles of relatively uniform composition, thicker particles appear darker than thinner ones (Figure 6.2(a)). Similarly, areas where elements of high atomic number are present are darker than those made up of elements with lower atomic number. This can be demonstrated for clay materials, when iron-

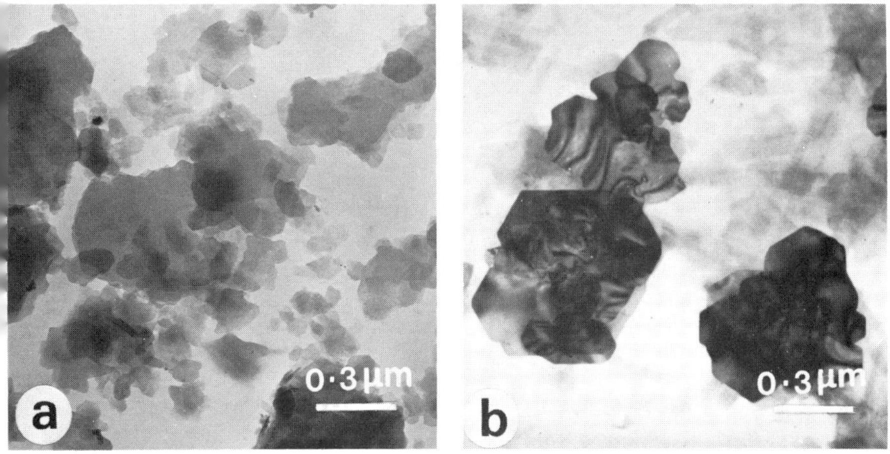

Figure 6.2 Micrographs demonstrating the effect of (a) particle thickness and (b) composition (iron oxide v. aluminosilicate) in image contrast

bearing phases, which appear darker, are present with aluminosilicates, which appear lighter (Figure 6.2(b)). Using this principle, shadow casting by evaporating metals of high atomic number, such as platinum or gold at low incident angles, has been developed to improve image contrast of thin clay particles (see section 6.4.3).

6.2.2 *Electron radiation and electromagnetic lenses*

Electrons are charged particles which have a rest mass of 9.107×10^{-28} grams. The interaction of scattered electrons, however, can be treated as a wave phenomenon, and de Broglie showed that the wavelength of the moving particles could be calculated by the following equation:

$$\lambda = \frac{h}{mv}$$

where λ = wavelength, h = Planck's constant (6.6×10^{-27}), m = mass of the particles, and v = velocity of the particles. The velocity of the electron beam in the TEM is controlled by the accelerating potential (50–400 kV). At 80 kV, the velocity exceeds half that of light, and the particles gain mass due to relativistic effects. Taking this into account, the wavelength can be calculated (Gard, 1971) from the accelerating voltage (in volts, V) from the relationship:

$$\lambda = 12.27/[V(1 + 0.978 \times 10^{-6}V)]^{1/2}$$

This relationship is shown in Figure 6.3.

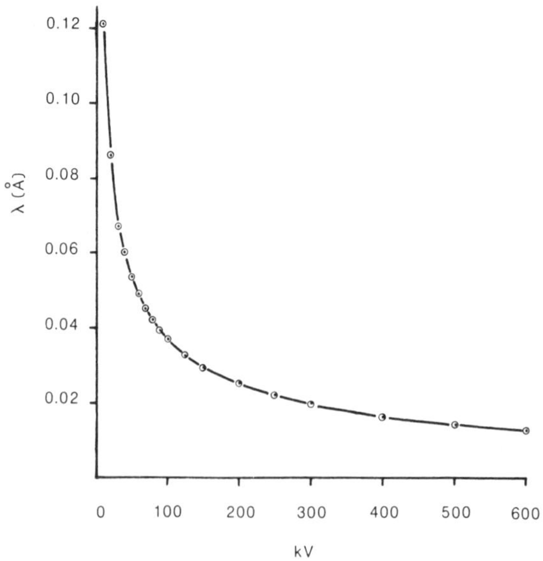

Figure 6.3 The relationship between the effective wavelength, λ, of generated electrons *v.* accelerating voltage (kV)

6.2.3 *Electron diffraction of small crystals*

Crystalline particles that are sufficiently thin to be transparent to electrons diffract a number of parallel electron beams, which form a primary diffraction spot pattern in the back focal plane of the objective lens. A geometric representation of this phenomenon is provided by the Bragg Law (see Chapter 2). Here the crystal is considered to consist of a number of parallel planes which repeat at regular intervals, d, and scatter the incident electron beam. At certain angles (Θ) between the electron beam and the atomic planes, the path difference between the scattered electrons is $n\lambda$, where n is an integer 1, 2, 3, etc., and λ is the wavelength of the electron beam such that:

$$n\lambda = 2d \sin \Theta$$

At these 'Bragg angles' diffraction maxima may occur, and the spacing between the planes, d, can be calculated where λ and Θ are known. The wavelength of electrons is very small, and so 2Θ does not exceed 5° over most of the diffraction pattern, and the approximation:

$$d = \lambda L/R$$

is often used, which is accurate to 0.1% for $d > 1$ Å, where L is the camera length (distance from specimen to the recording plate) and R is the distance from the diffraction maxima to the direct beam spot (000).

To determine the cell dimensions by ED it is desirable to orient the crystal so that a principal zone axis is obtained, and most instruments employ tilting stages for this purpose. For example, to obtain the a and b cell dimensions of most clay minerals, it is useful to find a [001] zone axis, the diffraction pattern of which can be indexed as hk0 (Figure 6.4). For most layer-silicate clay minerals the condition that $\mathrm{h} + \mathrm{k} = 2n$ applies, and maxima with indices 100, 010, 120, etc. are absent. Although it is convenient to index such a pattern as hk0, for a three-dimensional lattice, the monoclinic symmetry commonly applied to most layer silicates, requires that the spots (diffraction maxima) correspond with hk reciprocal lattice rows. To correct the a cell dimension derived from h00 ED maxima for monoclinic geometry, it should be multiplied by $\sin \beta$.

When the clay particle is oriented properly with respect to the electron beam, the symmetrically related diffraction maxima are uniform in intensity, for example, $020 \sim 0\bar{2}0$, $110 \sim \bar{1}\bar{1}0$, and $1\bar{1}0 \sim \bar{1}10$. The higher-order maxima are more sensitive to crystal orientation, and therefore the uniform intensity of the 060–$0\bar{6}0$, 330–$\bar{3}\bar{3}0$, and $3\bar{3}0$–$\bar{3}30$ pairs are better indicators of proper orientation. Fortunately, layer silicates have a β angle which is sufficiently close to 90°, and most clay particles are sufficiently thin, so that particles lying on 001 require little or no tilt to yield uniform hk0 patterns. Thicker particles generally require more precise orientation, and sometimes it is necessary to slightly tilt the specimen to obtain uniform [001] zone maxima.

The relative intensities of ED maxima from thin crystals composed of

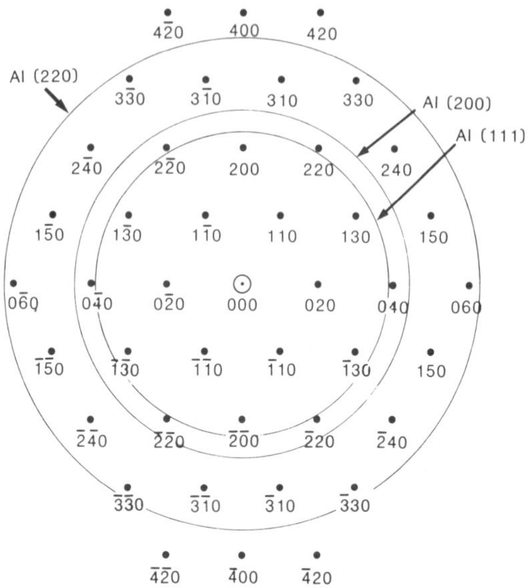

Figure 6.4 Indexing of an hk0 single-crystal spot pattern from a typical layer silicate with $b = 9.0$ Å. Also shown are diffraction rings from aluminium, commonly used as an internal standard for calculation of d-spacings

elements with moderate to low atomic number, where the intensity of the diffracted radiation is very small relative to the incident beam, can be treated as kinematic. Such kinematic intensity calculations can be used to determine the electron-density distribution of the specimen, and so provide good approximations for crystal structure analysis (Vainshtein, 1964).

Detailed discussion of this aspect is beyond the scope of this book, but in general, for thicker crystals the intensity of the diffracted radiation becomes large, and the effects of multiple diffraction (diffraction of diffracted radiation) produces complex dynamic interactions such as moiré fringes, Kikuchi line patterns (Hirsch *et al.*, 1965), and convergent beam electron-diffraction patterns (Steeds, 1979). For conventional ED studies of clay particles, the critical crystal thickness beyond which diffraction becomes dynamic has been estimated at 25–60 nm (Vainshtein, 1964), and specifically for muscovite as 36 nm—for the strongest diffraction maxima, 331—at an accelerating voltage of 80 kV (Grim and Güven, 1978). It should be noted that this critical thickness refers to crystals, not aggregates, and the electron diffraction of thick aggregates composed of thin crystals can still be treated as kinematic. Most clay crystals are thin enough to allow kinematic approximations to be used in the study of the relative intensities of single-crystal hk0 spot patterns and polycrystalline hk ring patterns.

Electrons interact with matter $\sim 10^4$ times more strongly than do X-rays, and it is therefore possible to obtain electron-diffraction patterns from microcrystalline particles with insufficient periodicity to produce useful X-ray maxima. However, the small volume element examined by ED always raises the question of whether an observed diffraction phenomenon is representative of the bulk specimen. Further developments in ED of clays could lie in structural analysis, using statistical studies of single-crystals from a large number of observations. Important contributions have already been made in this area (Méring and Oberlin, 1971; Grim and Güven, 1978), in addition to analysis from oblique-texture polycrystalline powder patterns (Zvyagin, 1967).

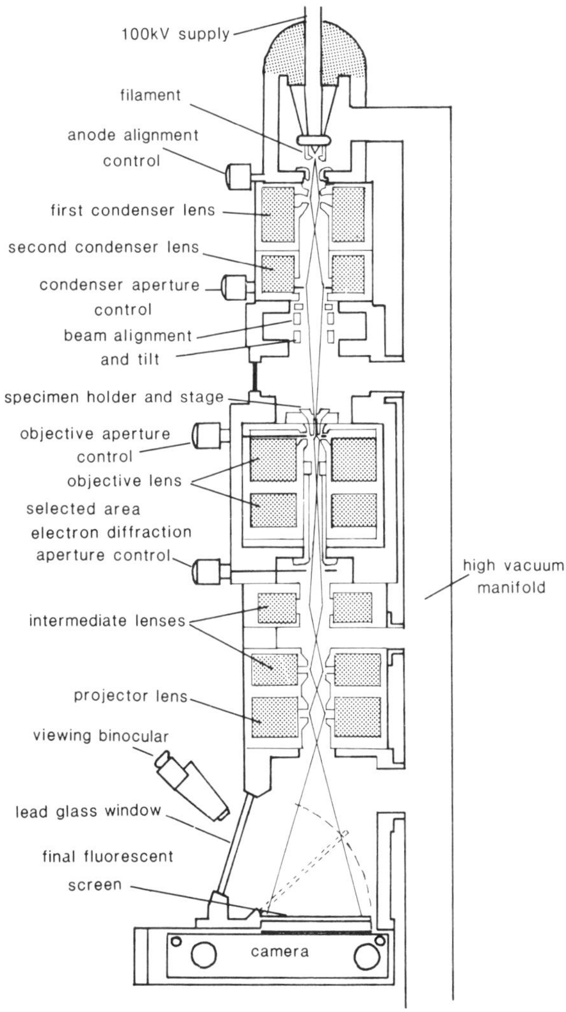

Figure 6.5 Schematic diagram of the configuration of a six-lens instrument

6.3 Instrumentation

6.3.1 *Instrument configuration*

Transmission geometry requires a linear arrangement, with the specimen lying between the electron source and observation screen. This is usually accomplished on an evacuated (10^{-4} torr) vertical column, with the electron gun at the head, followed by the first and second condenser lenses and condenser aperture, specimen chamber, objective aperture and lens(es), intermediate aperture and lens, projector lens(es), and fluorescent screen (Figure 6.5).

The instrument can be converted from imaging to diffraction mode by adjusting the intermediate lens current so that the primary electron-diffraction (ED) pattern is focused on the object plane of the projector lens and appears on the viewing screen enlarged several hundred times (Figure 6.6). The objective aperture must be withdrawn because the primary diffraction pattern is formed by electrons scattered beyond the edge of the aperture. For selected-area diffraction (SAD), see Figure 6.6(d), a diffraction aperture is placed below the position of the objective aperture, which allows diffraction from an area as small as 0.1 μm in diameter.

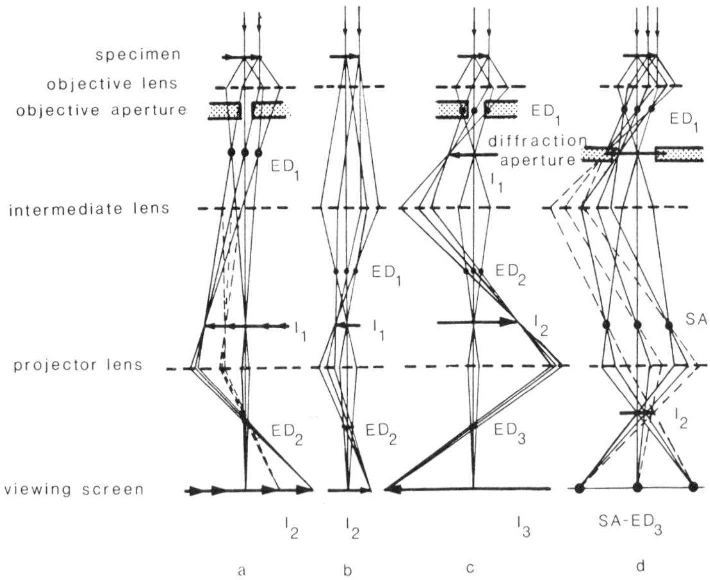

Figure 6.6 Ray diagrams for microscope with three imaging lenses: (a) Low magnification; the primary image is in the object plane of the projector, and its magnification can be reduced by the intermediate lens acting as an objective. (b) Very low magnification, with the objective lens switched off, and the intermediate lens acting as an objective. (c) High magnification; all three imaging lenses give magnified images. (d) Selected-area electron diffraction (SAD); the intermediate lens is adjusted to image the initial diffraction pattern (ED_1) in the object plane of the projector.

6.3.2 *Electron generation sources and lenses*

The electron source serves as the illumination system for the TEM imaging mode, as a monochromatic source of radiation in the diffraction mode and as a microprobe in the analytical mode. Many of these operations are performed at a variety of magnifications, and so the electron source must be versatile, and capable of high intensity. The accelerating voltages of most instruments range from 20 kV to 200 kV, and for most clay materials 100 kV is a reasonable compromise between performance and equipment costs. The source of electrons is commonly a hairpin tungsten filament.

The first condenser lens reduces the image of the source and condenses the beam, which is then focused on the specimen by the weaker second condenser lens. The area of illumination can be reduced to 1–2 μm, so that thc image is bright at high magnification. The second condenser is adjusted so that the field of illumination is uniform at various magnifications, this adjustment being automatic on some instruments.

6.3.3 *Specimen chamber*

The specimen is supported on a grid, usually 3.05 mm in diameter. The specimen grid is held in a cartridge and placed onto the specimen stage in the microscope, via an access port in the instrument column. The stage can be positioned relative to the electron beam by manual or automated micro-manipulators so that the entire grid can be examined. A goniometric specimen stage allows up to 60° tilt and can be used for ED studies, as well as for recording of stereomicrographs at lower angles ($\pm 5°$). Two types of tilt goniometric stages are the double-tilt stage (usually motor-driven using foot-pedal switches) and the rotation-tilt stage, where the specimen grid can be rotated, as well as titled in one direction. The determination of the effective tilt angle which can be obtained from these two types of stages, and the methods of obtaining symmetrical zone-axis diffraction patterns, are presented in Gard (1971).

6.3.4 *Information output devices*

Traditionally, the variable-contrast electron image and electron-diffraction pattern have been observed on a fluorescent screen near the base of the instrument column. Detailed examination is made possible by using a binocular telescope (usually 10 × magnification) on a portion of the observ-ation screen tilted into position for optimum viewing. Below the screen, most instruments have a camera which records the images and diffraction patterns on photographic film. Modern instruments may also be equipped with a scanning electron beam and detectors which produce an image on video output devices; these are designated scanning transmission electron microscopes (STEM). The advantages of these systems include improved microanalytical capabilities, video recording, image intensifying of weak signals (Anderson, 1968), digital image analysis, and digital analysis of diffraction patterns (Tokiwai *et al.*, 1983). Furthermore, the video images can be more readily combined with microanalysis data.

6.4 Specimen preparation

6.4.1 *General clay preparation techniques*

As most bulk clay materials are complex mixtures containing coarse grains and organic matter, it is usually necessary to separate the fine-grained clay particles for TEM study. The $< 2\,\mu m$ fraction is usually obtained using aqueous dispersion procedures: the method chosen greatly influences the material observed by TEM. Procedures should be adopted to maintain sample purity and avoid possible contaminants. The following can all contribute to sample purity: disposable pipettes and small specimen holders; ultrasonic baths for assisting dispersion and cleaning laboratory ware (ultrasonic probes may introduce fragments from titanium metal tips); microcentrifuges which can accommodate 1–2 ml volumes; a drying oven dedicated solely to microscope specimens; a dust-free laboratory environment; a readily available supply of freshly distilled water; and refrigeration for short-term storage of sample suspensions. The mildest possible procedure should be chosen in order to separate and purify the clay while avoiding the introduction of contaminants. In many cases the $< 2\,\mu m$ fraction can be easily obtained from mechanically disaggregated material in aqueous suspension, which is then sedimented. Grinding, when necessary, should be done under a liquid, preferably alcohol. In some cases, carbonate material may be removed by treatment with acetic acid (25%) until the residue no longer effervesces when tested with dilute HCl. Organic material can be removed using dilute H_2O_2, and amorphous iron, aluminium oxide, and silica materials can be extracted using 0.2 M ammonium oxalate solution (Smith, 1984). Retention of highly dispersed material can be ensured by using dialysis tubing at the final stages of washing or removal of soluble components.

Clay materials are often composed of aggregates of small particles that possess a large number of unsatisfied surface charges balanced by external cations or anions. Various dispersing agents and procedures may be necessary to disperse the aggregates, the particular procedure being determined by the type of clay present. For this reason it is desirable to characterize the specimen by XRD or infrared spectroscopy, so that the appropriate dispersion technique is applied. The clay material in the $< 2\,\mu m$ fraction is commonly in an aggregated state, and further dispersion is often required.

The dispersibility of a clay is determined in part by the type of ion on its surface. Na^+ and Li^+ are cations with high dispersing properties. Aggregated clays can be dispersed by first placing the material in a 1 M chloride salt solution, followed by agitation, centrifugation at high speed, decanting, and suspension in fresh salt solution. The procedure is repeated, so that the clay is washed three times, once overnight. After final centrifugation, the salt solution is decanted, the clay is suspended in distilled water, and the remaining excess ions are removed by dialysis against de-ionized or distilled water until free of chloride ion. The clay in suspension will now be highly dispersed, and can be further size-fractionated by centrifugation. The optimum size-fraction varies from sample to sample. The following fractions are suggested: $< 0.5\,\mu m$ for well-crystallized illites and chlorites, $< 0.2\,\mu m$ for poorly crystallized illites and interstratified clays with less than 40% smectite layers, and $< 0.1\,\mu m$ for smectites and interstratified clays with $> 40\%$ smectite layers. The suspension should be characterized by XRD of its sedimented aggregate, even when the suspension appears to be water-clear. Vermiculites and vermiculite interstratifications are difficult to study by TEM because they are often composed of large particles inherited from parent mica crystals. These clays can, however, be treated with large organic cations such as the *n*-butylammonium ion, so that they exfoliate in aqueous systems (Walker, 1967), so producing thin particles which can be more readily investigated. Kaolinitic clays (dickite, kaolinite and halloysite) often disperse well with small amounts of sodium pyrophosphate (1 mg/g clay) or sodium hexametaphosphate (1 ml of 0.1 M Calgon: 100 ml of water: 1–5 g of clay). Small amounts may be added to most clay suspensions to inhibit the reaggregation of clay particles after dispersion. The $< 1\,\mu m$ fraction is recommended for well-crystallized kaolinite. Unlike most clays, imogolite disperses in acid conditions. Farmer *et al.* (1980) recommended a pH of 3.5 (0.001 M HCl) with ultrasonic treatment. For TEM studies the $< 0.5\,\mu m$ fraction is recommended.

6.4.2 *Specimen supports*

A good general-purpose specimen support for clay materials is a thin 200-mesh copper grid, with a centre disc indication and a rectangular pattern made of different width axis bars to aid in orientation and location.

A carbon support film rests upon the grid. This film is made by evaporating a 2 nm carbon layer onto a freshly cleaved sheet of mica, using a vacuum-coating apparatus equipped with a thickness gauge. The carbon-coated mica sheet is then carefully immersed in distilled water, so that the carbon film separates and floats on the water surface (Figure 6.7). The copper grids (shiny side up) are then immersed in the water and placed on a wire gauze, which is then carefully positioned under the floating carbon film, and raised so that the film uniformly covers the grids. Following this treatment, the grids are placed on a clean piece of filter paper and allowed to dry. The grid is then placed shiny side down and attached to a support arm, and a small drop of the clay suspension is applied to it. The grid is then placed in a drying oven at 50°C.

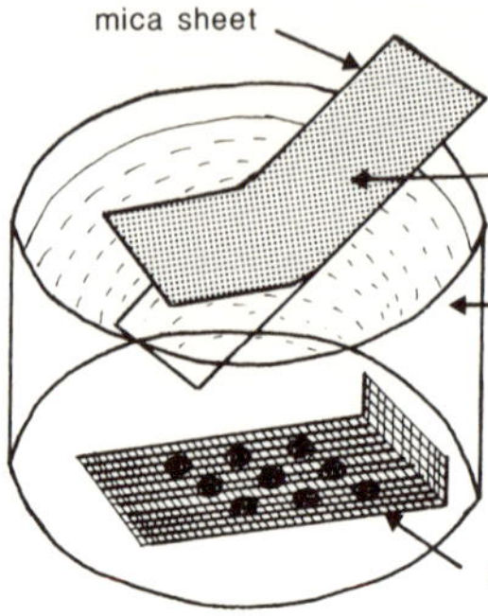

Figure 6.7 Flotation of carbon film from the mica sheet onto water with support grids on wire mesh in the bottom of the dish

Frequently, the concentration of clay particles is too high, and the film is cluttered with overlapping particles. In this case, the clay suspension should be diluted, further dispersed in an ultrasonic bath, and re-examined until an optimum dispersion is obtained. As a general rule it is necessary to dilute the clay suspension until only a faint scattering of light can be observed. For fine clay fractions, i.e. $< 0.2\,\mu m$ esd (equivalent spherical diameter), the suspension is diluted until water-clear.

6.4.3 *Shadow casting*

This is a useful technique for improving contrast, and for the quantitative determination of thickness. Very thin clay particles, such as smectites and interstratified clays, may be difficult to resolve. In such a case, a drop of the diluted clay suspension is spread over a sheet of freshly cleaved mica and allowed to dry. The mica sheet is then carefully attached to a shadowing stage with a projection of known height (such as a 2–3 nm screw head) and placed in a vacuum-coating apparatus equipped with a thickness gauge. The stage is set so that it is titled to $\sim 10^\circ$ with respect to the shadowing source, and Pt metal is evaporated onto the area, until the gauge registers 6 nm. The stage is then placed perpendicular to the shadowing direction and coated uniformly with 2 nm of carbon. Once removed, the projection on the stage should cast a visible shadow 10–15 mm in length, which can be measured ($\pm$ 0.05 mm) to determine accurately the shadowing angle, using the relationship:

$$\tan a = h/l,$$

where a is the shadowing angle, h is the height of the projection and l is the shadow length. The mica sheet is then carefully immersed in distilled water so that the carbon film with the Pt-shadowed clay particles floats and is picked up on the copper grids (Figure 6.7). TEM examination should show that the edges

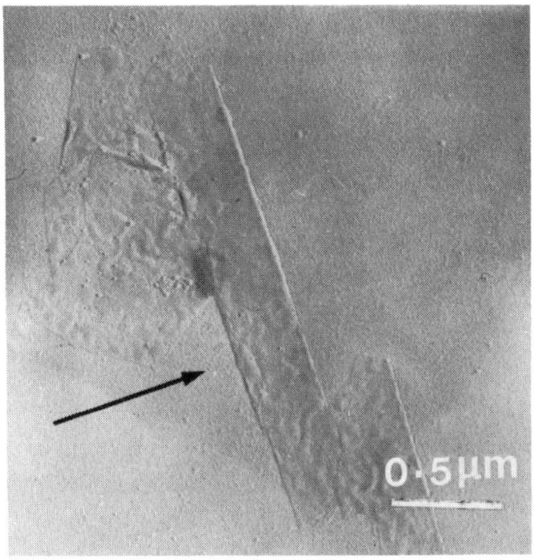

Figure 6.8 Shadow-cast preparation of beidellite; arrow indicates shadowing direction

of the particles lying towards the Pt source have accumulated fine-grained metal and appear dark, whereas the edges of the particles lying away from the platinum source have cast a shadow where no Pt grains are present and therefore appear light. The longer the shadow, the greater the thickness of the particle. In this way, the contrast of the particle edges is enhanced (Figure 6.8). The technique also allows determination of particle thickness. By measuring the length of the shadow parallel to the shadowing direction, the particle thickness (T) is given by:

$$T = (l \times \tan a)/m$$

where l is the shadow length, a is the shadowing angle, and m is the magnification of the micrograph. When the shadow length is not measured exactly perpendicular to the shadowing direction, the value T is divided by the cosine of the angle of deviation. This technique is accurate to ± 0.4 nm, and can measure particles as thin as 1.0 nm. More details on this technique and its uses are given in section 6.5.3.

6.4.4 *Replicas*

It is possible to study the orientations and configurations of clay particles from the surface features of a bulk sample, using replica techniques. A variety of techniques exist (Comer, 1971). The Pt/C-shadowed replica method is useful for clays. This technique is analogous to the procedure described in the previous section, except that, after coating, the specimen is placed in 48% hydrofluoric acid, so that it dissolves to leave the Pt/C replica. The replica film is then water washed, placed upon a copper grid and examined (Figure 6.9).

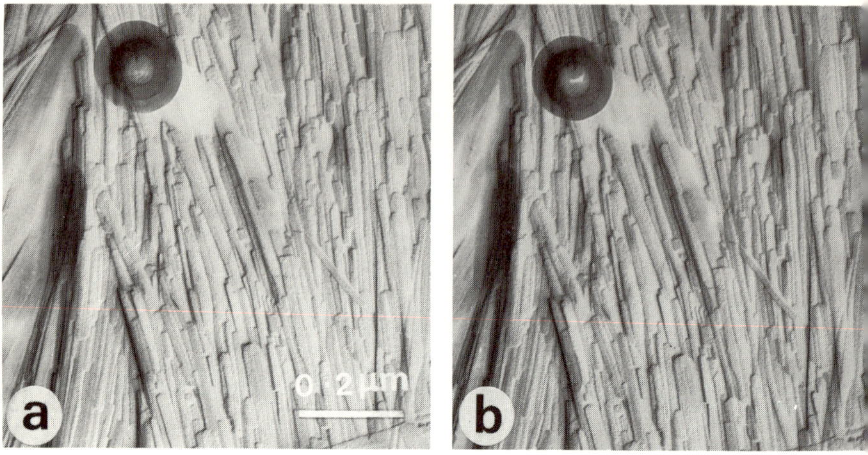

Figure 6.9 (a) and (b) stereopair of a replica of a fracture surface of sepiolite

Surface features up to 2.5 nm in relief are observable with this technique. Diffraction or microanalysis studies are not possible because of the acid dissolution of the specimen.

6.4.5 *Microtome sections*

Microtome sections are useful because they enable the orientation of clay particles with their *c*-axes perpendicular to the electron beam. This edge-on orientation is necessary for the study of local lattice images and 001 diffraction (section 6.6.1). Drops of clay suspensions can be dried on a cured epoxy surface, and the resulting sedimented aggregate can be 'sandwiched' by the addition of more epoxy (Spurr, 1969). Alternatively, fine-grained clay, such as that obtained from freeze-dried suspensions, can be gently mixed into the liquid epoxy and cured. Several varieties of epoxy and impregnation techniques are available, and some that form stable complexes with expansible clays (Yoshida, 1973) have been developed. The cured epoxy block is trimmed and mounted on an ultramicrotome (Reid, 1975). Sections less than 50 nm thick are then cut, preferably using a diamond knife, floated on distilled water, and picked up on copper support grids. Sections should be coated with 2 nm of carbon prior to TEM examination, to enhance their stability in the electron beam.

6.4.6 *Ion-beam thinning*

This is an alternative method of preparing specimens for lattice-fringe image study (Paulus *et al.*, 1975; Bresson, 1981). It is generally performed on small samples of bulk material impregnated with epoxy, which are ground to $\sim 30\,\mu m$ in thickness. The section is inspected using an optical microscope, and then mounted in a vacuum ion-beam milling apparatus, where an intense ion beam perforates the section. It is then mounted on a copper grid and examined by TEM. Images are observed at the thin edges of the section, near perforations. Ion-beam thinning is recommended for study of bulk specimens which contain coarse materials (such as quartz) that would otherwise damage the knife of a microtome. With heterogeneous specimens, however, it may not be clear whether the lattice images obtained originate from fine-grained clay materials, or from coarse-grained phyllosilicates.

6.5 Analytical procedures for clay materials

6.5.1 *Operating conditions*

The observation of clay materials can be adequately carried out at operating voltages of 100 kV. Higher voltages, however, offer greater resolution, and in general, slower rates of beam damage to sensitive materials. Furthermore, the sample penetration depth can be improved by a factor of ~ 1.7 when the operating voltage is increased from 100 kV to 200 kV.

Although most clay materials are relatively stable in the electron beam, radiation damage of the specimen is sometimes observed under the operating conditions for TEM. In some diffraction patterns and lattice-fringe images the effects can be seen as a gradual loss of detail, and in other instances, there is damage in the form of physical distortion and change in contrast. These effects are most severe at high magnification, and, therefore, appropriate operating procedures can be adopted to minimize exposure to high beam intensities until immediately before micrograph recording (Agar, 1978). These procedures include general observation at low magnification, prefocusing on a region of the field not including the area of interest, and avoiding unnecessary exposure to the beam.

6.5.2 *Specimen assessment*

When examining specimens that have been subjected to dispersion and size-fractionation procedures, the distinction between aggregates and particles should be borne in mind. (Yariv and Cross, 1979, refer to aggregates as secondary particles). Primary particles are considered to be free and distinct, whereas aggregates are complex arrangements of primary particles. Aggregates are commonly observed at low magnifications ($\times$ 10 K–20 K), whereas primary particles are commonly observed at higher magnifications. Sometimes primary particles are large and easy to resolve, as in kaolinite (Figure 6.14), but in smectites, the primary particles are very small and the distinction is more obscure. Shadow-casting techniques can be employed to improve contrast of small primary particles (section 6.4). Sometimes it may not be clear whether the observed aggregates derive from the bulk specimen, or if they have been formed in suspension or after drying.

Size-fractionated material can be very different from the bulk clay, particularly when the samples are composed of mixtures of clay minerals, and characterization of the various fractions by an independent method such as XRD is advised to guard against any bias introduced during sample preparation. Other minerals may also be present in the specimen; often, phases not detected by conventional methods are observed by TEM, including, for example, goethite and hematite in sediments, and lepidocrocite, goethite, gibbsite and boehmite in soils (Mackenzie *et al.*, 1971). Introduction of contaminants and spurious features by sample preparation methods must also be assessed. The presence of trace amounts of salt, resulting from incomplete removal of excess ions, is indicated by graininess on the specimen support film, and by the characteristic electron-diffraction pattern of the halides. Where concentration of the suspension has been achieved by evaporation, the precipitation of sparingly soluble microcrystalline phases such as baryte may occur. Concentration of fine-grained material by ultracentrifugation is preferable, followed by resuspension of the pellet in a smaller volume of distilled water.

6.5.3 *Data acquisition*

Examination of clays should proceed from low (10 K) to high magnification (100 K), encompassing as large an area of the specimen as possible, so as to obtain a representative assessment of the overall character of the material. After obtaining this overall impression, electron micrographs and electron-diffraction patterns should be acquired, to characterize the clay. The image should be focused using the telescope and tilting screen, to the optimum, slightly underfocused condition.

The morphological characteristics of clay particles and aggregates are often described in the following terms (examples are given in brackets).

Particles:

- plates: most clay particles can be regarded as plates with large lateral dimensions relative to thickness
- lath: a flat, elongated particle, (e.g. illite, hectorite, nontronite); curled and folded thin particles can also appear to be lath-like (e.g. beidellite)
- euhedral: particles showing straight edges, usually psuedo-hexagonal, indicative of crystal habit (e.g. kaolinite, illite, rectorite)
- subhedral: particles showing irregular edges (smectite, chlorite)
- equant: particles whose areal dimensions are approximately equal (kaolinite)
- subequant: particles whose lateral dimensions are unequal, but to a lesser degree than laths (illite, smectite)

fibres:	particles which exhibit extreme elongation, greater than that of laths (palygorskite, sepiolite, chrysotile)
tubes:	particles which have hollow cylindrical forms, (halloysite, imogolite, chrysotile)

These terms can be combined in certain ways, such as tubular fibres (chrysotile), and euhedral plates, (kaolinite). Particle morphology is not necessarily indicative of the mineral type, and some clay minerals, such as illite and rectorite, display a variety of forms, from euhedral laths to equant plates. On the other hand, some forms are completely characteristic, as in the case of imogolite.

Aggregates:

lamellar:	tabular, platy, flat material
foliated:	folded or crumpled material (smectites)
vermicular:	elongated worm-like material (kaolinite)
fibrous:	radiating or oriented arrangements of elongated particles
mosaic:	component particles appear to have crystallographic relationships
irregular:	composed of random arrangements of particles
polyphase:	composed of more than one mineral.

As is the case for particle terminology, combined terms can be used, such as mosaic-fibrous (exhibited by some illites) and foliated-lamellar (smectites).

New perspectives on clay morphology can be gained by studying stereoscopic images from micrograph pairs taken at different tilt orientations (using the goniometric specimen stage, Figure 6.9). With a tilt angle of 10°, (+ 5° for one micrograph, and − 5° for the other, across the tilt axis) resolution of features 200–500 nm in size can be achieved at magnifications of × 50 K to × 100 K. The micrograph pairs must be oriented correctly (N–S), with respect to the stereoscope viewer (E–W). Where this task is complicated by image rotation, a method for calculating the effective tilt-axis direction with respect to the micrographs can be used (Gard, 1971). Micrographs can usually be oriented and mounted properly by trial and error.

Statistical particle-size distribution analysis provides valuable information concerning the nature of clay material (Nadeau *et al.*, 1984b, 1985; Nadeau, 1985). These studies were conducted on Pt/C-shadowed specimens, usually from well-dispersed, size-fractionated suspensions. Conley (1966) and Bates (1971) reported measurements of particle length (L), width (W), and thickness (T), mean particle diameter, $(L \times W)^{1/2}$, and shape factor O $(T/(L \times W)^{1/2})$ for kaolinite. In such studies, for accurate thickness determination it is important that representative, flat-lying particles are measured, with no obstructions in the shadow area. Satisfactory preliminary results can be obtained from 10–30 particle measurements, and for detailed studies, 100–200 particles. The use of a digitizing tablet for micrograph analysis also allows the determination of particle area (A), and perimeter (P) which, when combined with the thickness

(T) of planar particles, allows the calculation of volume (V = T × A), basal surface area (A × 2), and edge surface area (T × P).

If the particle density is assumed (usually 2.6 g/cm^2) then the particle mass and surface area in m^2/g can also be determined. Nadeau (1985) showed that a useful approximation of surface area in m^2/g from the particle thickness is given by the expression:

$$S = 800/Tm$$

where S is the surface area, and Tm is the mean particle thickness in nm.* The particle surface area and mass data can be combined with independent cation-exchange capacity (CEC) data to determine the average surface charge density on the clay surface. For example, a smectite sample with a CEC of 103 meq/100 g, has particles with a mean surface area of 11.04×10^4 nm^2, mass of 167×10^{-18} g, and has 10.4×10^4 sites/particle, with a charge density of 1.06 nm^2/site.

Microchemical analysis of clay particles allows identification of different particle types, as well as the study of variation in clay crystal chemistry. For discrimination purposes, the ratio of the major elements is often used, such as Al/Si for kaolinite, K/Si for illite, or Mg/Si for chlorite. The analytical mode can be used for either individual particles, or for area scans which include a large number of particles. In specimens composed of very small particles, EDS systems may only provide useful information in the latter mode. Some problems with microanalysis of clays are their decomposition in the electron beam and the difficulties in obtaining reproducible determination of volatile elements such as potassium.

6.6 Electron diffraction of clay particles and aggregates

6.6.1 *General features of clay materials*

Platy clay particles and irregular lamellar aggregates commonly yield single-crystal spot patterns and polycrystalline ring patterns respectively. The single crystal pattern shows a pseudohexagonal symmetry, and can be indexed as hk0 (Figure 6.4) corresponding with an 001 section. Such patterns are observed from free and individual particles when examined by selected-area diffraction (SAD), but when a large area of the specimen is examined, many individual particles diffract and a polycrystalline ring pattern is observed. The pattern results from the hk0 maxima of randomly distributed individual particles. Similar patterns are produced by SAD of irregular lamellar

*This approximation is valid for clay materials, whose total surface area consists predominantly of basal surface area. For clays whose basal surface area accounts for only ~ 65–85% of the total surface area, such as kaolinites, the value S can be multiplied by a factor to account for the significant amount of edge surface area, which in the case of Tm = 50 nm, would be 1.24.

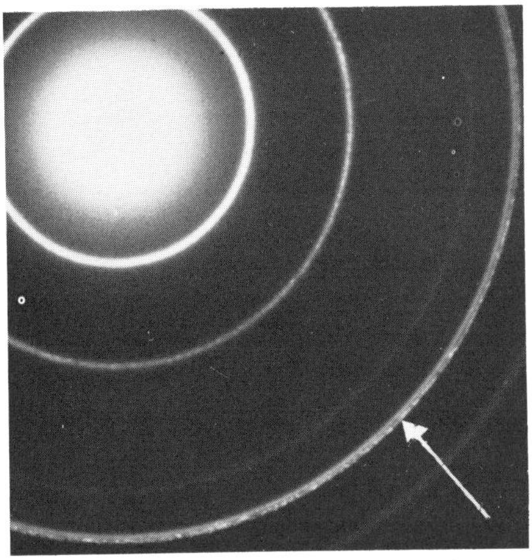

Figure 6.10 Part of a polycrystalline ring pattern with double 06 rings (arrowed) from the presence of both dioctahedral (outer ring) and trioctahedral components in the specimen

aggregates, in which the component particles are lying with their *c* axes approximately parallel to the beam but with no preferred orientation in the *ab* plane. Such patterns are often described as turbostratic. Polycrystalline patterns can be further designated as monomineralic and polymineralic, the latter consisting of sets of reflections from two or more phases. For example, the presence of both dioctahedral and trioctahedral clays can be determined by examination of the 06, 33 ring, which would have two intensity maxima (Figure 6.10).

In general, platy clay particles give very similar hk0 spot patterns, thus excluding the use of ED as a primary means of identification, this being a consequence of their similarities in structure. However, some clays, such as antigorite, tubular halloysite, sepiolite and palygorskite have characteristic diffraction features which allow unequivocal identification.

6.6.2 *Determination of* d_{hkl}

When d_{hkl} is determined using the formula $d = \lambda \mathrm{L}/\mathrm{R}$, (section 6.2.3), it is necessary to record the objective lens current (± 0.1 V) at the time when the diffraction patterns are recorded. In this way, a calibration curve of lens current against λL is obtained. Alternatively, an internal standard can be applied to the specimen, usually in the form of an evaporated metal such as Al or Au (Table 6.1). Using Al as an internal standard, enables determination of *d* up to 0.2% accuracy (Brindley and De Kimpe, 1961). The distance $2r_x$ between corresponding spots is measured across the direct beam spot from the negative, and the diameter $2r_s$ of the hkl ring from the standard is measured in

Table 6.1 Spacing of some internal standards for diffraction studies.

Element	*d*	*I*	hkl
Al	2.338	100	111
	2.024	47	200
	1.431	22	220
	1.221	24	311
Au	2.355	100	111
	2.039	52	200
	1.442	32	220
	1.230	36	311

a similar manner. The calculation is:

$$d_x = (2r_s)(d_{\mathrm{hkl}})/2r_x.$$

6.6.3 *Application of* hk0 *diffraction*

When a free and individual clay-mineral particle yields a single-crystal diffraction pattern, it is referred to as a fundamental particle. The crystallographic orientation of the particle can be established, and for euhedral particles the crystal forms can be identified. In the case of subequant, lath- and fibre-shaped particles, the direction of elongation can be established. Thin lath-like 1 M illite particles are commonly *a*-axis elongated, this being established by the strong (020) spots (Figure 6.11(a)). Conversely, diffraction patterns from large equant particles of 2 M_1 muscovites, show weak 020 spots (Figure 6.11(b)). The relative intensities of the hk0 spots are of particular interest in dioctahedral clays, because they are sensitive to the octahedral sites occupied. However, these considerations are beyond the scope of this book, and the interested reader should refer to papers by Méring and Oberlin (1971) and Grim and Güven (1978) for further details.

6.6.4 *Dark-field imaging of clay particles*

The diffraction maxima of the clay crystals can be used to create an image of the particle(s) from which the maxima originated. This is accomplished by deflecting the main beam, so allowing only the diffraction maxima to pass through the objective aperture. In this mode the background and amorphous material appear dark, and the crystalline material which contributes to the diffraction maxima appears light. This technique helps to resolve crystalline particles when they are associated with amorphous material, and also improves the contrast of very small crystalline particles.

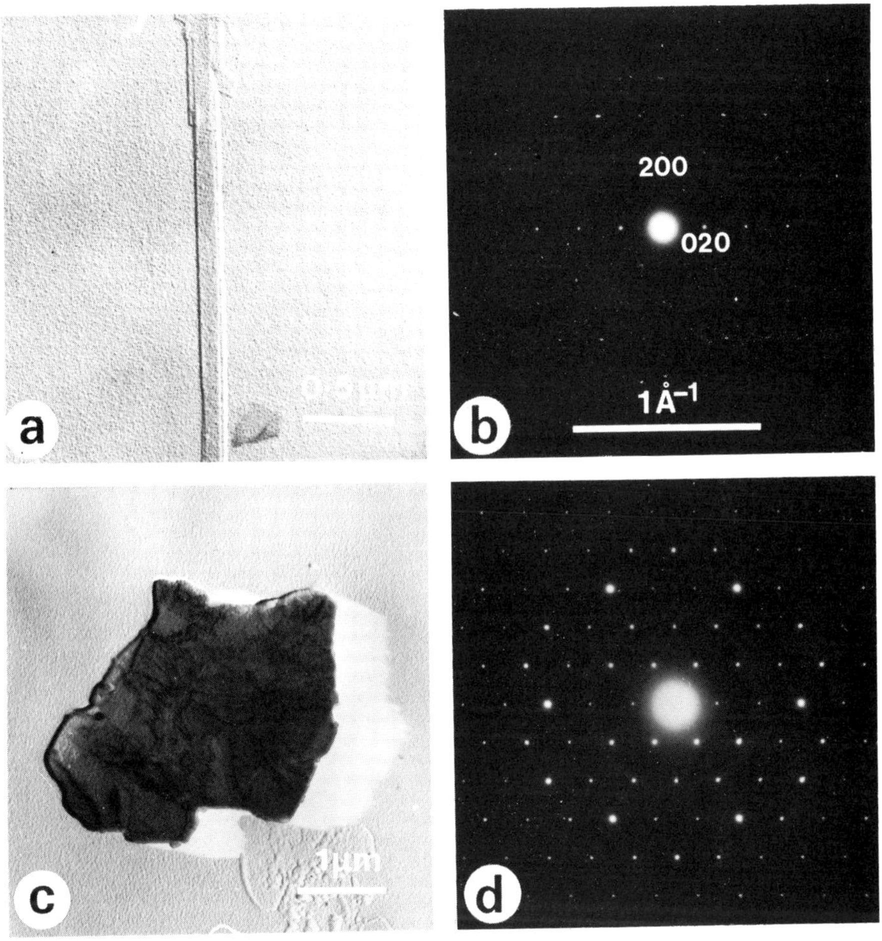

Figure 6.11 Pt/C-shadowed preparations of (a) 1M illite lath, and (c) equant $2M_1$ muscovite flake; (b) and (d) diffraction patterns from (a) and (c), respectively; diffuse rings are from Pt. The scale shown on (b) is the same for all subsequent diffraction patterns

6.6.5 *001 diffraction and lattice-fringe images*

Examination of 001 diffraction maxima requires that the *c*-axis of the clay particles should be perpendicular to the electron beam. Usually, this can only be obtained from microtome or ion-beam-thinned sections of epoxy-impregnated specimens, but sometimes particles which have curled up, or have twisted on-edge (Figure 6.12(a)), allow lattice-fringe images to be acquired in conventional preparations on a support film (Figure 6.12(b)). In clays, most lattice fringes are one-dimensional, in that only the c^* reciprocal lattice direction is normal to the electron beam. Crystals with three-dimensional

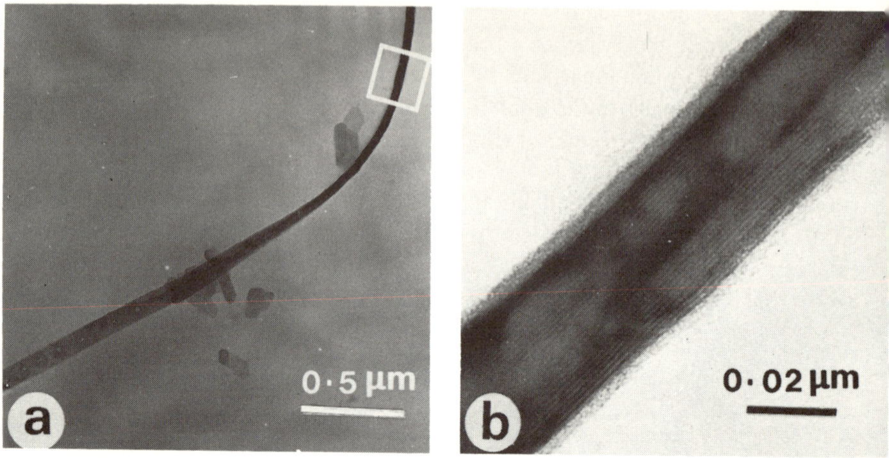

Figure 6.12 (a) Long thin illite lath which has twisted on-edge towards the top right-hand edge of the micrograph. (b) Lattice-fringe image from tilted portion of the lath with the *c*-axis perpendicular to the electron beam

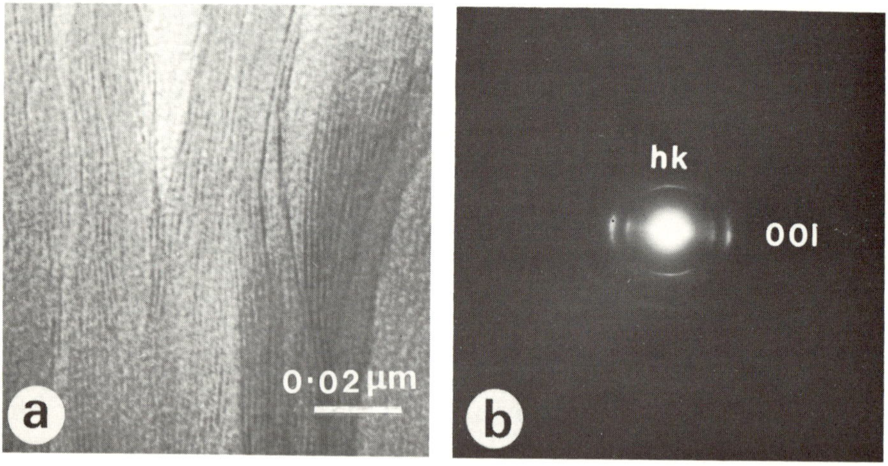

Figure 6.13 (a) Lattice-image of ultramicrotome section of montmorillonite embedded in resin. (b) Diffraction pattern showing 00l arcs and hk diffraction bands

periodicity can, however, be tilted such that a [hk0] zone axis, preferably [010] or [100], is obtained, and impressive two-dimensional lattice-fringe structure images have been obtained from large phyllosilicate crystals. Highly disordered clays, such as turbostratic smectites, have no three-dimensional periodicity and only yield one-dimensional lattice-fringe images (Figure 6.13(a)). Tilting of the specimen in the *ab* plane produces little, if any, change in the diffraction character. Instead of single-crystal spot patterns, 001 arcs and hk diffraction bands are obtained (Figure 6.13(b)). The arcs reflect the spacing between the adjacent silicate layers, and the bands represent the distribution of hk maxima from the individual layers, which are randomly arranged with respect to the *ab* plane. Lattice images may be obtained from specimens composed of aggregates of thin particles consisting of a small number of layers (3–10). A complex pattern results from a number of small single-crystals contributing to the 001 diffraction, but with no crystallographic continuity across the particle interfaces. Discrimination between particle interfaces and 'stacking faults' is difficult, and the microscopist must be aware that either can be present in the lattice-fringe images of clay materials.

6.7 Micrographs and electron-diffraction patterns of common clay minerals

The morphology of the clay minerals, as observed under TEM, has been illustrated at length in the atlases of Beutelspacher and van der Marel (1968) and Sudo *et al.* (1981). Structural information, as adduced from electron-diffraction patterns has been just as extensively discussed by contributors to the Electron Optical Monograph on Clay Minerals published by the Mineralogical Society (Gard, 1971). An illustrated overview is attempted here of the diagnostic TEM features of the main minerals occurring in clay materials. The discussion is anything but exhaustive and the reader should refer to the above publications for further details.

6.7.1 *Kaolinite minerals*

Well-crystallized kaolinite is generally unmistakeable under the TEM, and consists of relatively large, well-defined, hexagonal-shaped particles, bounded by prism and pinacoid faces (Figure 6.14(a)). The size of the particles varies widely, but often falls in the 0.5–2.0 μm range. The thickness of the kaolinite hexagonal plates usually exceeds 0.05 μm, and on average, the ratio of plate diameter to thickness (aspect ratio) is 6:1. Well-crystallized kaolinite always yields a well-defined, hexagonal, single-spot SAD pattern (Figure 6.14(b)).

Poorly ordered kaolinite is generally composed of subequant, platy particles with no particular well-developed form (Figure 6.14(c)). These particles tend to be thinner than those of well-crystallized kaolinite and yield somewhat poorer ED patterns (Figure 6.14(d)). Poorly crystallized kaolinite would be difficult to distinguish from other platy clay minerals, such as mica or chlorite, on the basis of TEM observations alone. There is no absolute correspondence between structural order and morphological perfection in kaolinites, and examples are known of poorly ordered, euhedral kaolinite and highly ordered anhedral kaolinite (Robertson *et al.*, 1954).

Dickite usually occurs as large, exceptionally well-formed, hexagonal

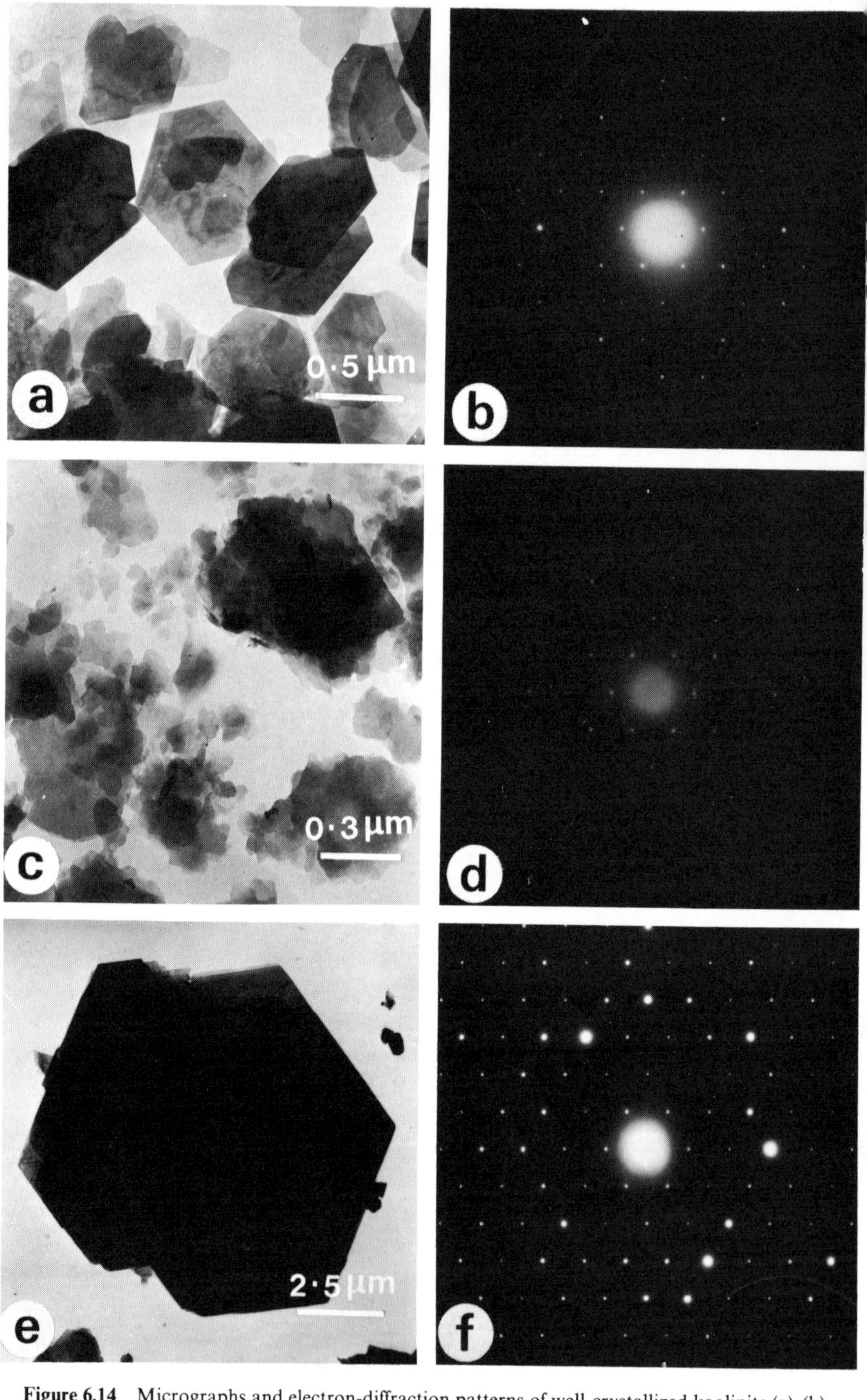

Figure 6.14 Micrographs and electron-diffraction patterns of well-crystallized kaolinite (a), (b), poorly ordered kaolinite (c), (d), and dickite (e), (f). SAD spot pattern (b) also shows rings from aluminium (Al) as on internal standard

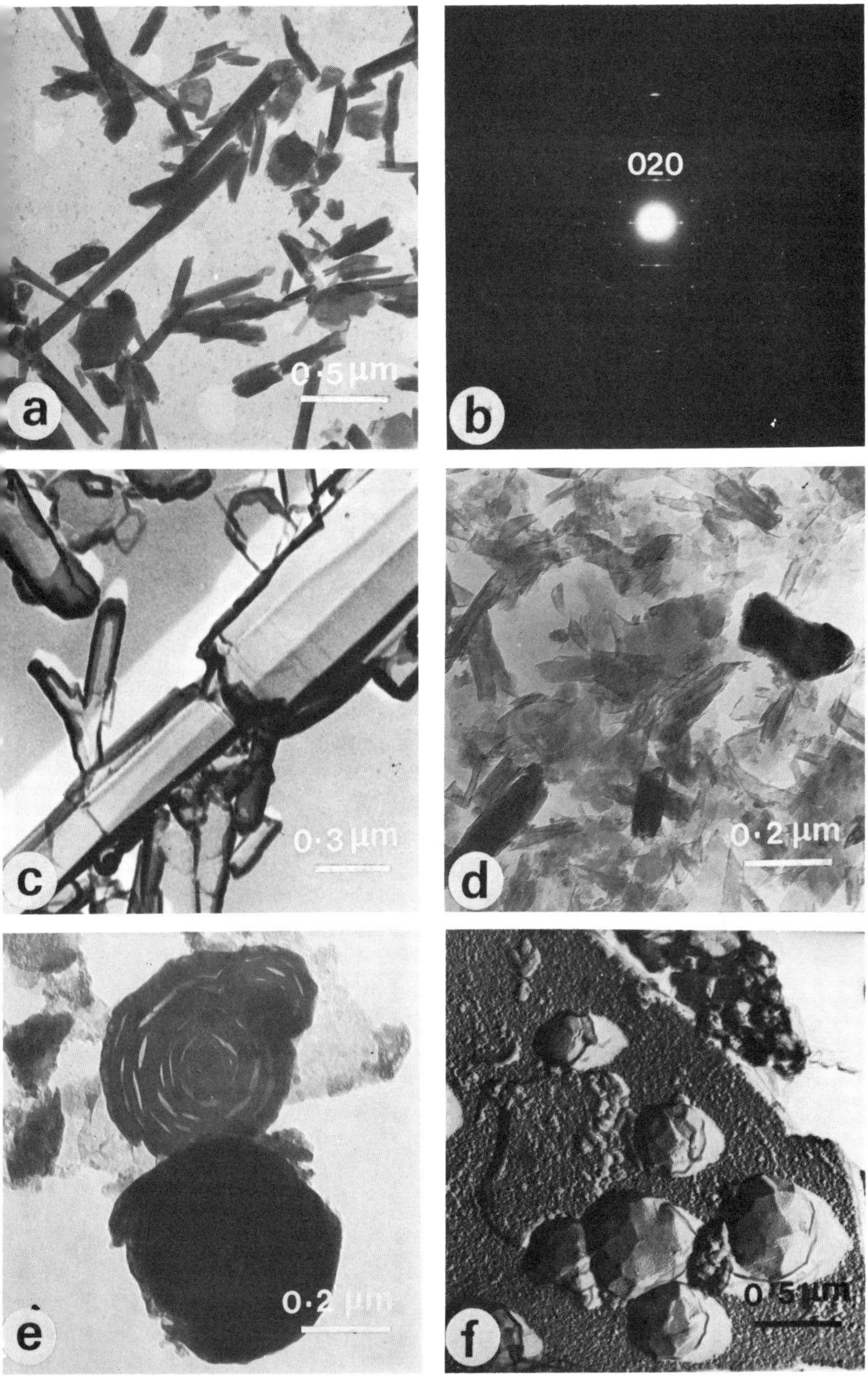

Figure 6.15 (a) Tubular halloysite: diffraction pattern (b) shows streaking along rows due to curvature of lattice in a tubular formation. Spacing between rows indicates *b*-axis elongation. Rings are from Al (c) Replica of a well-crystallized form of tubular halloysite. (d) Halloysite with poorly developed tubular morphology. (e) Replica of spheroidal halloysite showing well-developed crystal faces. (f) Thick and thin spheroids, the latter showing spiral structure

crystals (Figure 6.14(e)), showing the prism and pinacoid faces recognized in kaolinite, and sometimes even pyramids and domes. Particles up to 0.5 mm in diameter have been found, enabling optical and single-crystal X-ray studies to be made. Most commonly, dickite occurs in the 2–50 μm fraction and should certainly be suspected if large 'kaolinite-like' particles occur in non-clay fractions. As might be anticipated, dickite yields exceptionally clear, single-crystal-type, hexagonal ED patterns (Figure 6.14(f)).

Halloysite exhibits considerable morphological variability; in its commonest form, it consists of fine, tubular or lath-like particles (Figure 6.15(a)). Replica studies of halloysite sometimes reveal the development of crystallographic faces parallel to the tube axes (Figure 6.15(c)). In some soils, however, tubular morphology is extremely poorly developed, and may amount to little more than incipient curling of the long edges of lath-like particles (Figure 6.15(d)). Halloysites separated from volcanic soils may show a quite different morphology, and may occur as spheroids (Figure 6.15(f)), sometimes revealing an unusual spiral structure (Figure 6.15(e)). In general, the morphology of halloysite is sufficiently distinct from the other kaolinite minerals for TEM observations to be regarded as quite diagnostic.

6.7.2 *Serpentine minerals*

The most common varieties of serpentine minerals can be readily recognized under the TEM.

Chrysotile is unmistakable because of its finely fibrous form, the fibres themselves giving an impression of flexibility, as they are often curved or bent (Figure 6.16(a)). Ultrastructural studies show that chrysotile has a tubular morphology, consisting of spirally rolled layers, the fibres themselves often being composed of bundles of smaller fibrils. Typically, chrysotile fibres yield an electron-diffraction pattern consisting of streaked reflections arranged in layer lines. Indexing and measurement of the pattern shows that fibres are usually elongated along the *a*-axis (Figure 6.16(b)).

Antigorite is commonly made up of platy particles with distinct rectangular outlines (Figure 6.16(c)), or of short, stiff fibres. Such a morphology is not entirely diagnostic, but the ED pattern (which is relatively easily obtained) is quite specific. This is because the mineral has a corrugated sheet structure, which leads to a hexagonal pattern characterized by distinct clusters of subsidiary spots around the main reflections (Figure 6.16(d)).

Lizardite is composed of platy particles of indeterminate shape but with strong outlines (Figure 6.16(e)). A typical single-crystal type, hexagonal ED pattern is yielded by such plates (Figure 6.16(f)), there being no indication of the subsidiary reflections found with antigorite.

6.7.3 *Chlorite and illite*

These minerals are considered jointly, because in many clay materials it would be difficult to differentiate them on the basis of TEM observations alone.

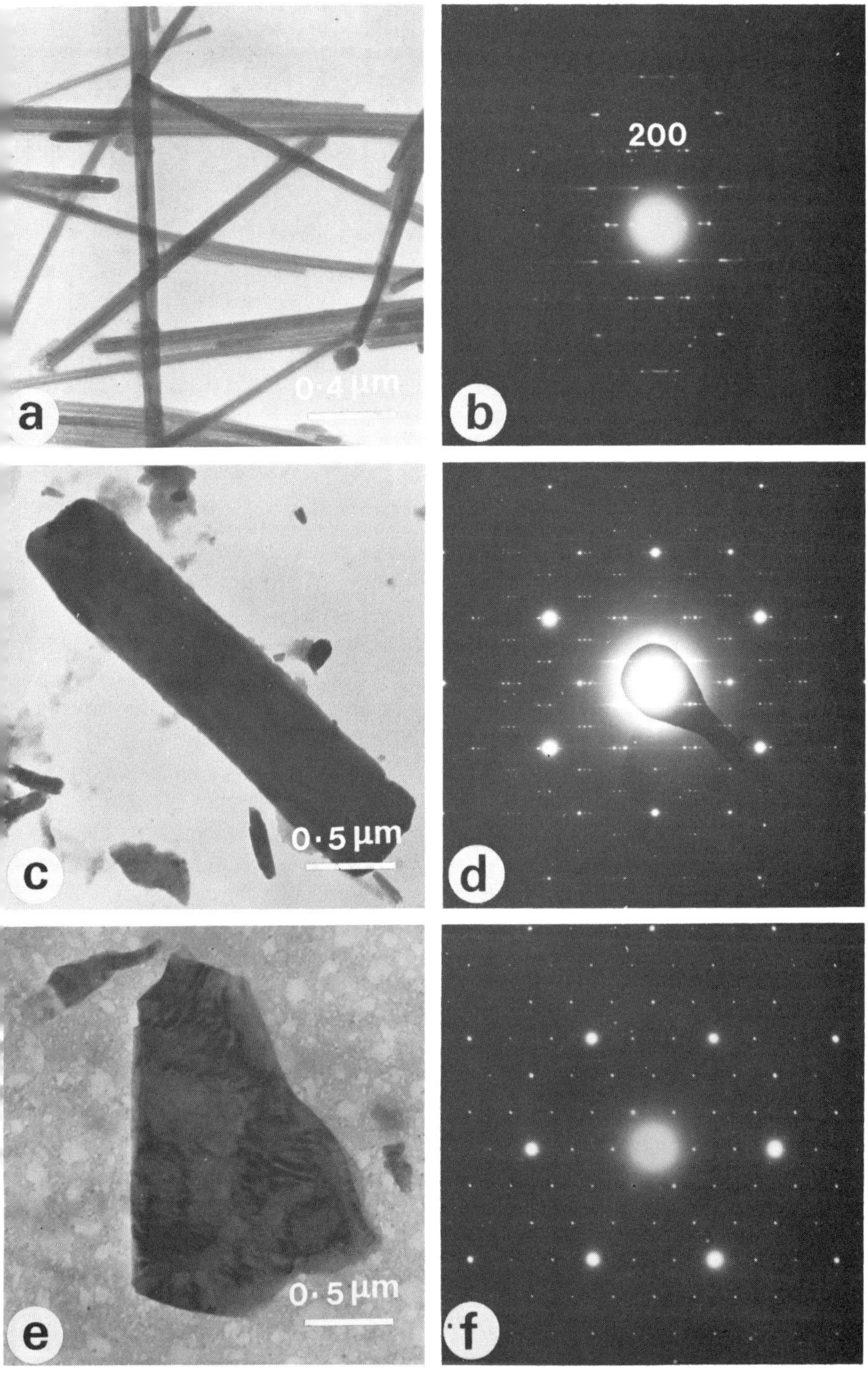

Figure 6.16 (a) Chrysotile, (b) diffraction from chrysotile indicates *a*-axis elongation. Note the spacing between layer lines compared with the halloysite pattern (Figure 6.15(b)) can be diagnostic. (c) Antigorite with characteristic diffraction pattern (d). (e) and (f) show lizardite

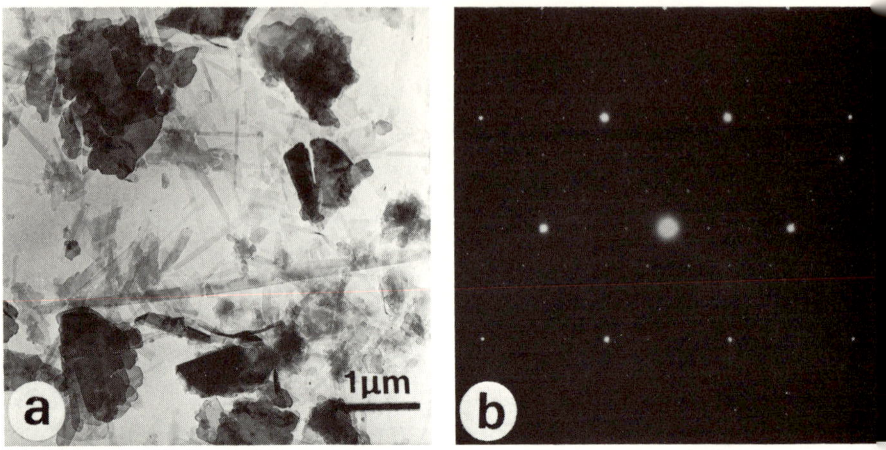

Figure 6.17 (a) Platy subhedral chlorite and thin, lath-like illite. (b) Diffraction pattern from chlorite plate. See Figure 6.11 (b) and (d) for examples of illite diffraction

Usually, they are composed of platy particles with indeterminate shapes yielding similar hexagonal electron-diffraction patterns. A distinction might be made by measuring the 060 reflections (in the case of trioctahedral chlorite and dioctahedral illite) but the difference is very small and cannot always be made with confidence. In some materials, however, chlorite and illite do have quite different morphologies. Figure 6.17 shows a clay fraction containing both minerals, which was separated from a reservoir sandstone. In this instance the illite is in a filamentous or lath-like form, whereas the chlorite occurs in platy, more electron-dense particles. Celadonite and glauconite also tend to occur in short lath-like particles.

6.7.4 *Smectites*

This group of minerals is quite varied morphologically, although most specimens are characterized by extremely fine-grained, poorly defined particles, with diffuse outlines and curled edges.

Montmorillonite typically occurs in this way (Figure 6.18(a)), and yields SAD patterns consisting of continuous or semicontinuous rings (Figure 6.18(b)). Such a pattern is indicative of turbostratic disorder, and shows that the constituent layers of the particle selected are randomly stacked one above the other. Many montmorillonite particles can therefore be regarded as aggregates, and indeed, a foliated, lamellar aggregate structure is often obvious.

Nontronite often shows a lath-like morphology, with the direction of elongation again being the *a*-axis (Figure 6.18(c)), but the mineral can also occur in poorly defined particles, as described at the beginning of this section. Hectorite is another smectite that occurs in *a*-axis-elongated laths.

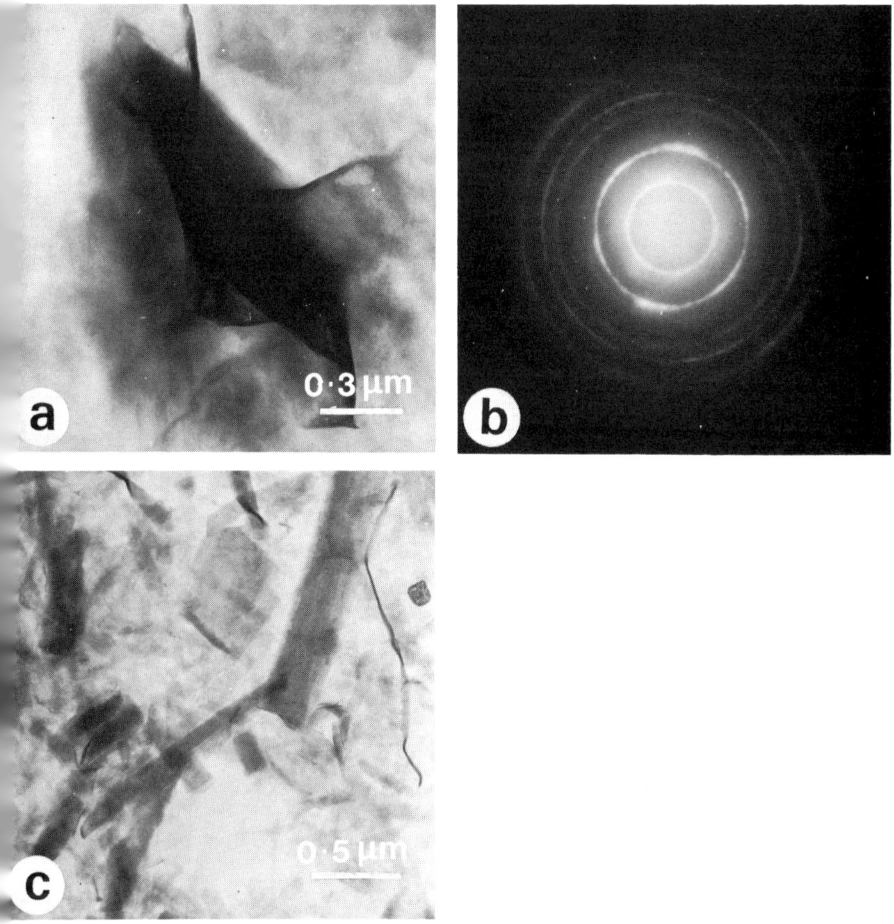

Figure 6.18 (a) Foliated montmorillonite aggregate giving: (b) a ring pattern indicative of turbostratic disorder in stacking of the layers. (c) Lath-like nontronite

Saponite occurs both in isometric, platy particles and in stubby laths aligned along *a* (Figure 6.19(a) and (b)). Electron diffraction often shows evidence of a higher degree of structural order than is found, for example, in montmorillonite. Thus, what appear to be relatively thick particles may yield spotty rather than continuous ring patterns, and single-crystal hexagonal spot patterns also occur (Figure 6.19(c) and (d)). Beidellite also appears to possess greater order than is the norm for smectites.

6.7.5 *Magnesium-rich clays*

In general, these minerals occur in either platy or fibrous forms, the latter being of unquestionable diagnostic value.

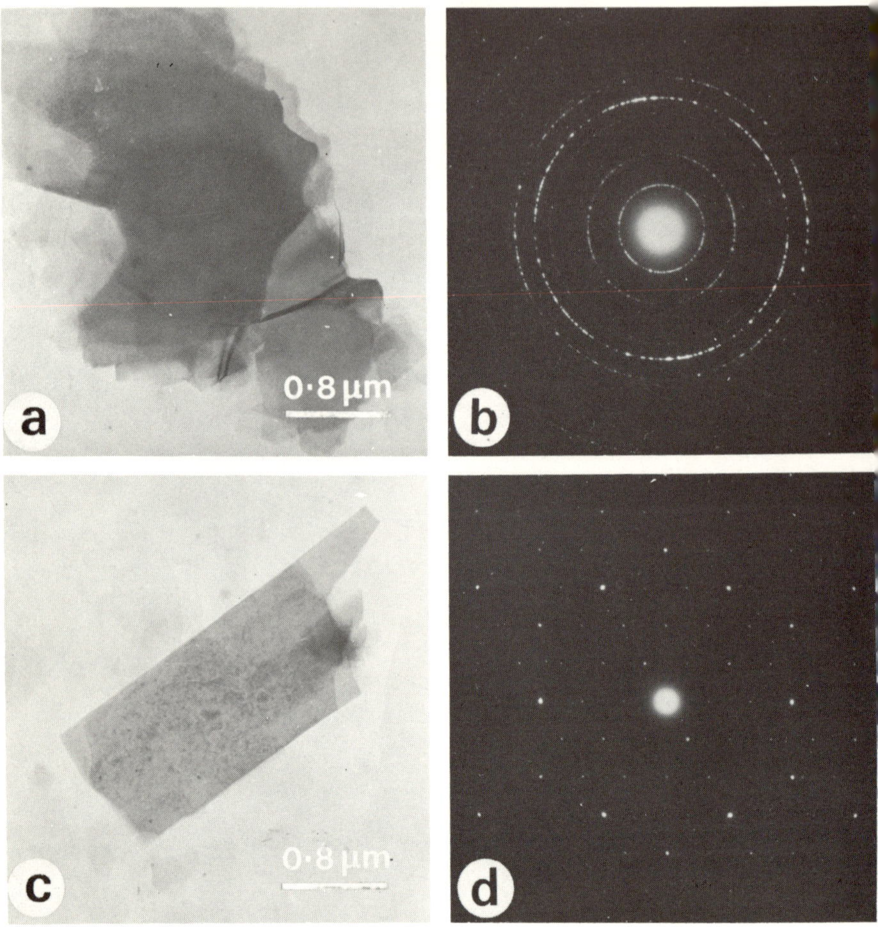

Figure 6.19 (a) and (b) platy saponite aggregate giving incomplete rings or arcs in diffraction. (c) and (d) lath-like saponite with single-crystal spot pattern

Talc is composed of rather thick, well-defined, platy particles that yield the usual, single-crystal hexagonal ED pattern (Figure 6.20(a) and (b)). In contrast, other magnesium-rich phyllosilicate clays such as kerolite, stevensite and swelling chlorite are characterized by a montmorillonite-like morphology and yield diffraction patterns typical of turbostratic disorder.

Palygorskite and sepiolite are both composed of *a*-axis-elongated fibres. The fibrous form in palygorskite is often better developed than that of sepiolite (Figure 6.20(c) and (e)), although this observation may be of little diagnostic value. Both minerals are sensitive to electron-beam damage and care is required in the taking of micrographs. Electron-diffraction patterns show a series of weak layer lines, and on the central line, the characteristic 10.8 and 12 Å spacings of palygorskite and sepiolite respectively can be determined (Figure 6.20(d) and (f)).

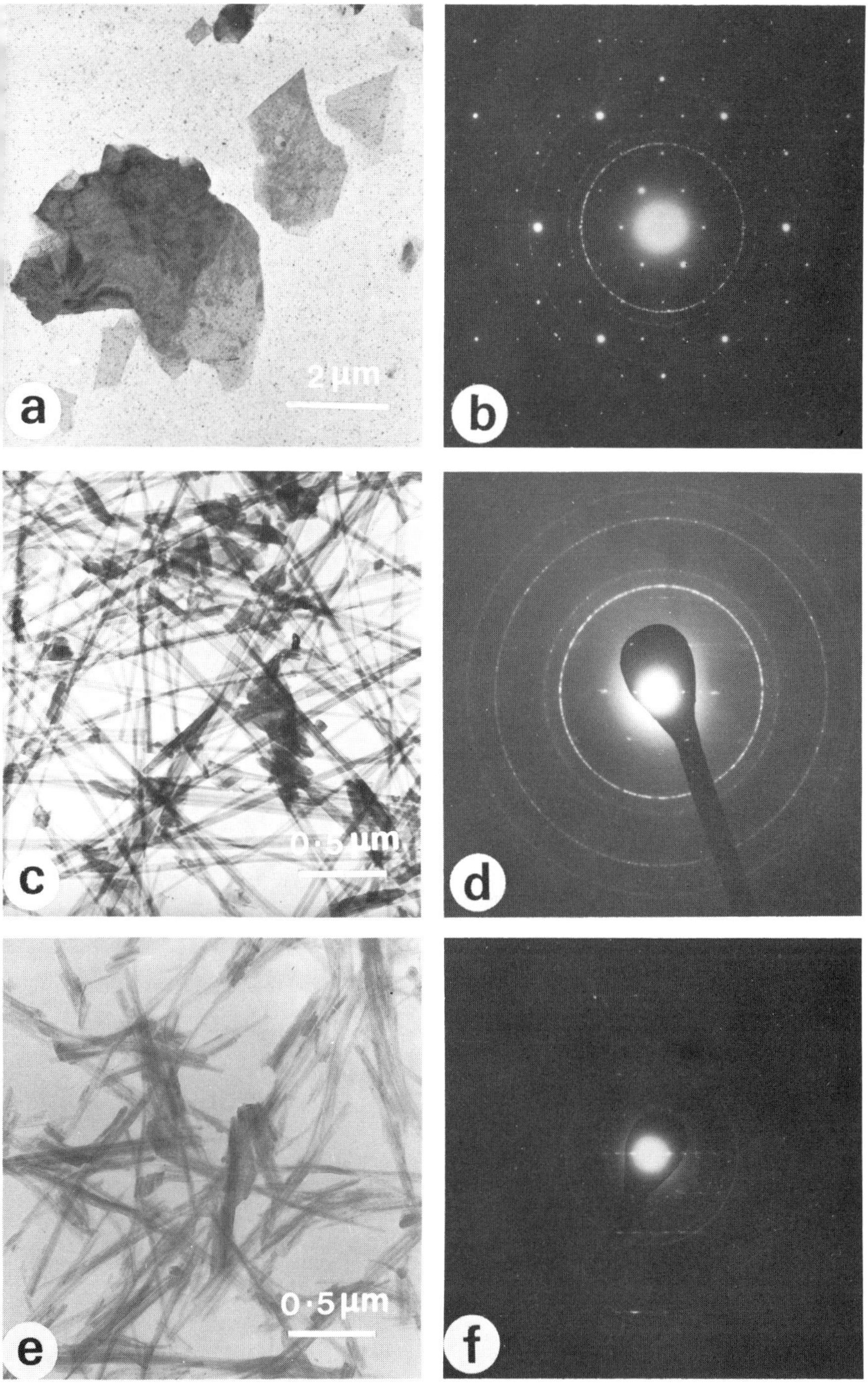

Figure 6.20 (a) Subhedral platy talc particles; (b) talc SAD spot pattern with rings from Al. (c) Palygorskite and (e) sepiolite fibres. Diffraction patterns (d) and (f), respectively, appear similar, apart from different spacing of spots along central row; 10.8 Å for palygorskite and 12 Å for sepiolite. Streaking along layer lines indicates structural disorder; rings are from Al

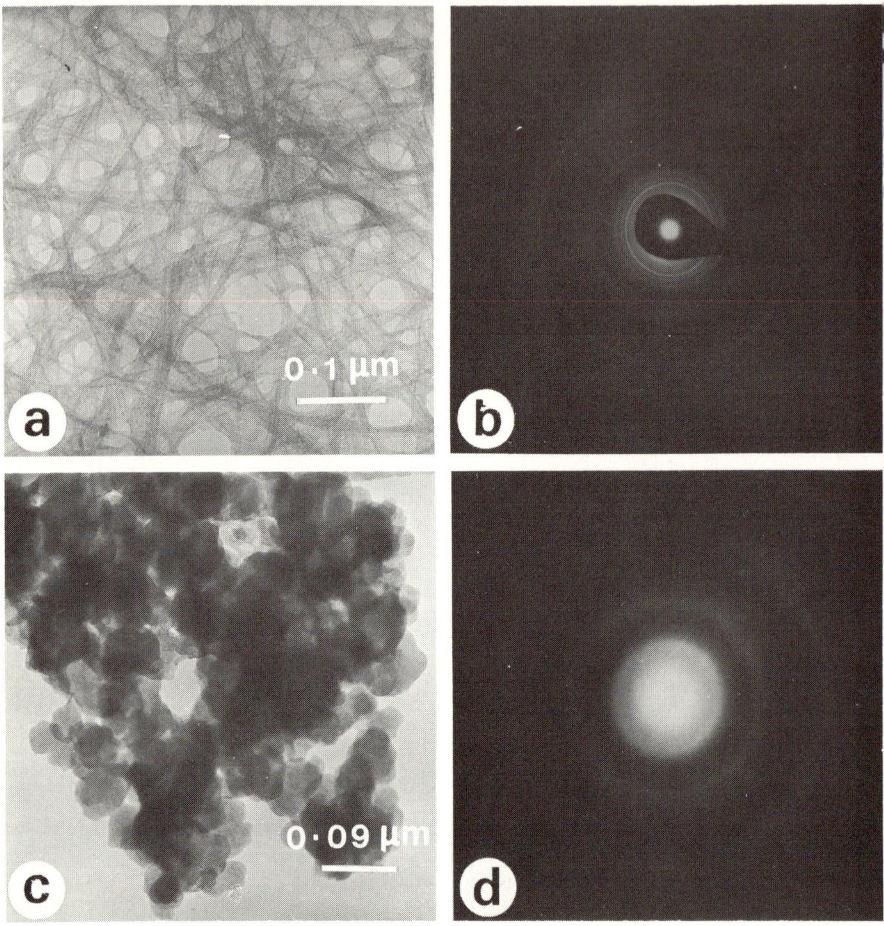

Figure 6.21 (a) and (b) show morphology and diffraction patterns which are quite diagnostic of imogolite; *d*-spacings of rings shown in (b) are from centre outwards: 4.1, 3.8, 3.3, 2.3, 2.1 and 1.4 Å. (c) and (d) allophane appears to be virtually amorphous showing only two very diffuse rings in SAD

6.7.6 *Imogolite and allophane*

Imogolite can be identified unequivocally by electron microscopy, which can be regarded as the most diagnostic method. The mineral consists of a mass of long and exceedingly fine fibres (Figure 6.21(a)). The fibres range in thickness from ~ 100 to 200 Å, but are some microns long. High-resolution micrographs show that the fibres are really bundles of tubes, and bring out a characteristic 'railway-track' appearance. The electron-diffraction patterns vary, according to whether the fibres are randomly oriented or in a parallel alignment, but they can be indexed by taking the fibre direction as the *c*-axis and the interfibre distance as the *b*-axis (Figure 6.21(b)). A reflection corresponding with the

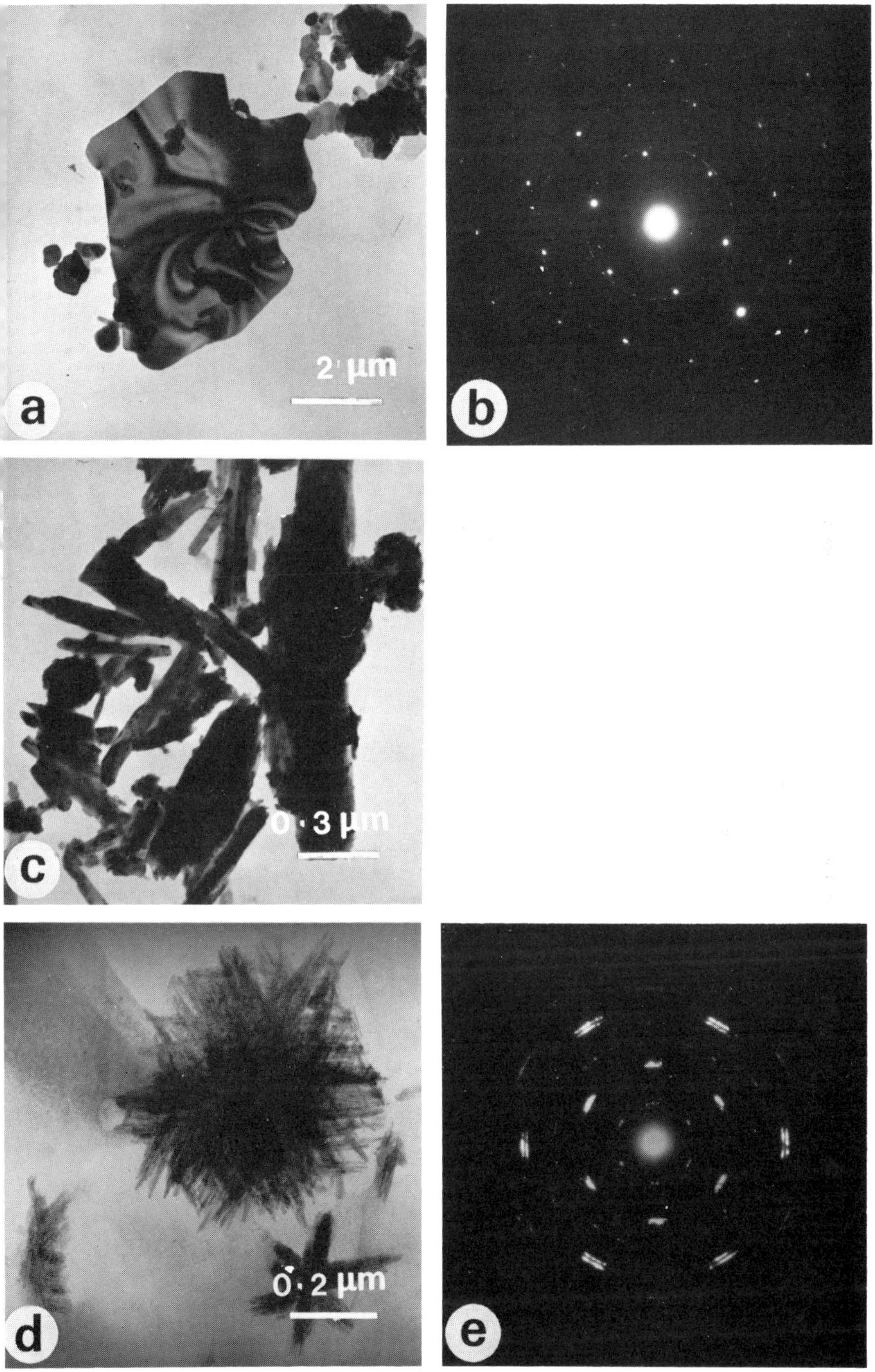

Figure 6.22 (a) Electron-dense platy hematite particle gives hexagonal spot pattern (b), with spacing of 2.5 Å; rings from Al. (c) Stubby lath-like goethite and (d) the more typical, acicular star-shaped goethite particles giving (e) a pseudohexagonal pattern of short arcs due to twinning

intertube distance (23 Å) can be observed, but only for short exposures.

Allophane consists of fine, rounded particles (Figure 6.21(c)), which are sometimes seen to have a ring structure at high magnification. Electron-diffraction patterns are poor, consisting essentially of only two broad ring reflections, yielding spacings at 3.3 and 2.25 Å (Figure 6.21(d)).

6.7.7 *Iron minerals*

These minerals show a range of morphological characteristics, that help to identify them even when they occur as minor components in complex mixtures.

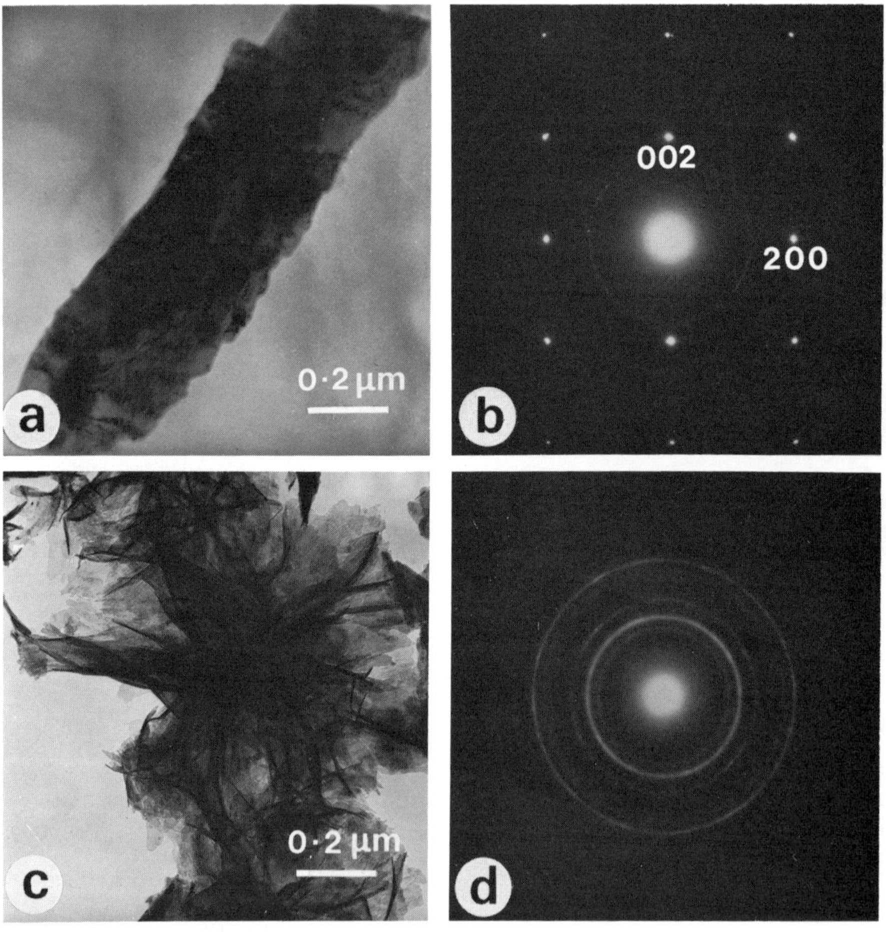

Figure 6.23 (a) Well-crystallized lepidocrocite lath, elongated along *c*-axis. (b) SAD pattern representing the (h0) plane of reflections; rings from Al. (c) Foliated aggregate of poorly crystallized lepidocrocite which can be identified from ring diffraction pattern (d), with *d*-spacings 6.3, 3.0, 2.5, 2.1, 1.6 and 1.5 Å

Hematite commonly occurs as small, electron-dense, platy particles, in a rough hexagonal or subequant shape and with well-defined outlines. Interference effects on the crystal surface are common (Figure 6.22(a)). The electron-diffraction pattern (Figure 6.22(b)) is a single-crystal, hexagonal type, similar to that of the layer silicates. However, measurement of the repeat distance enables hematite to be readily distinguished from all the major clay minerals yielding a spacing of 2.5 Å as opposed to 4.5 Å for the layer silicates.

Goethite tends to occur as poorly developed, lath-like particles elongated along the *c*-axis, and also as star-shaped twin crystals (Figure 6.22(c) and (d)). The ED patterns yielded by these twins are quite characteristic, with hexagonally arranged clusters of slightly arced spots (Figure 6.22(e)). Careful

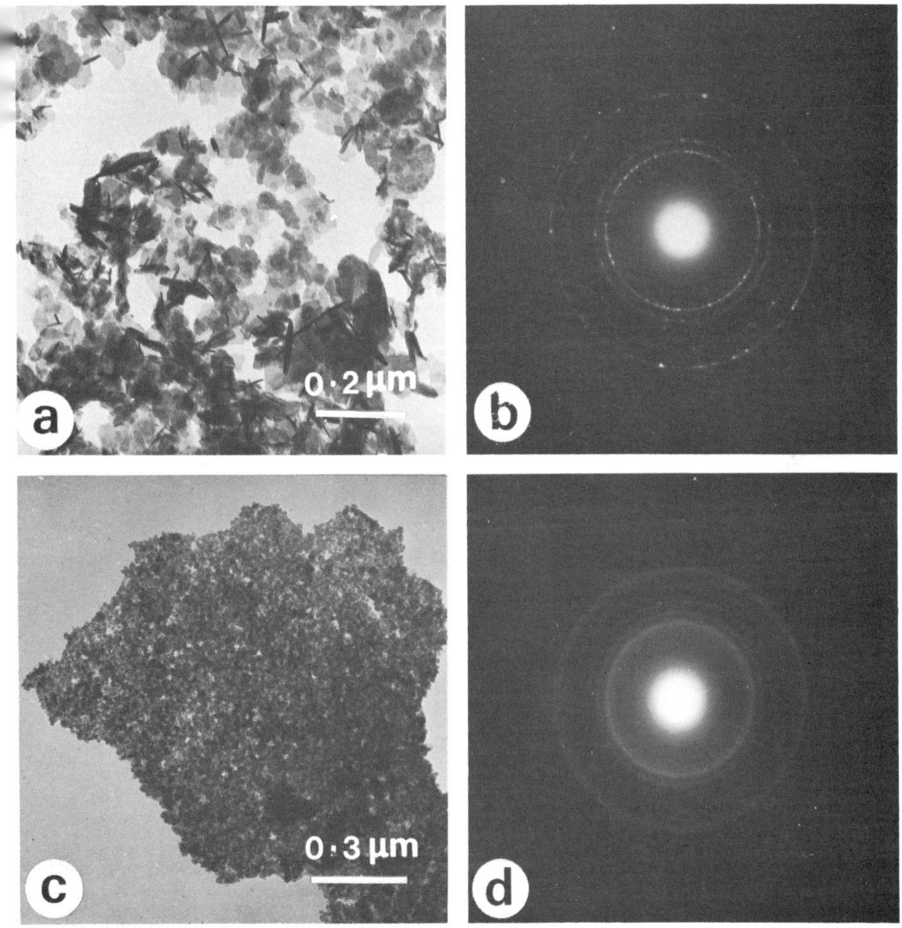

Figure 6.24 (a) and (b) Feroxyhyte. (*c*) and (*d*) ferrihydrite. Note that feroxyhyte usually gives a four-ring pattern with *d*-spacings 2.5, 2.2, 1.7 and 1.5 Å, while ferrihydrite gives a fifth ring at 1.9 Å

measurement of the 111 reflection can be used to determine the amount of Al substitution in the goethite structure.

Lepidocrocite often shows a stubby, *c*-axis-elongated, lath-like form (Figure 6.23(a)), and yields a diffraction pattern displaying a rectangular array of spots (Figure 6.23(b)) from the [010] zone axis, with $d_{200} = 1.55$ Å and $d_{002} = 1.96$ Å. Lepidocrocite also occurs in a poorly crystallized, crumpled sheet form, which shows a characteristic ring-like ED pattern (Figure 6.23(d) and (e)).

Ferrihydrite and feroxyhyte both show a fine-grained, indeterminate morphology, and yield continuous-ring ED patterns (Figure 6.24(a), (b), (c)

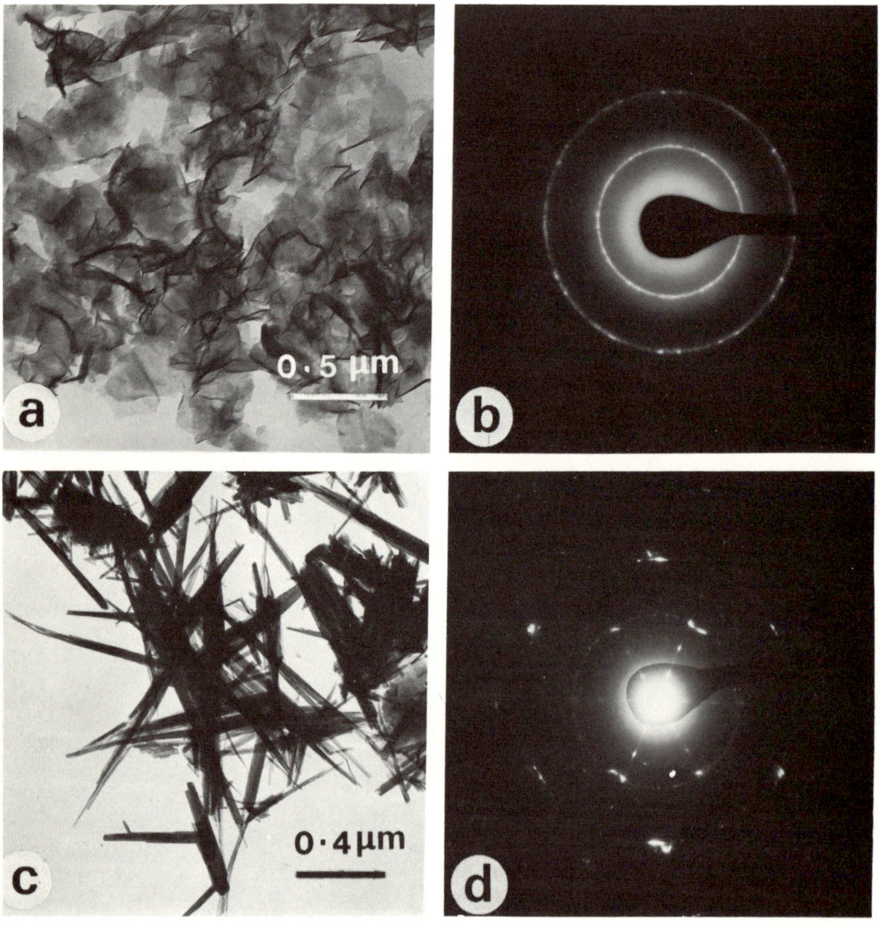

Figure 6.25 (a) and (b) Poorly crystallized vernadite giving only two diffraction rings at 2.5 and 1.4 Å. (c) Cryptomelane type of material, identified from diffraction pattern (d), with closest spaced row of spots of 6.9 Å

and (d)). Measurement of these rings yields the same series of spacings recorded by X-ray diffraction, so that the ED pattern is quite diagnostic.

6.7.8 *Manganese minerals*

These minerals may be very difficult to identify by XRD when they occur in clay materials. Their identification by electron microscopy is now a rapidly developing field, the two examples shown next merely illustrating the potential of the method.

Vernadite is an example of a very poorly crystalline phyllomanganate, and is characterized by thin, crumpled sheets which yield a ring ED pattern with characteristic spacings at 2.5 and 1.4 Å. (Figure 6.25(a) and (b)).

Cryptomelane by contrast, is a well-crystallized tectomanganate and shows a distinct acicular form. SAD patterns from a star-shaped twin show streaked rows of spots from each component crystal, with a *d*-spacing of 6.9 Å. (Figure 6.25(c) and (d)).

6.7.9 *Silica minerals*

Quartz is not easily studied by the TEM, since it rarely occurs in sufficiently small particles. In addition, the absence of any particular cleavage plane makes it difficult to characterize the mineral by electron diffraction.

The tridymite-like mineral that occurs in clays has a platy morphology (Figure 6.26(a)), with individual particles ranging from 0.1 μm to greater than 1 μm in diameter, depending on the nature of the occurrence. The SAD pattern

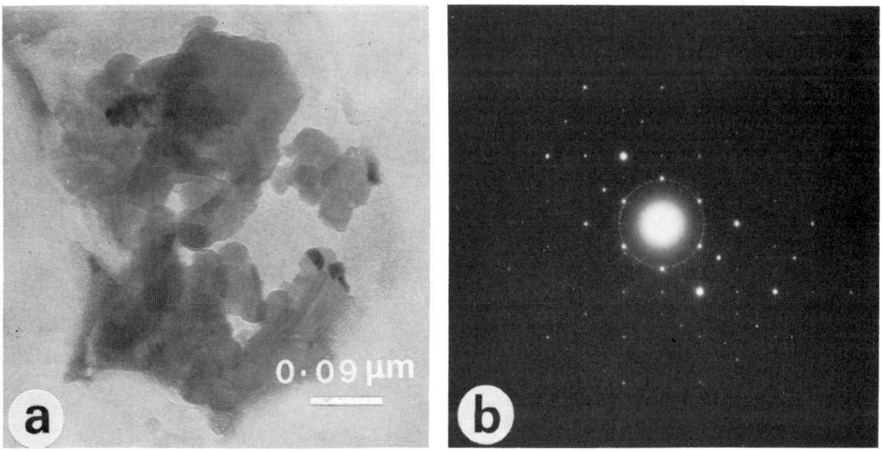

Figure 6.26 (a) Small platy tridymite crystals, (b) SAD pattern shows hexagonal array of spots with strong streaking between reflections, indicating disorder of the structure. Diffracting rings here are from montmorillonite

(Figure 6.26(b)) from individual flakes yields a hexagonal array of spots with *d*-spacing of 8.64 Å, and also shows strong streaking between the reflections. This pattern is indicative of disordered α-tridymite (Wilson *et al.*, 1974).

References

Agar, A.W. (1978) 'Operation of the electron microscope', in *Principles and Practice of Electron Microscope Operation*, (A.W. Agar, R.H. Alderson and D. Chescoe eds.), North-Holland, Amsterdam, 166–190.

Anderson, K. (1968) An image intensifier for the electron microscope, *J. Sci. Instr.* (*J. Phys. E*) **Ser. 2, 1**, 601–603.

Bates, T.F. (1971) 'The kaolin minerals', in *The Electron-Optical Investigation of Clays*, (J.A. Gard ed.) Mineralogical Society, London, 109–157.

Beutelspacher, H. and van der Marel, H.W. (1968) *Atlas of Electron Microscopy of Clay Minerals and their Admixtures*. Elsevier, Amsterdam, 333 pp.

Bresson, L.-M. (1981) Ion micromilling applied to the ultra-microscopic study of soils. *Soil Sci. Soc. Am. J.*, **45**, 568–573.

Brindley, G.W. and De Kimpe, C. (1961) Identification of clay minerals by single crystal diffraction. *Am. Mineral.*, **46**, 1005–1016.

Comer, J.J. (1971) 'Specimen preparation', in *The Electron-Optical Investigation of Clays*, (J.A. Gard ed.) Mineralogical Society, London, 79–108.

Conley, R.F. (1966) Statistical distribution of particle size and shape in the Georgia kaolins. *Clays & Clay Miner.*, **14**, 317–330.

Eitel, W. and Gotthardt, E. (1940) Der Einfluß von Elektroneninterferenzen auf die Abbildung von Kristallen im Ubermikroskop. *Naturwissenschaften*, **28**, 367.

Farmer, V.C., Russell, J.D. and Berrow, M.L. (1980) Imogolite and proto-imogolite allophane in spodic horizons: evidence from a mobile aluminium silicate complex in podzol formation. *J. Soil Sci.*, **80**, 673–684.

Gard, J.A. (1971) *The Electron-Optical Investigation of Clays*, Mineralogical Society, London, 383 pp.

Grim, R.E. and Güven, N. (1978) *Bentonites: Geology, Mineralogy, Properties and Uses* (*Dev. Sedimentol.*, **24**) Elsevier, Amsterdam, 256 pp.

Hirsch, P.B., Howie, A., Nicholson, R.B., Pashley, D.W. and Whelan, N.J. (1965) *Electron Microscopy of Thin Crystals*, Butterworth, London, 549 pp.

Iijima, S. and Buseck, P.R. (1978) Experimental study of disordered mica structures by high-resolution electron microscopy. *Acta Crystallog.*, **A34**, 709–719.

Mackenzie, R.C., Follett, E.A.C. and Meldau, R. (1971) 'The oxides of iron, aluminium, and manganese,' in *The Electron-Optical Investigation of Clays*, (J.A. Gard ed.), Mineralogical Society, London, 315–344.

Méring, J. and Oberlin, A. (1971) The smectites. In *The Electron-Optical Investigation of Clays*, (J.A. Gard ed.), Mineralogical Society, London, 193–229.

Nadeau, P.H. (1985) The physical dimensions of fundamental clay particles. *Clay Miner.*, **20**, 499–514.

Nadeau, P.H., Tait, J.M., McHardy, W.J. and Wilson, M.J. (1984a) Interstratified XRD characteristics of physical mixtures of elementary clay particles. *Clay Miner.*, **19**, 67–76.

Nadeau, P.H., Wilson, M.J., McHardy, W.J. and Tait, J.M. (1984b) Interparticle diffraction: a new concept for interstratified clays. *Clay Miner.*, **19**, 757–769.

Nadeau, P.H., Wilson, M.J., McHardy, W.J. and Tait, J.M. (1985) The conversion of smectite to illite during diagenesis: evidence from some bentonites and sandstones. *Miner. Mag.*, **49**, 393–400.

Paulus, M., Dubon, A. and Etienne, J. (1975) Application of ion-thinning to the study of the structure of argillaceous rocks by transmission electron microscopy. *Clay Miner.*, **10**, 417–426.

Reid, N. (1975) 'Ultramicrotomy', in *Practical Methods in Electron Microscopy*, 3. II (A.M. Glauert ed.) North-Holland, Amsterdam.

Robertson, R.H.S., Brindley, G.W., and Mackenzie, R.C. (1954) Mineralogy of the kaolin clays from Pugu, Tanganyika. *Am. Mineral.*, **39**, 118–139.

Ruska, H. (1943) Ubermikroskopische Untersuchungen an Asbeststaub und Asbestlungen, *Arch. Gewerbepath., Gewerbehyg.*, **11**, 575–578.

Smart, P. and Tovey, N.K. (1981) *Electron Microscopy of Soils and Sediments: Examples*, Clarendon Press, Oxford, 178 pp.

Smart, P. and Tovey, N.K. (1982) *Electron Microscopy of Soils and Sediments: Techniques*, Clarendon Press, Oxford, 400 pp.

Spurr, A.R. (1969) A low-viscosity epoxy resin embedding medium for electron microscopy. *J. Ultrastruct. Res.*, **26**, 31.

Smith, B.F.L. (1984) The determination of silicon in ammonium oxalate extract of soils. *Comm. Soil Sci. & Pl. Anal.*, **15**, 199–204.

Steeds, J.W. (1979) 'Convergent beam electron diffraction,' in *Introduction to Analytical Electron Microscopy*, (J.J. Hren, J.I. Goldstein, and D.C. Jay eds.), Plenum Press, New York, 387–422.

Sudo, T., Shimoda, S., Yotsumito, H. and Acta, S. (1981) *Electron Micrographs of Clay Minerals.* (*Dev. Sedimentol.*, **31**) Elsevier, Amsterdam, 203 pp.

Tokiwai, M., Moriguchi, S., Shinkawa, T., and Watanabe, E. (1983) Automatic computer analysis of electron diffraction patterns. *JEOL News*, **21E(2)**, 2–6.

Vainshtein, B.K. (1964) *Structure Analysis by Electron Diffraction*, Pergamon Press, Oxford, 420 pp.

Walker, G.F. (1967) Chemical exfoliation of vermiculite and the production of colloidal dispersions. *Science*, **156**, 385–387.

Wilson, M.J., Russell, J.D. and Tait, J.M. (1974) A new interpretation of the structure of disordered α-cristobalite. *Contrib. Miner. & Petrol.*, **47**, 1–6.

Yada, K. (1967) Study of chrysotile by a high resolution microscope. *Acta Crystallog.*, **23**, 704–707.

Yariv, S. and Cross, H. (1979) *Geochemistry of Colloidal Systems*, Springer-Verlag, Berlin, 450 pp.

Yoshida, T. (1973) Elementary layers in the interstratified clay minerals as revealed by electron microscopy. *Clays & Clay Miner.*, **21**, 413–420.

Zvyagin, B.B. (1967) *Electron-Diffraction Analysis of Clay Mineral Structures*, Plenum Press, New York, 364 pp.

7 Chemical analysis

D.C. BAIN AND B.F.L. SMITH

7.1 Introduction

Of the determinative methods used to study clay minerals, chemical analysis is the oldest and most established. Before the development of the earliest instrumental techniques, such as X-ray diffraction and thermal methods, the identification of clay-mineral phases was accomplished by chemical analysis, supplemented where possible by optical data and physical characteristics such as specific gravity, colour, hardness, etc. However, fine-grained materials cannot be characterized reliably using these methods alone, and misidentifications were common. With the routine application of infrared spectroscopy and electron microscopy, in addition to X-ray diffraction and thermal analysis, instrumental techniques are now unquestionably superior to chemical analysis for clay-mineral identification and a clay is often described and identified without recourse to chemical analysis. Nevertheless, for complete characterization, a chemical analysis is still essential, as this is the only way that a structural formula can be calculated and the distribution of cations in the structure defined directly. A full characterization should include the determination of all the major and minor elements (with the proportion of the iron present in the ferrous and ferric forms) and also the determination of the cation-exchange capacity and, if possible, the anion-exchange capacity.

Classification of clay minerals has always depended heavily on chemical data, and many clay-mineral groups, such as smectites, illites and chlorites, are subdivided according to composition. The characterization of a clay mineral as 'dioctahedral' or 'trioctahedral' can be determined from a *b*-dimension measurement using X-ray or electron-diffraction techniques, but these terms also have a chemical connotation, indicating the valency of the cations occupying the octahedral sheets. The type and proportions of cations, however, are best determined by chemical analysis. For example, a trioctahedral mineral contains divalent octahedral cations, which could be Fe^{2+}, Mg^{2+}, or both, in various proportions. Sometimes there is a complete gradation between two species of clays in the same group and these are best distinguished by their chemistry, e.g. montmorillonite and beidellite.

Thus, despite the many and sophisticated instrumental techniques available for studying clays, a good and complete chemical analysis is still necessary for the comprehensive study and characterization of any clay mineral. This chapter outlines the approach required to obtain such a chemical analysis, from the preparatory stages of sample handling, grinding and cation saturation, to the methods of analysis. Analysis of clay is, in effect, a silicate analysis, but there are special problems associated with clays that are not encountered with some other types of silicate materials, and particular attention is paid to overcoming these problems. The section on methods of analysis is not comprehensive, but draws attention to techniques of particular interest in the study of clays and lists references with more details of other relevant methods. The analysis of ferrous iron and determination of exchange capacity are dealt with in more practical detail, since these aspects of silicate analysis are often neglected. The final section details the procedure for calculating the structural formula from the chemical analysis of a clay mineral, indicating how impurities can affect the results obtained.

7.2 Sample preparation

No matter how much time is spent on a particular analytical method, or how accurate and precise the method may be, the quality of the final analysis is primarily dependent on the quality of the sample preparation; i.e. the analysis is only as good as the sample it represents. This fact cannot be overemphasized, yet it is too often ignored. First of all, the sample should be properly selected in the field, to ensure that it is representative of the total material. The analyst and the individual requiring the analysis should ensure that the sample is handled and prepared so that it is not altered by oxidation or contamination with the elements to be determined. Often an analyst is presented with a sample and a certain type of analysis is requested without prior consultation. To obtain the maximum benefit from a chemical analysis, the person requiring the analysis should consult the analyst at the earliest possible stage, so as to obtain guidance on the methods and equipment for separation, handling and grinding.

The many problems involved in sampling rocks and minerals, and the subsequent treatment and preparation of these samples for chemical analysis are described in detail by Johnson and Maxwell (1981). Fortunately for the clay mineralogist, many of these problems are minimized if truly clay-sized material (i.e. $< 2\,\mu m$) is under investigation. If the clay fraction has been separated by sedimentation or centrifugation, and has been disaggregated by gentle grinding (by hand in an agate mortar after drying) then the problems of grain size, homogeneity, representative subsampling, contamination and alteration by prolonged grinding are avoided. However, as this book is concerned with clay minerals in a broader sense, i.e. layer-silicate minerals, some of which may occur in particles considerably larger than clay size (e.g. mica, chlorite, vermiculite and serpentine) some details are presented relating to grinding and representative sampling.

7.2.1 *Grinding*

The use of a grinding mill to reduce the particle size of the sample inevitably causes contamination. The aim is to minimize this contamination, or to ensure that any elements introduced in the grinding process are not important in the analysis. One factor which affects the level of contamination is the length of the grinding time. With its high efficiency and consequent short grinding times (less than 10 minutes) the swing-mill or disc-mill is commonly used, with a lining material suitable for the analysis. A choice can be made between tungsten carbide, nickel–chromium alloy or agate. However, appreciable oxidation of chlorite and biotite can occur when using this method (Reay, 1981), and in heterogeneous materials some of the more resistant components may not be crushed to the same size as the less resistant components. Consequently, this method is not suitable for small quantities of material.

Particle shape can affect efficiency of grinding, and in a mixture, minerals of certain shapes will be ground more effectively than others. Platy minerals are notoriously difficult to grind and this can cause difficulties in the analysis. For example, in the analysis of pressed powder pellets of a rock containing biotite, using X-ray fluorescence spectrometry, Volborth (1964) found that the increase in the biotite surface relative to the other non-platy minerals gave serious inaccuracies in the analysis.

In the authors' experience, one of the most effective mills for grinding platy particles is the *McCrone Micronising Mill*. The grinding vessel is filled with an ordered array of some 48 identical, cylindrical grinding elements (corundum or agate), which on vibration rotate and grind the powder circulating between the grinding surfaces. To minimize oxidation of ferrous iron and sulphur, alcohol or acetone can be added to produce a slurry. Platy minerals can be reduced to less than 10 μm diameter in about 15–30 minutes by this method.

Health risks can result from inhalation of small fibres, and therefore it is advisable to wet grind any sample known to contain fibrous minerals.

Materials high in clay content can be difficult to grind in the dry state due to caking, but this can be alleviated by pre-drying at 140°C.

Whatever method is chosen, excessive grinding should always be avoided as it may cause changes in the mineral structure—important if the sample is to be examined by XRD, IR and EM methods—and contamination increases with grinding time. If it is necessary to sieve the sample at any stage, brass and stainless-steel sieves should be avoided, particularly if trace elements are to be determined. Nylon mesh sieves are recommended.

7.2.2 *Representative sampling*

If the sample of interest is a mineral separate, often the main problem is how to obtain a sufficient quantity of material for analysis. This does, however, eliminate the difficulty in acquiring a representative portion, as the total available material is analysed. Such a situation can arise with the clay fraction of a coarse-textured soil, where the clay fractions may comprise only a few percent of the total soil.

If, however, the sample of interest is available in bulk, then it must be reduced to a smaller representative amount which is statistically equivalent to the whole sample. The size of a representative sample is influenced by the size of the particles in the material. Jackson (1962) has shown that for samples containing particles up to 100 μm in diameter the optimum sample weight is about 5 mg, so that precautions are not necessary to ensure a representative sample for clay-size material ($< 2\,\mu$m). If, however, the sample contains very coarse particles, the sample must be split (after grinding if necessary) in an unbiased manner, using a sample splitter or riffler. This should be used, repeatedly if necessary, on the subdivided portions, until the desired quantity is obtained. Allen and Khan (1970) critically evaluated several sampling techniques, and concluded that the spinning riffler is by far the best technique for subsampling powders.

7.2.3 *Cation saturation*

Most clays, particularly those containing smectites and vermiculites, have exchangeable cations which will vary according to the environment in which the clay was formed or located. Sometimes all the exchange sites are occupied by the same cation, but this is often not the case, and varying proportions of Na^+, K^+, Mg^{2+} and Ca^{2+} commonly occur as exchangeable cations. If a chemical analysis is required for calculation of the structural formula of a particular clay mineral, it is advisable to saturate the clay with an index cation not present in the sample. Most clays contain some non-clay impurities or small amounts of another clay mineral, and these other phases may contain some of the elements which are possible exchangeable cations. For example, a smectite-rich clay may contain impurities such as carbonates (calcite and dolomite) or feldspars, and as a result, allocation of elements such as Na, K, Mg and Ca to the smectite formula becomes difficult, if not impossible. An additional complication is that Mg^{2+} occurs in octahedral sites as well as in exchangeable sites.

Thus, saturating the clay sample with an index cation simplifies calculation of the structural formula, and, by choosing an appropriate cation, correction of the final analysis for the presence of impurities is also simplified. Sometimes a less common element such as Ba or Sr should be chosen, if it is known that the impurities contain all the common saturating cations (Na^+, K^+, Ca^{2+}, Mg^{2+}). In the analysis of bentonites in which interstratified illite-smectite was the dominant clay mineral, and contained K, Mg and possibly Na, with calcite as the suspected impurity, Nadeau and Bain (1986) saturated the clays with Sr^{2+}.

Saturation with the chosen cation is performed by treating the clay with a

1 M salt solution of a cation which will not enter the clay-mineral structure. After centrifuging and decanting off the excess salt solution, the treatment is repeated, and the clay is shaken overnight in fresh salt solution, to ensure that all naturally occurring exchangeable cations have been replaced with the chosen cation. Finally, the sample is washed free of excess salt; however, the sample should not be washed unnecessarily, because desaturation can occur, along with alteration by hydrolysis (Chaussidon, 1982). If chloride solutions are employed, removal of chloride may be conveniently monitored by testing with silver nitrate. Washing with alcohol is advisable in the latter stages to remove the excess exchange solution, and if the clay is very finely dispersed (particularly if Na^+ or Li^+ is the exchangeable cation), dialysis is strongly recommended to prevent loss of the very fine particulate material.

7.3 Methods of major elemental analysis

Major elemental analysis methods have changed dramatically over the last fifty years. Initially, the 'Classical Scheme' of analysis was adopted, as devised by Lundell and Hoffman (1938), Groves (1951) and Hillebrand *et al.* (1953). These methods consisted of a decomposition step, followed by gravimetric determination of the constituent elements, and although recognized by many analysts to be accurate, the procedure is tedious, very time-consuming and requires quite a large sample ($\sim$ 1 g), which in many cases is not available. The development of colorimetric methods has led to their introduction into more rapid analytical schemes, whereby much smaller sample weights ($\sim$ 0.1 g) can be utilized (Shapiro and Brannock, 1956).

More recently, techniques such as flame photometry, atomic emission and absorption, and X-ray fluorescence spectroscopy have resulted in greatly reduced analysis times and sample weights, at the expense of a small reduction in accuracy which could be minimized by replication.

The elements normally analysed and, by convention, reported in the form of percentage oxides, are: SiO_2, Al_2O_3, Fe_2O_3, FeO, CaO, MgO, K_2O, Na_2O, TiO_2, MnO, H_2O^+, H_2O^-, P_2O_5 and CO_2. The summation should result in a total in the range 99.75–100.50% for 'classical analyses' and 99.0–101.0% for 'rapid' techniques for an acceptable analysis. Chalmers and Page (1957) recommended that reports of elemental analyses should include an estimate of the precision and accuracy of the results. The latter should always be monitored by carrying out a complete analysis of a mineral or rock of similar known composition.

A quick qualitative analysis is a very useful precursor to any quantitative determination. This was previously achieved by standard qualitative analysis, but microprobe analysis or X-ray fluorescence now gives not only a rapid indication of the range of elements present, but also a measure of their relative amounts.

Two forms of analytical techniques are available for the measurement of the elemental composition of clay minerals:

(i) Non-destructive: where no further sample preparation is necessary and the sample can be re-used for other analyses,

(ii) Destructive: where the sample must undergo decomposition into a suitable form for analysis.

7.3.1 *Sample decomposition*

Few of the wide range of decomposition methods available can cause complete dissolution of all types of clay minerals. Dolezal *et al.* (1968) comprehensively reviewed these, and their book (translated into English) is invaluable for any analyst. They described three main decomposition methods: (a) acid, (b) fusion and (c) sintering, but only techniques (a) and (b) are in current use.

(a) *Acid decomposition.* Samples are dissolved in hydrochloric, hydrofluoric, nitric, sulphuric, perchloric or phosphoric acid or combinations of these. The most widely used acid-dissolution technique is decomposition with hydrofluoric acid, either alone or in conjunction with nitric or sulphuric acid, in special Teflon decomposition vessels. These allow rapid decomposition of the sample without loss of silica by volatilization of SiF_4, essential if silicon is one of the required elements (Bernas, 1968, Lim and Jackson, 1982). Previously, a separate decomposition technique was necessary for the determination of silica (Shapiro and Brannock, 1956).

(b) *Fusion.* Decomposition by fusion at high temperature, with alkali carbonates, hydroxides, borates or pyrosulphates, usually in the sample: flux ratio of 1:7, results in a melt which can then be dissolved in water or acid to give a solution containing the constituent elements.

7.3.1.1 *Decomposition by hydrofluoric acid (Bernas, 1968).* Fifty ($\pm$ 0.01) mg of a prepared, representative sample (< 200 mesh) are placed in a Teflon decomposition vessel with 0.5 ml aqua regia. The latter is particularly necessary if any organic material is present. Three ml of HF (48%) are added and the vessel is tightly sealed and placed in an oven at 110°C for 1 hour. After cooling, the dissolution products are quantitatively transferred to a 50 ml plastic beaker containing 2.8 g of HBO_3. 10 ml distilled water are added and the mixture stirred magnetically to dissolve any insoluble fluorides. When clear, the solution is diluted to 100 ml and stored in a plastic bottle ready for major elemental analysis.

7.3.1.2 *Fusion with lithium metaborate (see Ingamells, 1966, Shapiro, 1967, and Verbeek* et al.*, 1982).* 100 ($\pm$ 0.01) mg of the prepared representative sample (< 200 mesh) are weighed into a small platinum crucible, 700 mg $LiBO_3$ added and thoroughly mixed, before heating gently over a Bunsen burner. The temperature is gradually increased to around 900°C, and the heating is continued for a further 10 min to ensure complete decomposition. The crucible and melt are allowed to cool and then placed in a 50 ml beaker. 35 ml distilled water and 4 ml nitric acid (conc.) are added to the crucible, and the contents stirred magnetically, using a Teflon follower, for $1\frac{1}{2}$ h to ensure complete dissolution. The solution is then quantitatively transferred to a 100 ml standard flask, diluted to the mark, and is reserved for major elemental analysis.

Table 7.1 Spectrochemical instrumental methods.

Technique	References
UV/visual	Shapiro and Brannock (1956), Langmyhr and Graff (1965), Shapiro (1967), Jeffery (1981)
Atomic absorption Flame emission	Shapiro (1967), Bernas (1968), Verbeek *et al.* (1982), Jeffery (1981), Baker and Suhr (1982), Ure (1983)
Inductively coupled plasma	Thomson and Walsh (1983)
X-ray fluorescence	Norrish and Chappell (1977), Jones (1982),
Neutron activation analysis	De Soete *et al.* (1972), Helmke (1982)

7.3.2 *Analytical determinations*

A wide range of analytical techniques is currently available to quantify the elemental composition of clays. The methods are too numerous and diverse to be detailed in this handbook, but the main methods in use are listed in Table 7.1, with references to more detailed information.

The choice of method depends on a number of factors, such as availability of equipment, size of sample, adequate sensitivity and accuracy of the required analysis. UV/visual methods are cheap and fairly simple to use, but suffer from interference effects. Conventional atomic-absorption and flame-emission spectrometry have the disadvantage of being single-element techniques but the advantages of easy sample presentation and low cost. Inductively coupled plasma (ICP) has several advantages when dealing with clay samples: it has a linear dynamic range over several orders of magnitude and can analyse small samples. However, the inherently high sensitivity of ICP is lost in the dilution process. All these techniques are solution-analysis methods, and consequently, are prone to losses and contamination during the process of clay dissolution. X-ray fluorescence (XRF) and neutron activation analysis (NAA) are multi-element techniques for solid samples. XRF can be non-destructive, but for major elemental analysis it is usually advisable to fuse the sample and dilute with flux to reduce interelement effects. It is very accurate, but cannot differentiate between ferrous and ferric iron. NAA can also be non-destructive, has high sensitivity and accuracy, but suffers the major drawback of requiring a nuclear reactor.

7.4 Analysis for ferrous iron

Ferrous iron frequently forms the major part of the iron content of rocks and minerals. Measurement of its content with regard to the environment in which these materials have formed can give valuable information, and can, in some cases, differentiate between members of an isomorphous series, e.g. micas. It is very important to know the proportion of iron in the ferrous form when the structural formulae of clay minerals are calculated. If all the iron present is assumed to be in the ferric state, misleading estimations of the layer charge can occur. Atmospheric oxidation rapidly changes ferrous iron to the ferric state, a process which is well illustrated in the case of poorly drained gley soils, where the blue-grey colour, due to ferrous iron, changes with improved drainage conditions, e.g. along root channels, to rusty brown ferric stains.

7.4.1 *Sample preparation*

Great care must be taken in preparing small representative samples for the determination of ferrous iron. Prolonged grinding can result in the oxidation of ferrous to ferric iron, and to overcome this, Hillebrand (1919) recommended very gentle grinding to <70 mesh. Kolthoff and Sandell (1956) suggested gentle grinding under alcohol, and this is currently the adopted procedure.

7.4.2 *Decomposition*

At this stage too, precautions must be taken to prevent oxidation of the ferrous iron components, by carrying out the decomposition under an inert atmosphere or liquid. In practice, this is accomplished by decomposing the material in a boiling mixture of hydrofluoric and sulphuric acids; the steam so generated prevents any aerial oxidation. The decomposition, however, must not be prolonged, because ferrous iron can be oxidized by concentrated sulphuric acid.

An alternative technique is to carry out the decomposition in the presence of an oxidant such as dichromate or vanadate (Wilson, 1960) which will immediately oxidize any ferrous iron to ferric. This can then be reversed stoichiometrically (see following equation) by altering the pH, and the regenerated ferrous iron can then be determined.

$$V^{5+} + Fe^{2+} \rightleftharpoons V^{4+} + Fe^{3+}$$

7.4.3 *Sources of interference*

The presence of sulphides (pyrrhotite, pyrites), organic matter and oxidants (manganates, vanadates) can result in substantial errors in the determination of ferrous iron. Graphite itself does not interfere, but other forms of organic matter can cause poor results.

7.4.4 *Procedures*

The determination of ferrous iron in rocks and minerals is carried out on separate subsamples of the material used for total elemental analysis. The samples are decomposed by hydrofluoric or hydrofluoric/sulphuric acid mixtures and the ferrous iron is measured in the resulting solution. Titrimetric and spectrophotometric methods have been used to quantify the ferrous iron released. Titrations with oxidants such as permanganate, dichromate or vanadate, or back titrations of excess oxidants have been utilized. A spectrophotometric method for the determination of ferrous and ferric iron in minerals as *ortho*-phenanthroline complexes has been extensively studied by Stucki (1981). He determined Fe^{2+} conventionally, using a fluorescent lamp for the total iron and the Fe^{3+} by difference. Hey (1941) suggested the use of iodine monochloride as an oxidant, in a medium consisting of hydrochloric and hydrofluoric acids, measuring the ferrous iron by titrating the liberated iodine with iodate, or by back titrating the excess iodate with thiosulphate (Van Loon, 1965). This technique has been used by Nicholls (1960) to determine ferrous iron in samples containing organic matter.

Other instrumentally orientated techniques for measuring ferrous iron are: potentiometric titrations, polarography and NMR or Mössbauer spectroscopy. The latter is used mainly for unstable compounds and, as the measurements are very time-consuming, it is used only for special cases (Goodman, 1980).

7.4.5 *Analytical methods*

Two main methods in current use for the determination of ferrous iron in rocks and minerals are detailed:

(a) *Shapiro and Brannock (1952) method.*
The mineral sample (< 70 mesh) is decomposed in a platinum crucible using a mixture of hydrofluoric and sulphuric acids at boiling point. The oxidation of any ferrous iron present is minimized because air is excluded by the steam that is generated. When decomposition is complete, the crucible and contents are plunged into a boric acid solution and the ferrous iron liberated is then measured quantitatively by titration with an oxidant.

Reagents:
Sulphuric acid—4.5 M.
Diphenylamine sulphonate indicator: dissolve 0.2 g sodium diphenylamine sulphonate in 1 litre distilled water, then add 1 litre 85% phosphoric acid (analytical reagent: AR) and mix well.
Standard dichromate solution (0.0167 M).

Decomposition and determination:
Place 0.500 g sample (< 70 mesh) in a platinum crucible (100 ml capacity), add 20 ml 4.5 M H_2SO_4, cover, and carefully boil over a gas burner until the steam generated excludes atmospheric oxygen. Carefully add 10 ml hydrofluoric acid (AR) to the crucible, cover and continue the boiling for a further 15 min. Plunge the crucible and contents into 400 ml of 2.5% boric acid in a 800 ml beaker, add 10 ml diphenylamine sulphonate indicator, and titrate immediately with standard dichromate solution to a purple end-point. Determine a reagent blank.

$$1 \text{ ml } 0.0167 \text{ M } K_2Cr_2O_7 \equiv 5.58 \text{ mg } Fe^{2+}$$

(b) *Wilson (1960) method.*
In this method, the decomposition of the sample takes place in the presence of an oxidant—vanadate at room temperature. Two methods of determining the amount of ferrous iron released will be described: titrimetric and spectrophotometric.

Decomposition:
3–20 mg of ground sample (< 70 mesh) are weighed into a 40 ml polystyrene vial, and 0.100 ml of 0.139 M ammonium vanadate solution are added using a microsyringe burette. If the FeO content is > 1 mg, more vanadate solution must be used. 1 ml of hydrofluoric acid is added and the vial stoppered and allowed to stand until complete decomposition is achieved; for most minerals overnight (16 hours) is sufficient. The resulting solution is often cloudy, due to the presence of insoluble sulphates and fluorides.

(*c*) *Titrimetric method.*

Reagents:
Ammonium vanadate 0.139 M—dissolve 1.6260 g ammonium vanadate (NH_4VO_3) in 20 ml 5 M H_2SO_4, and dilute to a final volume of 100 ml.
Ammonium ferrous sulphate 0.07 M—dissolve 2.75 g $(NH_4)_2Fe(SO_4)_2 \cdot 6H_2O$ in 100 ml 0.5 M H_2SO_4 (1 ml ≡ 10 mg FeO).
Barium diphenylamine sulphate indicator—0.01% w/v aqueous.
Hydrofluoric acid—(AR) 40% w/v.
Sulphuric acid (5 M).

Determination:
3 ml 5 M H_2SO_4 are added to the solution obtained after the decomposition procedure, and the mixture stirred magnetically to dissolve any previously insoluble products. 5 drops of barium diphenylamine sulphonate and 3 ml distilled water are then added, and the mixture immediately titrated to a pale green end-point with 0.07 M ammonium ferrous sulphate, using an microburette for increased accuracy. Blank titrations should always be carried out on the reagents, and the difference between the blank titration and the sample titration is related to the ferrous iron content of the sample.

(d) *Spectrophotometric method.*

The decomposition of the sample is carried out in a similar manner to that used for the titrimetric determination. Ferrous iron released is immediately oxidized to the ferric state by the vanadate oxidizing reagent, the latter being reduced to the vanadous form. The reaction is reversible, stoichiometric and pH-dependent. Wilson (1960) used a solution of beryllium sulphate to complex any excess hydrofluoric acid. An increase in pH causes the vanadous form to be oxidized to vanadate, and the appropriate amount of ferric iron to be reduced to ferrous iron which can then be determined by forming a coloured complex with dipyridyl.

Shapiro (1960) preferred to use a solution of boric acid, instead of the highly carcinogenic beryllium sulphate, to complex excess fluoride ions. Organic matter and sulphide minerals do not interfere in the determination of ferrous iron by this method.

Reagents:
2:2′ Dipyridyl—0.15% w/v aqueous.
Beryllium sulphate—50% w/v aqueous (*N.B. carcinogenic*).
Ammonium acetate buffer—50% w/v aqueous.
Standard iron solution (stock)—dissolve 0.5458 ferrous ammonium sulphate (AR) in 100 ml water and add 5 ml 3.5 M sulphuric acid. Slowly add a dilute solution of potassium permanganate until the mixture is slightly pink, then dilute to 1 litre. (1 ml ≡ 0.1 mg FeO.)
Standard iron solution (working)—dilute 100 ml stock solution to 1 litre. (1 ml ≡ 10 μg FeO.)
Preparation of standard ferrous iron calibration curve—pipette a range (0–80 μg FeO) into a series of standard 100 ml standard flasks, add ~ 100 mg ascorbic acid to each flask to reduce the iron to the ferrous state, and after 10 min develop the dipyridyl complex as in the spectrophotometric determination.

Determination:
When the decomposition is complete, add 5 ml 50% beryllium sulphate solution *carefully* (by safety pipette) to the decomposed mixture in the polystyrene vial, and stir till any insoluble fluorides dissolve completely. Then add 5 ml 2:2 dipyridyl solution and 10 ml 50% ammonium acetate buffer to bring the pH to ~ 5. Quantitatively transfer the solutions to 100 ml standard flasks and make up to the mark with distilled water. Measure the

absorptions of the red ferrous dipyridyl complex at 525 nm, and compare with the standard calibration curve. A standard rock (e.g. US Geological Survey: Flanagan, 1973) of similar composition—for comparison—and a reagent blank, should be carried through the full procedure.

7.5 Exchange capacity

Clay minerals can adsorb cations and anions and retain them in an exchangeable form. Exchangeable ions normally associated with clay minerals are: the cations Ca^{2+}, Mg^{2+}, H^+, K^+, NH_4^+, Na^+ and Al^{3+} and the anions SO_4^{2-}, Cl^-, PO_4^{3-} and NO_3^-. Exchange capacity can be defined as the total ionic charge of the adsorption complex active in the adsorption of ions. It is normally measured at pH 7 and reported as milli-equivalents per 100 g (meq/100 g).

The exchange reaction of minerals is of great importance in agriculture, plant-nutrient availability, soil structure, rock weathering processes, ceramic and concrete work, and in the oil industry, where bentonites are widely used in drilling.

The exchange capacity of minerals can be considered from two main viewpoints: cation-exchange capacity (CEC) and anion-exchange capacity (AEC).

7.5.1 *Cation-exchange capacity*

CEC can be defined as the sum of the exchangeable cations that a mineral can adsorb at a specific pH, i.e. a measurement of the negative charges carried by the mineral. These negative charges in clay minerals can be derived from:

(i) Isomorphous substitution within the structure
(ii) Broken bonds at edges and external surfaces
(iii) The dissociation of accessible hydroxyl groups.

Negative charges, due to isomorphous substitutions (i) within mineral structures, are known as permanent charge, and are independent of conditions such as pH and ion activity, whereas those from sources (ii) and (iii) are so dependent. For this reason, the type of CEC methods used should always be reported with the results. The high levels of CEC for clay minerals (Table 7.2) can, to some extent, be attributed to their large areas of external surface accessible to hydrated cations.

Kelley (1948) and Thomas (1977) have comprehensively reviewed the background of CEC measurements, Metson (1956) and Grim (1968) reviewed the methods for clay minerals, and Chapman (1965), and Rhoades (1982) those for soils.

Rhoades (1982) suggested four types of method:

(i) summation method: measurement, and summation of, displaced ex-

Table 7.2 Cation-exchange capacities of clay minerals (Grim, 1968).

Mineral	CEC (meq/100 g)
Kaolinite	3–15
Halloysite	5–50
Illite	10–40
Chlorite	10–40
Allophane	25–50
Montmorillonite	60–150
Vermiculite	100–150

changeable cations, i.e. exchangeable Na + Ca + Mg + K + exchangeable acidity

(ii) direct displacement of saturating index cation

(iii) displacement of saturating index cation after washing free from saturating salt

(iv) radioactive tracer measurement of index cations (^{45}Ca: Blume and Smith, 1954; ^{133}Ba: Bache, 1970).

For most minerals, the measurement of CEC is not dependent on the pH of saturation. Methods used for soils are mostly based on saturation, with ammonium as the index cation, but for some minerals low results are obtained (Chapman, 1965). For 1:1 minerals (kaolinite, halloysite) ammonium cannot completely replace aluminium and hydrogen, and for 2:1 minerals (vermiculite, micas, illite) the ammonium index ion becomes fixed and is not exchangeable. Sodium and barium have been suggested as more suitable saturating ions because they can be easily exchanged and determined. The use of divalent barium as a saturating cation is considered by many workers to be more effective than that of monovalent ions.

No general method exists that can be reliably used for all clay minerals, and the method chosen should therefore be appropriate to the type of mineral to be analysed. Details of three methods in current use are given:

(a) *Ammonium saturation (Chapman, 1965, modified).*

Reagents:
Ammonium acetate (1 M) pH 7—58 ml acetic acid (conc.) and 55 ml ammonium hydroxide diluted to 1 litre and the pH adjusted to 7 + 0.01.
Isopropyl alcohol—99%.
Sodium chloride—aqueous w/v solution acidified to be 0.005 M in HCl

Methods:
Weigh accurately 10–100 mg clay mineral into a 15 ml centrifuge tube. Add 10 ml molar ammonium acetate solution (pH 7), shake and leave overnight. Centrifuge and remove the clear supernatant liquid by suction. Add a further 10 ml ammonium acetate solution, and repeat the saturation process another four times. Wash free from excess ammonium salts by centrifuging with six 10 ml portions of isopropyl alcohol.

Shake the ammonium-saturated clay with 10 ml NaCl (acidified) solution, centrifuge and siphon off the supernatant liquid into a 100 ml standard flask. Repeat the process another six times and dilute to 100 ml with distilled water. The ammonium ion replaced can then be measured by micro-Kjeldahl distillation, spectrophotometrically by the isocyanurate method (Fraser and Russell, 1969), or by using a selective-ion electrode (Busenberg and Clemency, 1973).

(b) *Sodium saturation.*

Reagents:
Sodium acetate (1 M)—dissolve 136 g $NaOAc \cdot 3H_2O$ in 1 litre distilled water, and adjust the pH to 8.2 with HOAc or NaOH. Other reagents as in (a).

Method:
The saturation technique is the same as described above, using sodium acetate as the saturating salt and ammonium acetate to displace the index sodium ion. This can then be determined easily by flame photometry or atomic-emission spectroscopy. The results are normally reported as meq/100 g oven-dry clay.

(c) *Barium saturation (Gillman, 1979).*

Reagents:
Barium saturating solution (0.1 M)—dissolve 24.4 g $BaCl_2 \cdot 2H_2O$ in 1 litre distilled water.
Barium equilibrating solution (0.002 M)—dissolve 0.4889 g $BaCl_2 \cdot 2H_2O$ in 1 litre distilled water and adjust the pH to 7 with $Ba(OH)_2$ or HCl.
Magnesium displacing solution (0.005 M).

Method:
Place 10–100 mg (accurately weighed) clay mineral in a 15 ml (weighed) centrifuge tube, add 10 ml 0.1 M barium chloride and shake for 2 hours. Centrifuge, siphon off the supernatant liquid and equilibrate the clay with 3 × 10 ml portions of 0.002 M barium chloride, shaking for 1 hour between centrifugations and discarding the supernatants. Then weigh the tube plus wet clay to measure the amount of entrained barium chloride solution. Add 10 ml 0.005 M magnesium sulphate solution and shake the suspension gently for 1 hour. Adjust the conductivity of the suspension to the conductivity of the 0.0015 M magnesium-sulphate reference solution by adding 0.005 M magnesium sulphate or distilled water. Shake the suspension gently for a further 16 hours (overnight) and adjust the conductivity again if necessary. Then note the weight of the centrifuge tube plus the suspension, to estimate the volume of magnesium sulphate or distilled water added. Centrifuge the suspension and determine the magnesium and chloride contents of the supernatant liquid by atomic-absorption spectrometry (Mg) and spectrophotometry or ion chromatography (Cl).

Calculation:
If V_1 = volume of entrained $BaCl_2$ (ml), V_2 = volume of added $MgSO_4$ (ml), V_3 = final supernatant volume, and C_1 = Mg concentration of supernatant in meq/ml.
Then

$$(MgSO_4\ \text{added}) \qquad \text{CEC (meq/100 g)} = \frac{100\ (0.01\ V_2 - C_1V_3)}{\text{wt of sample (g)}}$$

$$(\text{if water added}) \qquad \text{CEC (meq/100 g)} = \frac{100(0.1 - C_1V_3)}{\text{wt of sample (g)}}$$

Measurement of the chloride content of the supernatant also allows the estimation of AEC. If C_2 = chloride concentration of 0.002 M $BaCl_2$ and C_3 = chloride concentration of supernatant in meq/ml then:

$$\text{AEC (meq/100 g)} = \frac{100\ (C_3V_3 - C_2V_1)}{\text{wt of sample (g)}}$$

For minerals with higher CEC values, lower sample weights and higher concentrations of magnesium sulphate solutions (e.g. 0.01 M) should be considered.

7.5.2 *Anion-exchange capacity*

Anion-exchange capacity can be defined as the sum of exchangeable anions that a clay mineral can adsorb, usually expressed as meq per 100 g clay. It is considered to be a measure of the positive charges carried by a clay mineral. Grim (1968) suggested three types of anion exchange in clay minerals:

(i) Replacement of hydroxyl groups—a surface reaction, e.g. adsorption of phosphate by kaolinite

(ii) Anions of similar geometric size to silica tetrahedra (e.g. phosphate, arsenate, borate) may be adsorbed at the edges of clay minerals—a surface reaction

(iii) Free exchangeable aluminium and iron may form insoluble salts with exchanging anions, making it uncertain whether the exchange is due to this effect or to the clay mineral.

In the replacement of surface hydroxyls some destruction of the lattice may occur, resulting in inaccurate measurements. This can, however, be minimized by using low concentrations of saturating salts. AEC has been shown to be proportional to the surface area and to vary with the degree of crystallinity of the clay mineral. The range of anion-exchange values obtained for the common clay minerals is illustrated in Table 7.3.

Parfitt (1978) has reviewed the mechanisms of anion exchange for a wide range of clay minerals. He concluded that:

(i) Minerals with surface hydroxyls associated with aluminium and iron would adsorb anions

(ii) The adsorption of anions is dependent on pH, with maximum adsorption occurring at lower pH

(iii) The order of adsorption of anions by minerals can generally be considered to be phosphate > arsenate > selenite > molybdate > sulphate = fluoride > chloride > nitrate.

(iv) The presence of other anions can cause competition for adsorption sites, and exchangeable cations such as calcium and aluminium can, in some cases, result in the formation of insoluble byproducts.

Investigations of the anion-exchange capacity of clay minerals have been concerned mainly with phosphate adsorption. The release of hydroxyl ions by fluoride ion has also been utilized by Weiss *et al.* (1956), and Wada and

Table 7.3 Anion-exchange capacities of clay minerals (Grim, 1968).

Mineral	AEC (meq/100 g)
Vermiculite	4
Illite	4–17
Chlorite	5–20
Kaolinite	7–20
Montmorillonite	20–30

Okamura (1977) used ammonium chloride as the saturating salt to measure the CEC and AEC for clay minerals and soils. They showed that there are differences between soils and clays in the dependence of CEC and AEC on the pH and ion concentrations, these being most marked in the presence of allophane and imogolite.

Methods for the determination of the anion-exchange capacity of minerals usually involve saturation with an index anion and measurement of the amount adsorbed. Details of two methods are given:

(a) *Phosphate: Mehlich (1948).*

Reagents:

Triethanolamine solution—dilute 22.5 ml triethanolamine to 250 ml with distilled water and adjust the pH to 8.1 with HCl. Mix with 500 ml of 5% $BaCl_2 \cdot 2H_2O$ (w/v) and dilute to 1 litre.

Calcium chloride solution—dissolve 50 g $CaCl_2 \cdot 2H_2O$ in 1 litre distilled water and adjust the pH to 8.0 with $Ca(OH)_2$ solution.

Phosphate saturating solution—0.001 M H_3PO_4.

Molybdate reagent (Murphy and Riley, 1962)—dissolve 12 g ammonium molybdate in 200 ml distilled water. Dissolve 0.2670 g antimony-potassium tartrate in 500 ml 5M H_2SO_4. Mix thoroughly, dilute to 1 litre and store in a dark bottle.

Working reagent—dissolve 1.056 g ascorbic acid in 100 ml ammonium molybdate reagent and mix well (this is stable for 24 hours).

Methods:

Accurately weigh 10–100 mg clay mineral into a 15 ml centrifuge tube. Add 10 ml triethanolamine solution and shake for 30 min. Centrifuge and wash five times with 95% ethanol, shake with 10 ml calcium chloride solution for 30 min, centrifuge and wash again five times with 25% ethanol. Add 10 ml phosphate saturating solution (0.001 M), shake for 30 min and allow to stand for 24 hours. Shake again, centrifuge and reserve the supernatant liquid for the determination of adsorbed phosphate (i.e. concentration in saturating solution – concentration in supernatant). Extractable phosphate is determined on a separate sample with 0.03 M ammonium fluoride in 0.1 M HCl, using a 1:20 mineral–solution ratio.

Phosphate determination:

Pipette a suitable aliquot into a 100 ml standard flask, dilute to 80 ml with distilled water, add 8 ml working reagent and make up to 100 ml. Shake well and measure the absorption at 880 nm after 10 min, comparing results with those obtained for a series of phosphate standards in the range 0–400 μg P.

Then AEC in meq/100 g clay = (extractable P + P adsorbed) meq/100 g.

(b) *Chloride: Wada and Okamura (1977)*

A 0.3–0.5 g sample (accurately weighed in a weighed 15 ml centrifuge tube) is first saturated with Cl^- index anion, using 1 M NH_4Cl (10 ml) at pH 6, and is shaken for 30 min. The suspension is centrifuged, the supernatant discarded, and the residue equilibrated by washing 5 × 10 ml portions 0.1 M NH_4Cl. The residual excess saturating Cl^- ion is estimated from the increase in weight (Cl_{res}). The Cl^- ion is then replaced by washing with (5 × 10 ml portions) 1 M KNO_3 and the Cl^- released (Cl_{rel}) is measured by selective-ion electrode potentiometry or by spectrophotometry.

AEC (meq/100 g) clay = (Cl_{rel}–Cl_{res}) in meq/100 g.

7.6 Determination of a structural formula

7.6.1 *Correction for impurities*

Before calculating the structural formula of a clay mineral, it is essential to ensure that the chemical analysis represents only the mineral of interest. Clay samples are rarely monomineralic, and several impurities are normally present, which may be crystalline, amorphous, or both (the latter is particularly common in soil clays). Sometimes authors simply note that certain impurities are present and make no allowance in the formula calculations, but this can only lead to a formula of dubious validity.

If all the impurities have been identified, corrections can be made to the analysis before calculating the structural formula of the studied mineral, provided that:

(i) the impurities are present only in relatively small amounts
(ii) the amounts of the impurities can be determined
(iii) the chemical composition of each impurity is known with some certainty.

If the impurities are present in substantial quantities, then the chemical composition and amount of each impurity must be known with a very high degree of accuracy, and this is not always possible. Large corrections are best avoided, as the resulting formula—and any interpretations based on it—are open to question. For example, Evans and Adams (1975) corrected for up to 40% of impurities (consisting of quartz, albite, orthoclase and rutile) in the analysis of mudstones, to obtain illite and chlorite structural formulae. These were queried by Bjørlykke and Jorgensen (1976), on the basis that feldspar contents had been underestimated.

7.6.1.1 *Identification of impurities.* Once the decision has been made that a structural formula is to be calculated for a clay mineral, it is necessary to detect and identify all the other phases present in the sample. All available techniques should be employed, as no single technique is suitable for detection and identification of the wide range of possible impurities. X-ray diffraction is clearly suitable for most crystalline components, but detection limits are not as low as IR for kaolinite and quartz for example, or as low as DTA for gibbsite and goethite. Mössbauer spectrometry is particularly useful for identifying and quantifying the amounts of any iron phases present.

7.6.1.2 *Removal of impurities.* Ideally, it is desirable to eliminate impurities from the sample before the chemical analysis is performed. This is almost always preferable to applying corrections to the analysis, and should be pursued if at all possible. Several methods can be used to purify the phase to be analysed. For smectite-rich samples, impurities can often be eliminated by size fractionation, as smectite is usually concentrated in the very finest fraction. However, the smectite in the finest fractions does not necessarily have a chemical composition identical to that in the coarser fractions—a beidellite described by Nadeau *et al.* (1985) was found to be more montmorillonitic in the $< 1.4\,\mu m$ fraction than in the $> 3\,\mu m$ fraction. Nevertheless, size fractionation is particularly useful in removing the non-phyllosilicates such as quartz, feldspars, carbonates, zeolites and amphiboles.

Another physical method of separating phases associated with clays is magnetic separation, which can be used either to remove iron-containing impurities, or to concentrate an iron-containing phyllosilicate such as chlorite. Magnetic separation rarely produces a single pure phase, but has been successfully applied to clay suspensions to reduce the level of impurities (Schultz and Dixon 1979, Russell *et al.*, 1984).

Chemical methods can also be used to remove impurities, but these must be monitored carefully, as the reagent used may attack the mineral of interest. Carbonate impurities can be eliminated by acid dissolution, but this is not to be recommended if trioctahedral clays are present, since they are very susceptible to acid attack. A greatly superior method of removing carbonate is to pass a stream of carbon dioxide through a suspension of the clay, thereby converting the carbonate to bicarbonate, which is soluble in water. This method is slow and probably unsuitable for routine use, but it ensures freedom from acid attack on clays. It was successfully used to remove calcite from a Jordanian clay, thereby allowing the structural formula to be calculated more accurately for the chromium-rich smectite-volkonskoite (Khoury *et al.*, 1984).

If X-ray-amorphous materials are present, chemical dissolution methods are necessary for their removal. The most common amorphous phases are iron oxides and hydroxides, which can be removed (along with crystalline forms such as goethite and hematite) by treatment with sodium dithionite or ammonium oxalate, and aluminosilicate phases which can be dissolved with sodium carbonate solutions. The use of these selective chemical treatments for dissolving or characterizing poorly ordered phases is discussed in detail in Chapter 8. The effects of these treatments on the crystalline clays should always be monitored by XRD, IR or EM if possible, as changes can be induced in the clays, e.g. Mössbauer spectrometry has revealed that deferration treatment can alter the Fe^{2+}/Fe^{3+} ratio in vermiculite and smectite (Ericsson *et al.*, 1984).

7.6.1.3 *Application of corrections for impurities.* Once the chemical composition has been obtained, any impurities that have not been removed prior to analysis must be corrected for before the analysis is used to calculate a structural formula. To do this, the chemical composition and the quantity of each impurity must be known. For many minerals, the composition is close to the theoretical composition, e.g. for quartz, cristobalite, carbonates, kaolinite, gibbsite, goethite, hematite, lepidocrocite, anatase and ilmenite. Difficulties arise, however, with minerals in which isomorphous substitution is common. Any feldspar present is unlikely to be a pure orthoclase, albite or anorthite end-member, and XRD or IR should be employed to identify the type of feldspar. Chlorite composition is also variable, but the relative intensities of the basal reflections on the XRD trace should give some indication of the chlorite species present. Smectites vary from highly aluminous types to iron-rich varieties, and an 060 measurement by XRD usually indicates whether the species is di- or trioctahedral. Illite is probably the most difficult mineral to correct for, as the K content varies widely. Use of the theoretical muscovite formula would allow for the minimum amount of illite present, but the K_2O content of illites can vary from 11.8% (as in muscovite) to as low as 4.8%.

The lower the quantity of any one impurity the less important it is to know its composition accurately. Provided that the amount of the impurity is less than 5% (for example), approximate compositions should not produce appreciable errors in the subsequent calculation.

Any Ti present in the analyses of clays is most likely to occur as anatase or possibly ilmenite. There have been some reports of Ti^{3+} occurring in the structure of kaolinite (Dolcater *et al.*, 1970; Rengasamy, 1976) and in octahedral sites in biotite and phlogopite (Evans and Raftery, 1980). In the absence of direct evidence of a particular Ti-mineral, the quantity of TiO_2 (normally < 2%) is simply subtracted. Titanium can occur in clays in an X-ray-amorphous form of anatase, in which case its presence is only revealed by chemical analysis (Bain, 1976).

Small amounts of P_2O_5, SO_3 and CO_2 are commonly deducted from the analysis, on the assumption that they are present in unidentified impurities.

An example of corrections applied to the analysis of a clay is presented in Table 7.4. The sample is a bentonite containing an interstratified illite-smectite, and the purpose of the analysis was to calculate a structural formula for this mineral. To obtain as accurate a formula as possible, the following steps were undertaken:

(i) A fine fraction (< 0.5 μm) was separated by centrifugation to concentrate the illite-smectite.

(ii) As calcite and gypsum were possible impurities, the sample was saturated with Sr^{2+}, so that any Ca detected in the analysis would indicate the presence of these phases.

(iii) An examination by Mössbauer spectroscopy indicated 1.63% goethite, and allowed the structural iron in the illite-smectite to be partitioned accurately between ferrous and ferric.

(iv) An examination by IR spectroscopy revealed the presence of kaolinite and quartz, estimated at 1.5% and 0.5% respectively.

The percentages of Fe_2O_3, SiO_2 and Al_2O_3 corresponding to the amounts of goethite, kaolinite and quartz determined, were subtracted from the chemical analysis (performed on the sample previously ignited at 1000°C). The TiO_2 percentage (0.36%) was assumed to be due to anatase, and was subtracted along with the small amounts of MnO, CaO, P_2O_5 and SO_3, which were presumed to be present in impurities at levels below the detection limit. The sum of the impurities was 4.3%, which is fairly small, but still produces a significant difference in the calculated structural formula (see section 7.6.2.1). The normalized analysis is also listed, so that the illite-smectite composition can be compared with others.

Table 7.4 Corrections for impurities in $< 0.5\,\mu m$ fraction of a bentonite, Sr^{2+}-saturated.

Analysis		Corrections						Corrected analysis	Normalized analysis
		Goethite (1.63%)	Kaolinite (1.5%)	Quartz (0.5%)	Anatase	Others	Sum		
SiO_2	53.23	—	0.81	0.5	—	—	1.31	51.92	53.93
TiO_2	0.36	—	—	—	0.36	—	0.36	—	—
Al_2O_3	31.87	—	0.69	—	—	—	0.69	31.18	32.39
Fe_2O_3	3.97	1.63	—	—	—	—	1.63	2.34	2.43
FeO	0.20	—	—	—	—	—	—	0.20	0.21
MnO	0.07	—	—	—	—	0.07	0.07	—	—
MgO	1.81	—	—	—	—	—	—	1.81	1.88
CaO	0.01	—	—	—	—	0.01	0.01	—	—
Na_2O	N.D.*	—	—	—	—	—	—	—	—
K_2O	5.92	—	—	—	—	—	—	5.92	6.15
P_2O_5	0.17	—	—	—	—	0.17	0.17	—	—
SO_3	0.06	—	—	—	—	0.06	0.06	—	—
SrO	2.90	—	—	—	—	—	—	2.90	3.01
Total	100.57						4.30	96.27	100.00

* N.D. = not determined.

7.6.2 *Calculation of formulae*

Ideally, the structural formula of a mineral should be determined by first calculating the unit-cell volume from the unit-cell dimensions, then calculating the mass of the unit cell after determination of the density of the mineral, and finally apportioning the elements from the chemical analysis, according to the percentage of each present (Brown *et al.*, 1978).

If the type of clay mineral has been established from X-ray diffraction, its structural formula is calculated by assuming the anion composition of the mineral, e.g. if the mineral is identified as a chlorite, its anion composition is $O_{20}(OH)_{16}$ with an anion framework charge of 56. On the other hand, if the mineral is a smectite, the calculation is based on a unit cell with $O_{20}(OH)_4$ and 44 negative charges.

The amount of water associated with clays varies according to humidity, and so it is best to express the analysis of a clay on an ignited basis, either by analysing the clay after ignition (at say 1000°C) or by recalculating the analysis to an ignited basis using the figure for weight loss at 1000°C.

The procedure for calculating the structural formula is as follows:

(i) Divide the weight percentage of each oxide by its molecular weight—this yields 'moles' oxide

(ii) Multiply 'moles' for each oxide by the number of positive charges per oxide to produce the number of gram equivalents of the cation

(iii) Divide the charge from the anion framework by the total of the gram equivalents of all the cations to produce a proportionality factor

(iv) Multiply the gram equivalent of each cation by the proportionality factor, and divide by the valency of the cation to produce the number of cations in the unit cell.

An example of this procedure is illustrated in Table 7.5. The calculation is performed as for a smectite, i.e. based on a unit cell with 44 negative charges, and the corrected analysis normalized to 100% is used. However, because of the introduction of the proportionality factor, the corrected analysis could be used without normalization, producing the same structural formula.

7.6.2.1 *Assignment of cations to structural sites.*

In order to obtain the structural formula, the numbers of cations have to be allocated to tetrahedral, octahedral and exchangeable positions in the silicate structure. Based on considerations of ionic radii, it is assumed that all cations having a coordination number of six are in octahedral sites, and those cations having coordination numbers greater than six are considered to be in sites between the aluminosilicate layers. The procedure is as follows:

(i) All Si^{4+} atoms are assigned to tetrahedral sites, and the remainder of the eight sites are filled with Al^{3+} atoms, which can have coordination numbers of four or six. For the illite-smectite in Table 7.5, there are 6.72 Si^{4+} atoms and (8–6.72), i.e. 1.28 Al^{3+} in the tetrahedral sheet.

Table 7.5 Calculation of structural formula of illite-smectite, from its chemical analysis, after corrections for impurities.

Oxide	Weight %		Molecular weight		Gram equiv. of cation		Proportionality factor		Gram equiv. cation per unit cell		Cation valence		Cations per unit cell
SiO_2	53.93	÷	60.06	× 4 =	3.5917	×	7.4867	=	26.8899	÷	4	=	6.72
Al_2O_3	32.39	÷	101.94	× 6 =	1.9064	×		=	14.2726	÷	3	=	4.76
Fe_2O_3	2.43	÷	159.70	× 6 =	0.0913	×		=	0.6835	÷	3	=	0.23
FeO	0.21	÷	71.85	× 2 =	0.0058	×		=	0.0434	÷	2	=	0.02
MgO	1.88	÷	40.32	× 2 =	0.0932	×		=	0.6978	÷	2	=	0.35
K_2O	6.15	÷	94.19	× 2 =	0.1306	×		=	0.9778	÷	1	=	0.98
SrO	3.01	÷	103.62	× 2 =	0.0581	×		=	0.4350	÷	2	=	0.22
					5.8771								

(ii) The remaining Al^{3+} atoms, and all Fe^{3+}, Fe^{2+}, and Mg^{2+} are assigned to the octahedral sites. Thus, for the illite-smectite, there are (4.76–1.28) = 3.48 Al^{3+} octahedral atoms, and the octahedral sheet is $(Al^{3+}_{3.48}\,Fe^{3+}_{0.23}\,Fe^{2+}_{0.02}\,Mg^{2+}_{0.35})$.

(iii) Any Ca, Na and K, and any other cations present, known to have coordination numbers greater than six are assigned to sites between the aluminosilicate layers. For the illite-smectite there are 0.98 K^+ atoms and 0.22 Sr^{2+} atoms.

It will be clear that unequivocal knowledge of the nature of the exchangeable cation obtained by saturating with a suitable cation is a distinct advantage at the cation assignment stage. In the natural state, Na, K, Mg, Ca, and even Al, can occur in exchangeable positions, and because Mg and Al usually occur in octahedral sites, this makes the assignment impossible. Saturating with a cation which is too large to fit in an octahedral site enables a confident assignment to be made. For the illite-smectite, Sr^{2+} was the saturating cation, which enabled the octahedral assignment to be completed, and also revealed that K^+ is present as an interlayer cation in twelve coordination sites. The final structural formula of the illite-smectite is:

$$K_{0.98}Sr_{0.22}(Si_{6.72}Al_{1.28})(Al_{3.48}Fe^{3+}_{0.23}Fe^{2+}_{0.02}Mg_{0.35})O_{20}(OH)_4$$

The total number of octahedral cations is 4.08, thereby indicating that the mineral is dioctahedral.

To enable the structural formula to be calculated, several assumptions are necessary. One assumption is that all the ferric iron is situated in the octahedral sheet, and this will usually be valid, although there have been reports (based mainly on Mössbauer studies) that iron can occur in interlayer positions (Cardile and Johnston, 1985) and Fe^{3+} can occur in tetrahedral sites (Sanz *et al.*, 1978).

If the formula had been calculated from the chemical analysis uncorrected for impurities, the resultant formula would have been:

$$K_{0.95}Sr_{0.21}(Si_{6.67}Al_{1.33})(Al_{3.38}Fe^{3+}_{0.37}Fe^{2+}_{0.02}Mg_{0.34})O_{20}(OH)_4$$

This is significantly different from the formula obtained from the corrected analysis, indicating different degrees of isomorphous substitution, and affecting the layer-charge calculations. This illustrates the importance of correction for impurities, even when their total is less than 5%.

7.6.2.2 *Calculation of layer charge.* Once the structural formula has been obtained, the charge distribution within the structure can easily be calculated. In an 'ideal' structure (such as pyrophyllite or talc) the tetrahedral sheet would have a positive charge of 32 (eight quadrivalent silicon cations) and the octahedral sheet a positive charge of 12 (either six divalent cations or four trivalent cations), the combined positive charge of 44 balancing the 44 negative charges on the anion framework. However, because of ionic substitutions this 'ideal' arrangement is rarely attained.

In the illite-smectite, the tetrahedral charge is $(6.72 \times 4) + (1.28 \times 3) = 30.72$, and so the tetrahedral sheet has a net negative charge of $(32-30.72) = 1.28$. The octahedral charge is $(3.48 \times 3) + (0.23 \times 3) + (0.02 \times 2) + (0.35 \times 2) = 11.87$ and the octahedral sheet has a net negative charge of $(12-11.87) = 0.13$. Consequently, the total charge deficit on the unit cell (tetrahedral + octahedral sheets) is $1.28 + 0.13 = 1.41$. This negative charge is balanced by the interlayer and exchangeable cations—in the illite-smectite this is $(0.98 \times 1) + (0.22 \times 2) = 1.42$. Considering the assumptions involved in the structural formula calculation, together with the unavoidable experimental errors in the analytical methods, a difference of 0.01 between layer charge and interlayer charge is not significant, and is less than is often encountered.

7.6.2.3 *Calculation of cation-exchange capacity.* As the exchange is known to be represented by the strontium content (3.01% SrO) in the illite-smectite, the cation-exchange capacity is:
$(1000 \times 3.01)/(103.62 \div 2) = 58$ meq Sr per 100 g ignited clay.

7.6.2.4 *Examples of calculated structural formulae.* There are numerous examples in the literature of structural formulae of clay minerals calculated from chemical analyses. A useful compilation of such analyses and formulae for a wide range of different clay groups is given in Weaver and Pollard (1973), and includes a discussion of the chemistry of each group. A selection of papers with analyses and formulae of particular types of clay is listed in Table 7.6;

Table 7.6 Examples in the literature of clay minerals with calculated structural formulae.

Mineral type	Reference
Saponite	Kohyama *et al.* (1973)
Saponite	Cowking *et al.* (1983)
Saponite	Ianovici *et al.* (1983)
Beidellite	Chen *et al.* (1976)
Ferruginous beidellite	Ozkan and Ross (1979)
Ferriferous beidellite	Wilson and Mitchell (1979)
Smectites	Curtin and Smillie (1981)
Iron-rich smectites	Brigatti (1983)
Volkonskoites	Mackenzie (1984)
Chlorites	Foster (1962)
Chlorite	Bain (1977)
Chlorite	Proust (1982)
Swelling chlorite	Macneill (1978)
Swelling chlorite	Bain and Russell (1981)
Swelling chlorite	Curtis *et al.* (1984)
Vermiculite	Ildefonse *et al.* (1979)
Vermiculite-smectite	Craw (1984)
Chlorite and illite	Bodine and Standaert (1977)

many of those included give details of impurities present in the sample and explain how the appropriate corrections were applied.

The importance of obtaining good, accurate structural formulae for a mineral before embarking on interpretations and speculations, cannot be overemphasized. This was highlighted by Schultz (1982), who critically examined several papers in which illite or highly illitic interstratified clay was deduced to have a much greater octahedral than tetrahedral charge. He doubted the existence of such phases, because of: 'inappropriate basic assumptions on calculation of structural formulas, failure to correct for impurities, and/or possible inaccuracies in chemical analyses'.

References

Allen, J. and Khan, A.A. (1970) Critical evaluation of powder sampling procedures. *Chem. Engineer*, May 1970, CE108–CE112.

Bache, B.W. (1970) Barium isotope method for measuring cation-exchange capacity of soils and clays. *J. Sci. Food & Agric.*, **22**, 169–171.

Bain, D.C. (1976) A titanium-rich clay. *J. Soil Sci.*, **27**, 68–70.

Bain, D.C. (1977) The weathering of ferruginous chlorite in a podzol from Argyllshire, Scotland. *Geoderma*, **17**, 193–208.

Bain, D.C. and Russell, J.D. (1981) Swelling minerals in a basalt and its weathering product from Morvern, Scotland: II. Swelling chlorite. *Clay Miner.*, **16**, 203–212.

Baker, D.E. and Suhr, N.H., (1982) Atomic absorption and flame emission spectrometry. *Agronomy*, **9**, 13–27.

Bernas, B. (1968) A new method for decomposition and comprehensive analysis of silicates by atomic spectrometry. *Anal. Chem.*, **40**, 1682–1686.

Bjørlykke, K. and Jorgensen, P. (1976) Comments on the paper: Chlorite and illite in some Lower Palaeozoic mudstones of mid-Wales. *Clay Miner.*, **11**, 165–167.

Blume, J.M. and Smith, E. (1954) Equilibration of exchangeable calcium and cation exchange capacity by equilibration with ^{45}Ca. *Soil Sci.*, **77**, 9–17.

Bodine, M.W. and Standaert, R.R. (1977) Chlorite and illite compositions from Upper Silurian rock salts, Retsof, New York. *Clays & Clay Miner.*, **25**, 57–71.

Brigatti, M.F. (1983) Relationships between composition and structure in Fe-rich smectites. *Clay Miner.*, **18**, 177–186.

Brown, G., Newman, A.C.D., Rayner, J.H. and Weir, A.H. (1978) 'The structure and chemistry of soil clay minerals,' in *The Chemistry of Soil Constituents*, (D.J. Greenland and M.H.B. Hayes eds.). Wiley, New York, 29–178.

Busenberg, E. and Clemency, C.V. (1973) Determination of cation exchange capacity of clays and soils using an ammonia electrode. *Clays & Clay Miner.*, **21**, 213–217.

Cardile, C.M. and Johnston, J.H. (1985) Structural studies of nontronites with different iron contents by ^{57}Fe Mössbauer spectroscopy. *Clays & Clay Miner.*, **33**, 295–300.

Chalmers, R.A. and Page, E.S. (1957) The reporting of chemical analyses of silicate rocks. *Geochim. Cosmochim. Acta*, **11**, 247–251.

Chapman, H.D. (1965) 'Cation exchange capacity,' in *Methods of Soil Analysis*. (*Agronomy*, **9**) (C.A. Black *et al.* eds.) American Society of Agronomy, Madison, Wisconsin, 891–901.

Chaussidon, J. (1982) 'The exchange complex: the general laws of anion and cation exchange,' in *Constituents and Properties of Soils*, (M. Bonneau and B. Souchier eds.). Academic Press, London, 377–398.

Chen, P.-Y., Wan, H.-M. and Brindley, G.W. (1976) Beidellite clay from Chang-Yuan, Taiwan; geology and mineralogy. *Clay Miner.*, **11**, 221–233.

Cowking, A., Wilson, M.J., Tait, J.M. and Robertson, R.H.S. (1983) Structure and swelling of fibrous and granular saponitic clay from Orrock quarry, Fife, Scotland. *Clay Miner.*, **18**, 49–64.

Craw, D. (1984) Ferrous-iron-bearing vermiculite-smectite series formed during alteration of chlorite to kaolinite, Otago schist, New Zealand. *Clay Miner.*, **19**, 509–520.

Curtin, D. and Smillie, G.W. (1981) Composition and origin of smectite in soils derived from basalt in Northern Ireland. *Clays and Clay Miner.*, **29**, 277–284.

Curtis, C.D., Ireland, B.J., Whiteman, J.A., Mulvaney, R. and Whittle, C.K. (1984) Authigenic chlorites: problems with chemical analysis and structural formulae calculation. *Clay Miner.*, **19**, 471–481.

De Soete, D.G.R., Gijbels, R. and Hoste, J. (1972) *Neutron Activation Analysis.* Wiley-Interscience, New York.

Dolcater, D.L., Syers, J.K. and Jackson, M.L. (1970) Titanium as free oxide and substituted forms in kaolinites and other soil minerals. *Clays & Clay Miner.*, **18**, 71–79.

Dolezal, J., Povondra, P. and Sulick, Z. (1968) *Decomposition Techniques in Inorganic Analysis.* Iliffe Books, London.

Ericsson, T., Linares, J., and Lotse, E. (1984) A Mössbauer study of the effect of dithionite/citrate/bicarbonate treatment on a vermiculite, a smectite and a soil. *Clay Miner.*, **19**, 85–91.

Evans, L.J. and Adams, W.A. (1975) Chlorite and illite in some Lower Palaeozoic mudstones of mid-Wales. *Clay Miner.*, **10**, 387–397.

Evans, S. and Raftery, E. (1980) X-ray photoelectron studies of titanium in biotite and phlogopite. *Clay Miner.*, **15**, 209–218.

Flanagan, F.J. (1973) 1972 values for international geochemical reference samples. *Geochim. Cosmochim. Acta*, **37**, 1189–1200.

Foster, M.D. (1962) Interpretation of the composition and a classification of the chlorites. *U.S. Geol. Surv. Prof. Pap.*, **414A**, 1–33.

Fraser, A.R. and Russell, J.D. (1969) A spectrophotometric method for determination of cation exchange capacity of clay minerals. *Clay Miner.*, **8**, 229–230.

Gillman, G.P. (1979) A proposed method for the measurement of exchange properties in highly weathered soils. *Australian J. Soil Res.*, **17**, 129–139.

Goodman, B.A. (1980) 'Mössbauer spectroscopy,' in *Advanced Chemical Methods for Soil and Clay Research*, (J.W. Stucki and W.L. Banwart eds.), D. Reidel Publishing Co., Dordrecht, Holland.

Grim, R.E. (1968) *Clay Mineralogy.* McGraw-Hill, New York, 185–233.

Groves, A.W. (1951) *Silicate Analysis.* Allen and Unwin, London.

Helmke, P.A. (1982) Neutron activation analysis. *Agronomy*, **9**, 67–84.

Hey, M.H. (1941) The determination of ferrous iron in resistant silicates. *Miner. Mag.*, **26**, 116–118.

Hillebrand, W.F. (1919) The analysis of silicate and carbonate rocks. *Bull. U.S. Geol. Surv.*, 700.

Hillebrand, W.F., Lundell, G.E.F., Bright, H.A. and Hoffman, J.I. (1953) *Applied Inorganic Analysis.* John Wiley, New York.

Ianovici, V., Neacsu, G. and Neacsu, V. (1983) Saponite from Moldova Noua, Romania. *Clay Miner.*, **18**, 205–207.

Ildefonse, P., Capin, E. and Velde, B. (1979) A soil vermiculite formed from a meta-gabbro, Loire-Atlantique, France. *Clay Miner.*, **14**, 201–210.

Ingamells, C.O. (1966) Absorptiometric methods in rapid silicate analysis. *Anal. Chem.*, **38**, 1228–1234.

Jackson, M.L. (1962) *Soil Chemical Analysis*, Taylor, Garnett, Evans and Co., Watford, 33.

Jeffery, P.G. (1981) *Chemical Methods of Rock Analysis.* Pergamon Press, Oxford.

Johnson, W.M. and Maxwell, J.A. (1981) *Rock and Mineral Analysis.* John Wiley, New York.

Jones, A.A. (1982) X-ray fluorescence spectrometry. *Agronomy*, **9**, 85–121.

Kelley, W.P. (1948) *Cation Exchange in Soils.* Rheinhold Publishing Corporation, New York.

Khoury, H.N., Mackenzie, R.C., Russell, J.D. and Tait, J.M. (1984) An iron-free volkonskoite. *Clay Miner.*, **19**, 43–57.

Kohyama, N., Shimoda, S. and Sudo, T. (1973) Iron-rich saponite (ferrous and ferric forms). *Clays & Clay Miner.*, **21**, 229–237.

Kolthoff, I.M. and Sandell, E.B. (1956) *Text book of Quantitative Inorganic Analysis.* Macmillan, New York.

Langmyhr, F.G. and Graff, P.R. (1965) A contribution to analytical chemistry of silicate rocks; a scheme of analysis for eleven main constituents based on decomposition by hydrofluoric acid. *Norges Geologiske Undersolkelse*, **230**, 1–128.

Lim, C.H. and Jackson, M.L. (1982) Dissolution for total elemental analysis. *Agronomy*, **9**, 1–11.

Lundell, G.E.F. and Hoffman, J.I. (1938) *Outlines of Methods of Chemical Analysis.* Wiley, New York.

Mackenzie, R.C. (1984) Constitution of and relationships among volkonskoites. *Clay Miner.*, **19**, 669–671.
Macneill, S. (1978) A chemical investigation of a chlorite intergrade mineral in the Keuper Marl. *Clay Miner.*, **13**, 357–365.
Mehlich, A. (1948) Determination of cation- and anion-exchange properties of soils. *Soil Sci.*, **66**, 429–445.
Metson, A.J. (1956) Methods of chemical analysis for soil survey samples. *New Zealand DSIR*, **12**, 83–90.
Murphy, J. and Riley, J.P. (1962) A modified single solution method for the determination of phosphate in natural waters. *Anal. Chim. Acta*, **27**, 31–36.
Nadeau, P.H. and Bain, D.C. (1986) Composition of some smectites and diagenetic illitic clays and implications for their origin. *Clays & Clay Miner.*, **34**, 455–464.
Nadeau, P.H., Farmer, V.C., McHardy, W.J. and Bain, D.C. (1985) Compositional variations of the Unterrupsroth beidellite. *Am. Mineral.*, **70**, 1004–1010.
Nicholls, G.D. (1960) Techniques in sedimentary geochemistry. Determination of ferrous iron contents of carbonaceous shales. *J. Sedim. Petrol.*, **30**, 603–612.
Norrish, K. and Chappell, B.W. (1977) 'X-ray fluorescence spectrometry,' in *Physical Methods in Determinative Mineralogy*, (J. Zussman ed.), Academic Press, London, 201–272.
Özkan, A.I. and Ross, G.J. (1979) Ferruginous beidellites in Turkish soils. *Proc. Soil Sci. Soc. Am.*, **43**, 1242–1248.
Parfitt, R.L. (1978) Anion adsorption by soils and soil material. *Adv. Agronomy*, **30**, 1–50.
Proust, D. (1982) Supergene alteration of metamorphic chlorite in an amphibolite from the Massif Central, France. *Clay Miner.*, **17**, 159–173.
Reay, A. (1981) The effect of disc mill grinding on some rock-forming minerals. *Miner. Mag.*, **44**, 179–182.
Rengasamy, P. (1976) Substitution of iron and titanium in kaolinites. *Clays & Clay Miner.*, **24**, 265–266.
Rhoades, J.D. (1982) 'Cation exchange capacity,' in *Methods of Soil Analysis*, (*Agronomy*, **9**), (A.L. Page *et al.* eds), American Society of Agronomy, Madison, Wisconsin, 149–157.
Russell, J.D., Birnie, A. and Fraser, A.R. (1984) High-gradient magnetic separation (HGMS) in soil clay mineral studies. *Clay Miner.*, **19**, 771–778.
Sanz, J., Meyers, J., Vielvoye, L. and Stone, W.E.E. (1978) The location and content of iron in natural biotites and phlogopites: a comparison of several methods. *Clay Miner.*, **13**, 45–52.
Schultz, D.G. and Dixon, J.B. (1979) High-gradient magnetic separation of iron oxides and other magnetic minerals from soil clays. *Proc. Soil Sci. Soc. Am.*, **43**, 793–799.
Schultz, L.G. (1982) Chemical characteristics and origin of Ordovician K-bentonites along the Cincinnati Arch: a discussion. *Clays & Clay Miner.*, **30**, 445–446.
Shapiro, L. (1960) A spectrophotometric method for determination of FeO in rocks. *Geological Survey Research*, B496–B497.
Shapiro, L. (1967) Rapid analysis of rocks and minerals by a single-solution method. *U.S. Geol. Surv. Prof. Pap.*, **575B**, B187–191.
Shapiro, L. and Brannock, W.W. (1952) Rapid analysis of silicate rocks. *Bull. U.S. Geol. Surv.*, **165**.
Shapiro, L. and Brannock, W.W. (1956) Rapid analysis of silicate rocks. *Bull. U.S. Geol. Surv.*, **165**.
Stucki, J.W. (1981) The quantitative assay of minerals for Fe^{2+} and Fe^{3+} using, 1, 10-phenanthroline. II a photochemical method. *Soil Sci. Soc. Am. J.*, **45**, 638–641.
Thomas, G.W. (1977) Historical developments in soils chemistry: ion exchange. *Soil Sci. Soc. Am. J.*, **41**, 230–238.
Thomson, M. and Walsh, J.N. (1983) *Handbook of Inductively Coupled Plasma Spectrometry*. Blackie, Glasgow and London.
Ure, A.M. (1983) 'Atomic absorption and flame emission spectrometry,' in *Soil Analysis*, (K.A. Smith ed.) Marcel Dekker, New York.
Van Loon. J.C. (1965) Titrimetric determination of the iron (II) oxide content of silicates using potassium iodate. *Talanta*, **12**, 599–603.
Verbeek, A.A., Mitchell, M.C. and Ure, A.M. (1982) The analysis of small samples of rock and soil by atomic absorption and emission spectrometry after a lithium metaborate fusion/nitric acid dissolution procedure. *Anal. Chim. Acta*, **135**, 215–228.
Volborth, A. (1964) Biotite mica effect in X-ray spectrographic analysis of pressed rock powders. *Am. Mineral.*, **49**, 634–643.

Wada. K. and Okamura, Y. (1977) in *Proc., Intern. Seminar on Soil Environ. & Fertility Management in Intensive Agric., Tokyo, Japan,* 811–815.

Weaver, C.E. and Pollard, L.D. (1973) *The Chemistry of Clay Minerals.* Elsevier, Amsterdam.

Weiss, A., Mehler, A., Koch, G. and Hoffman, U. (1956) Anion exchange capacity of clay minerals. *Zeit. Allemange Chem.,* **284**, 247–271.

Wilson, A.D. (1960) The micro-determination of ferrous iron in silicate minerals by a volumetric and a colorimetric method. *Analyst,* **85**, 823–827.

Wilson, M.J. and Mitchell, B.D. (1979) Comparative study of a vertisol and an entisol from the Blue Nile plains of Sudan. *Egypt. J. Soil Sci.,* **19**, 207–220.

8 Characterization of poorly ordered minerals by selective chemical methods

B.F.L. SMITH AND B.D. MITCHELL

8.1 Introduction

At the beginning of this century, ideas on the composition of clay-size material were greatly influenced by Van Bemmelen (1910) who suggested that this material existed as a completely amorphous colloidal complex. However, many mineralogists thought that clays could be a mixture, containing varying proportions of crystalline material (Mackenzie, 1963). The mixed-oxide hypothesis of Van Bemmelen (1910), associating the colloidal state with noncrystallinity persisted until Hendricks and Fry (1930), using X-ray methods showed the presence of large amounts of crystalline material in soil clays. This resulted in the virtual rejection of ideas advocating an amorphous nature for clays. Also, researches were directed almost exclusively towards elucidation of the physicochemical reactions within the crystal structure of clay minerals. Despite this emphasis on crystallinity, Mattson, in a series of papers on the colloidal chemistry of soil, continued to recognize the importance of inorganic, poorly ordered components in soils—a balanced view which has now become generally accepted (Rich and Thomas, 1960; Mitchell *et al.*, 1964). Soil clay can, therefore, best be regarded as consisting of both crystalline and noncrystalline minerals. Noncrystalline material is referred to by a number of names, such as X-ray-amorphous oxides and hydrous oxides, allophane, poorly ordered inorganic material and inorganic gel material. In 1972, Cradwick *et al.*, showed that imogolite, a poorly ordered, gel-like, hydrous aluminium silicate, first identified by Yoshinaga and Aomine (1962), actually had a well-defined crystalline structure which they characterized by electron diffraction and trimethylsilylation. This study and others like it have led to a reappraisal of the 'amorphous' material in soils.

8.2 Principles

A wide range of methods exist for the identification of well-defined crystalline clay minerals in soils, as described in previous chapters. However, until recently, definitive information on the nature of poorly ordered components in soils has been rather sparse. This very reactive material, often occurring in small amounts, is difficult to characterize. For some time, it was thought to be mainly confined to soils in regions of recent volcanic activity, (Fieldes, 1955; Yoshinaga, 1966) but now it has been shown to be an important constituent of temperate soils (Mitchell and Farmer, 1962, McKeague and Day, 1966, and Tait *et al.*, 1978). In soils, it often occurs in association with the surface of primary soil particles (Follett *et al.*, 1965; Jones and Uehara, 1973; and Farmer *et al.*, 1983) and can, therefore, have an influence incommensurate with the amount present (Kirkman *et al.*, 1966).

Poorly ordered material in soils consists mainly of oxides and/or hydroxides of silicon, aluminium and iron (Guillet and Souchier, 1982), either discrete or in various combinations. The occurrence of noncrystalline forms of silica, usually termed opaline silica, is frequently associated with immature soils of volcanic origin, although instances have been reported in Spanish soils, on granite, (Hoyos and Pino, 1958) and in organic surface horizons of Scottish soils (Mitchell and Farmer, 1962). This type of material is derived by weathering of primary silicate minerals, or by recycling vegetation where it occurs in humus-rich horizons in the form of phytoliths.

Poorly ordered oxides and hydroxides of aluminium in soils are not as widespread as those of silicon, being noted to a small extent in the early weathering stage of New Zealand soils (Yoshinaga *et al.*, 1973) and in B horizons of Scottish soils (Tait *et al.*, 1978). Noncrystalline iron oxides are generally stabilized by organic and silicon complexes, the rate of crystallization to goethite and hematite being dependent on environmental factors (Schwertmann, 1978). Combinations of the three components are found in many soils. Amorphous aluminosilicates include gel-like compounds, containing mainly aluminium, silicon, oxygen and water, but usually also some iron, magnesium, hydroxyl, and occasionally phosphate. Allophane is a general term for amorphous aluminosilicate gels, where the Al_2O_3/SiO_2 mol ratio falls in the range 0.5–2.0 (Ross and Kerr, 1934). Farmer *et al.* (1980) have shown that the composition of allophanic material in soils varies according to the pH of formation. Earlier, Yoshinaga and Aomine (1962) separated two mineral colloids from a soil clay fraction. One colloid remained dispersed in both acid and alkali, while the other flocculated in alkali. The former was amorphous to X-rays, and was considered to be allophane, and the latter, which possessed a low degree of crystallinity, was termed imogolite.

The conventional methods described in previous chapters can also provide some indication of noncrystalline inorganic material in soils. It is not always clear, however, whether there is a sharp division between the highly ordered crystalline core and the disordered surface material, or if there is a transition zone of increasing order inwards, as envisaged by Mattson in the 1930s. The only methods of obtaining even a semiquantitative estimation of highly disordered (X-ray amorphous) inorganic material in soils are those based on selective chemical dissolution.

8.3 Selective chemical methods

Poorly ordered material in soils is normally characterized by ascertaining its relative resistance to dissolution in various chemical reagents. Aluminosilicate gels form soluble silicates and aluminates in alkaline solutions. Both sodium hydroxide and sodium carbonate have been used for many years to dissolve siliceous and aluminous cements, which frequently bind clay particles into aggregates. Treatment with these reagents not only improves clay dispersion but also 'cleans up' the crystalline components, thereby assisting in their identification by instrumental techniques, particularly XRD. These selective chemical methods depend on high specific surface area of poorly ordered material in soils, which results in a higher dissolution rate than for crystalline minerals. A high ratio of dissolution reagent to sample weight ensures much better dissolution, and avoids saturation of the solution with silicon or aluminium.

A range of specified selective chemical methods will be given for estimating amounts of poorly ordered material in soils. Also, detailed analytical information will be provided on the methods in current use. Quantitative measurement of extracted aluminium, iron and silicon can be ascertained by either spectrophotometric or atomic-absorption methods. The latter, however, are not sensitive enough for the low amounts of silicon frequently obtained when extracting poorly ordered material from soils. Details of atomic-absorption methods are not given here, but these can be successfully used for most of the procedures, provided that the solutions are diluted to reduce interference from very high levels of sodium in the extractants.

8.3.1 *Alkali dissolution*

Boiling with 2% Na_2CO_3 has been employed (Jackson *et al.*, 1950) to remove the finely divided siliceous material that cements clay particles. Hashimoto and Jackson (1960) showed that digestion with Na_2CO_3 solution may not completely remove poorly ordered aluminosilicates, perhaps because of the slow rate of reaction and saturation of the extracting solution. The effect of hot and cold 0.5 M Na_2CO_3 solutions on pure clay minerals, accessory minerals, gels and soil clays (Follett *et al.*, 1965) showed:

(i) little reaction with the crystalline minerals,
(ii) appreciable solubility of silica gel and allophane,
(iii) increased dispersion.

Considerable amounts of SiO_2 and Al_2O_3 were extracted from the soil clays examined. The effects of hot and cold 0.5 M Na_2CO_3 treatment were compared with that of 0.5 M NaOH (Hashimoto and Jackson, 1960) and it was concluded that the amounts of Si and Al extracted vary with the reagent in the sequence 0.5 M NaOH > hot 0.5 M Na_2CO_3 > cold 0.5 M Na_2CO_3. This result is in agreement with the calculated reactivity of the reagents.

Treatment with hot 0.5 M Na_2CO_3 not only removes 'less readily soluble' aluminosilica gel material, but also some crystalline gibbsite (Jorgensen *et al.*, 1970) and can result in the formation of zeolite-like aluminosilicate products (Farmer *et al.*, 1977). Hashimoto and Jackson (1960) recommended 2.5 min of boiling with 0.5 M NaOH for the removal of allophane, free silicon and aluminium from soils. The reaction time was curtailed because prolonged treatment has been shown to attack crystalline components. Hashimoto and Jackson also pointed out that dehydroxylated kaolinite and halloysite were dissolved by this technique, whereas chlorite and montmorillonite, heated to 500°C, were virtually unaffected. A NaOH solution buffered at pH 9.7 has been used by Bruckert and Souchier (1975) to extract the more mobile organic complexes of iron and aluminium.

The two commonly used methods of alkali dissolution, namely:

(i) a 2.5 min boil with 0.5 M NaOH (Hashimoto and Jackson, 1960) and,

(ii) cold extraction with 0.5 M Na_2CO_3 (Follett *et al.*, 1965) will now be described:

(a) *2.5 min boil with 0.5 M NaOH (Hashimoto and Jackson, 1960).*

Reagents:
Sodium hydroxide (0.5 M)—dissolve 20 g NaOH (AR) pellets in 1 litre distilled water and store in a polyethylene bottle.

Dissolution procedure:
Place 50 mg sample (< 200 mesh) in a nickel crucible, add 50 ml 0.5 M NaOH and boil the suspension for 2.5 min. After cooling, the latter is quantitatively transferred to a 100 ml polypropylene centrifuge tube and centrifuged at 2200 g for 10 min. The supernatant solution is then siphoned off into a 200 ml standard flask, the residue is washed with distilled water, recentrifuged, and the wash water added to the solution in the 200 ml flask, diluted to the mark with distilled water and retained for the determination of aluminium and silicon.

(b) *Aluminium determination—Aluminon (Robertson, 1950).*

Reagents:
Ammonium acetate buffer, pH 4.5—dilute 230 ml glacial acetic acid (AR) and 216 ml ammonium hydroxide (AR) to 1 litre with distilled water
Dissolve 10 g Acacia in 100 ml distilled water.
Dissolve 107 g ammonium chloride AR in 150 ml conc. hydrochloric acid.
Mix the above solutions, dilute to 2 litres and adjust the pH to 4.5 ± 0.01
Aluminon (0.1%)—dissolve 1 g Aurine tricarboxylic acid in 150 ml distilled water to which 5 drops of ammonia have been added. Boil the solution to expel excess ammonia, cool and dilute to 1 litre with distilled water.
Aluminium standard solution (stock)—dissolve 1.7584 g potassium-aluminium sulphate in distilled water containing 50 ml 5 M hydrochloric acid, and dilute to 1 litre with distilled water.
Aluminium working standard—dilute 50 ml of stock Al standard to 1 litre with distilled water, giving a solution of concentration 10 μg/ml Al.
Thioglycollic acid.

Procedure:
Pipette a suitable aliquot (0–30 μg Al) into a 100 ml conical flask, adjust the pH to within the range 4–5, then dilute with distilled water to 50 ml total volume.

Add 10 ml M NH_4OAc buffer, then 1 drop of thioglycollic acid, to reduce any ferric iron present to the ferrous form—which does not form a complex with Aluminon. Then add 2 ml Aluminon reagent and place in a boiling water-bath for exactly 30 min. Cool and dilute to 100 ml with distilled water and measure the absorption on a spectrophotometer at 525 nm using distilled water in the reference cell. The absorptions are then compared with a standard aluminium curve prepared in the same manner for each set of determinations, in the range 0–30 μg Al. A blank determination should always be carried out.

(c) *Silicon determination—molybdate reaction (Webber and Wilson, 1965).*

Ammonium molybdate at ~pH 1.2 reacts with silica and any phosphate present to produce heteropoly acids. Tartaric acid is added to destroy the phosphomolybdic complex, thus allowing silicon to be determined as the silicomolybdic complex in the presence of phosphate. The sensitivity of this method can be greatly increased by measuring this complex as the reduced molybdenum-blue form, which can be achieved by the addition of an amino acid.

Reagents:
Ammonium molybdate solution:
Solution A: dissolve 89 g ammonium molybdate $(NH_4)_6Mo_7O_{24}\cdot4H_2O$ (AR) in ~800 ml distilled water.
Solution B: add 63 ml sulphuric acid (d_{20} 1.84) (AR) cautiously with stirring, to ~100 ml distilled water and cool to room temperature.

Add solution B to A, cool to room temperature and dilute to 1 litre with distilled water and store in a polyethylene bottle. This solution is stable for at least six months.
Tartaric acid—dissolve 280 g tartaric acid (AR) in 1 litre distilled water and store in a polyethylene bottle. This solution is stable for at least six months.
Reducing solution—dissolve 2.8 g of sodium sulphite $(Na_2SO_3\cdot7H_2O)$ and 0.6 g 1-amino-2-naphthol-4-sulphonic acid in ~100 ml distilled water. Add 30 g of sodium metabisulphite $(Na_2S_2O_5)$, dilute to 400 ml with distilled water and store in a polyethylene bottle. This solution is best kept in the dark and should be prepared fresh every month.
Silicon stock solution—weigh 1 g pure, dried silica into a platinum crucible, add 5 g anhydrous sodium carbonate (AR) and heat slowly over a Meker burner, gradually increasing the temperature until red heat, and then continue heating for a further 10 min, until the melt becomes clear. Allow the melt to cool before dissolving it in distilled water and diluting to 1 litre. The solution, stored in a polyethylene bottle, is stable for at least two years. Silica concentration ≡ 1000 μg/ml SiO_2.
Silicon working standard—dilute 10 ml of the stock solution to 1 litre with distilled water. Store in a polyethylene bottle. The solution is stable for at least 3 months. Silica concentration ≡ 10 μg/ml SiO_2.

Silicon flasks—for accurate and reproducible silicon determinations, a set of 100 ml standard flasks should be specially reserved. These flasks should initially be cleaned carefully with chromic acid (*hazardous oxidizing reagent*) by soaking overnight, then washed with 1:1 hydrochloric acid and washed with distilled water, and should be kept full of distilled water after each determination to prevent the formation of siliceous coatings on the flasks.

Procedure:
Pipette a suitable aliquot (5–80 μg/SiO_2) into the 'silicon' flasks, adjust the pH to 3–4, and dilute to ~80 ml with distilled water. Add 2.5 ml ammonium molybdate reagent and mix immediately. After 10 min add 5 ml tartaric acid, and after a further 5 min add 2.5 ml reducing solution. Dilute to 100 ml with distilled water. Stand the solution for one hour and measure the absorption at 820 nm in a 4 cm cell using distilled water as a reference. A wavelength of 650 nm may be used, with a 40% reduction in sensitivity over that obtained at 820 nm.

The results obtained are compared with a standard graph, prepared for a range of silicon solutions (0–80 μg SiO_2). A blank determination should be carried out on the reagents.

(d) *0.5 M sodium-carbonate extraction (cold)—Follett* et al. *(1965).*

Reagents:
Sodium carbonate (0.5 M)—dissolve 106 g Na_2CO_3 anhydrous (AR) in 2 litres of distilled water and store in a polyethylene bottle.

Dissolution procedure:
Weigh 100 mg sample (< 200 mesh) into a 100 ml polypropylene centrifuge tube, add 80 ml 0.5 M Na_2CO_3 solution, stopper well and place in an end-over-end shaker overnight (16 hours). Centrifuge for 10 min at 2200 *g*, and siphon off the supernatant solution into a 200 ml standard flask. Make up to the mark with distilled water and reserve the solution for the determination of aluminium and silicon. For soils, a 100 ml flask can be used, as the amounts of aluminium and silicon extracted are much smaller than from clay samples.

Aluminium and silicon contents are then determined as in the procedures already discussed.

8.3.2 *Sodium dithionite*

Free iron oxides and crystalline iron oxides are frequently removed from soil to improve mineral identification and quantification. Most methods employed to remove these oxides involve the reduction and mobilization of iron. Deb (1950) was among the first to use sodium dithionite to measure the hydrous iron content of soils. This method was modified and improved by Mitchell and Mackenzie (1954) to measure 'free' iron, by Mehra and Jackson (1960)—to determine the release of iron and aluminium, and more recently by Mitchell *et al.* (1971) to determine three major components of inorganic gel material in soils and soil clays, namely, silicon, aluminium and ferric iron.

Mehra and Jackson buffered the dithionite extraction of soils at pH 7 with molar sodium bicarbonate, and used sodium citrate to chelate the iron and aluminium released. However, the presence of citrate and dithionite in the extract causes difficulty in the determination of its silicon and aluminium content. Weaver *et al.* (1968) improved the determination of silicon by using a much higher concentration of ammonium molybdate reagent. Mitchell *et al.*, (1971) modified the above procedure to ensure a more reliable determination of silica and alumina in the dithionite extract of soils. They did not use sodium citrate, but instead followed the buffered dithionite treatment with a dilute acid wash to dissolve the precipitated iron sulphide. Then, aliquots from the extracts were treated with alkaline hypobromite, to destroy the dithionite and allow easy and rapid spectrophotometric determination of silicon and aluminium.

(a) *Dithionite–citrate–bicarbonate method—Mehra and Jackson (1960).*

Reagents:
Sodium dithionite ($Na_2S_2O_4$)
Sodium citrate (0.3 M)—88 g $Na_3C_6H_5O_7 \cdot 2H_2O$ (AR) 1 litre
Sodium bicarbonate (1 M)—42 g $NaHCO_3$ (AR) 1 litre
Sodium chloride (AR) saturated solution
Acetone (AR)

Dissolution procedure:
Place a suitable weight of sample (100 mg clay or 400 mg soil) in a 100 ml polypropylene centrifuge tube. Add 40 ml 0.3 M sodium citrate and 5 ml 1 M sodium bicarbonate and heat the mixture on a water-bath at 80°C for 30 min.

0.5 g of sodium dithionite is then added to the hot suspension, stirring frequently (or agitating by means of an ultrasonic probe) for 10 min. The mixture is then allowed to cool, before adding 10 ml saturated sodium chloride and 10 ml acetone (to encourage flocculation). After cooling to room temperature, the mixture is centrifuged at 2200 *g* for 10 min, and then the supernatant is siphoned off into a 200 ml volumetric flask, the residue washed with 50 ml distilled water, recentrifuged and the wash liquid added to the previous supernatant. The solution is then made up to the mark with distilled water and reserved for iron, aluminium and silicon determinations.

(b) *Iron determination—1, 10-phenanthroline (de Endredy, 1963).*

Reagents:
1, 10-phenanthroline (0.25%) solution in water
Sodium citrate (AR) (20%) solution in water
Ascorbic acid (L)
Hydrochloric acid (2 M)—dilute 172 ml conc. HCl (AR) to 1 litre with distilled water
Iron standard (stock)—dissolve 0.7022 g ferrous ammonium sulphate in 100 ml distilled water, containing 5 ml of 6 M sulphuric acid, add a dilute solution of potassium permanganate dropwise until the mixture is just pink in colour, then dilute to 1 litre with distilled water (1 ml = 100 μg Fe)
Iron standard (working)—dilute 100 ml of iron stock solution to 1 litre (1 ml = 10 μg Fe).

Procedure:
Pipette a suitable aliquot (0–60 μg Fe) into a 50 ml volumetric flask, and add 2 ml 2 M hydrochloric acid. Dilute the solution to about 25 ml with distilled water, add 100 $\pm$ 10 mg of ascorbic acid, mix thoroughly and allow the mixture to stand for 10 min. Then add 2 ml 0.25% 1, 10-phenanthroline solution and 5 ml 20% sodium citrate buffer, and after an hour measure the colour of the ferrous iron–phenanthroline complex at 525 nm. Compare this with a standard iron calibration curve prepared in a similar manner in the range (0–80 μg Fe) for the 4 cm cell, and (50–400 μg Fe) when a 1 cm cell is used. N.B. A reagent-blank determination should be carried through the complete extraction process.

(c) *Aluminium and silicon determinations.*

If aluminium and silicon must be measured in the sodium dithionite extracts, then pretreatment must be carried out to overcome possible interference from citrate and dithionite. Weaver *et al.* (1968) proposed a method for the determination of silica in citrate–bicarbonate–dithionite extracts of soils by destroying the dithionite, and oxidizing the extract by bubbling air through it. They also eliminated the interference of citrate by utilizing an excess of molybdate in the determination of silica. Mitchell *et al.* (1971) employed a different approach—citrate was not utilized in the extracting solution, and the dithionite was destroyed by alkaline hypobromite oxidation; any excess bromine was removed by acidification and heat treatment. An added bonus to this approach is that any organic colour developed in the extract is also destroyed, thus eliminating a possible source of interference in spectrophotometric determinations of aluminium (Mitchell and Smith, 1974).
Aluminium and silicon can then be determined, after pH adjustment of the treated solutions, by the procedures detailed in this section.

8.3.3 *Ammonium oxalate—Smith (1984)*

Tamm (1922) was the first to use oxalate at pH 3.3 to extract amorphous oxides and hydrous oxides from soils, showing that it has only a minimal effect on crystalline oxides and clay minerals. The method depends essentially on the pH and complexing ability of oxalate to extract colloid complexes of aluminium, iron, silicon and manganese. Schwertmann (1964) reassessed this reagent, finding that when used in the dark it dissolved only amorphous iron oxides. Oxalate-extractable iron and aluminium are used as a criterion in the classification of soils (McKeague and Day, 1966) and as a means of monitoring the mobility of iron and aluminium within the soil profiles (Schwertmann, 1973). In recent years, following the widespread recognition of allophanic materials in soils, (Childs *et al.*, 1983) it has also become necessary to determine silicon in oxalate extracts. When pure allophane (Al/Si:2/1) is treated with oxalate, a value of 13.3% is obtained for extractable silicon (Si_{ox}). Estimation of allophane using oxalate and pyrophosphate extractants depends on their similar chemical behaviour. An approximate estimation of allophane can be obtained (Parfitt and Henmi, 1982)—when its presence is indicated by the ratio of inorganic-extractable aluminium ($Al_{ox} - Al_{py}$) to oxalate-extractable silicon (Si_{ox}) falling within the range 1.5–2.6—by multiplying the value Si_{ox} by 7.5, i.e.

$$\% \text{ allophane} = \frac{100 \times \% \, Si_{ox}}{13.3}$$

$$= 7.5 \times \% \, Si_{ox}$$

where Al_{py} = pyrophosphate-extractable Al, Al_{ox} = oxalate-extractable Al, and Si_{ox} = oxalate-extractable Si.

(a) *Ammonium oxalate extraction*

Reagents:
Ammonium oxalate (0.2 M) pH 3—mix 700 ml ammonium oxalate solution (28.3 g/litre) with 535 ml 0.2 M oxalic acid (25.2 g/litre) and adjust to pH 3, storing the solution in the polyethylene bottle.

Procedure:
Place 200 mg soil (< 100 mesh) in a 100 ml polypropylene centrifuge tube, add 80 ml 0.2 M ammonium oxalate at pH 3 and shake the mixture for 4 hours in the dark. The supernatant solution, obtained after centrifuging at 2000 *g* for 10 min, is siphoned off into a 200 ml volumetric flask and diluted to the mark with distilled water. The solution is then reserved for the determination of iron, aluminium and silicon.

Pretreatment:
This is only necessary when aluminium has to be measured. Iron and silicon can be determined in the untreated extract using the methods detailed previously. For the silicon determination, excess ammonium molybdate reagent is utilized (Smith, 1984).

Aluminium:
Treat a suitable aliquot (0–30 μg Al) with 2 ml nitric acid (conc.), 1 drop sulphuric acid (conc.) and 20 ml distilled water, and then evaporate slowly on a hot-plate (~ 200°C) to destroy the oxalate. Then add 2 ml 5 M hydrochloric acid to the residue, evaporate to dryness and redissolve the residue in 40 ml warm distilled water. Adjust the pH of the solution with dilute NH_4OH to within the range 4–6 before determining the aluminium content spectrophotometrically, using the 'Aluminon' method previously described.

8.3.4 *Hydroxyl release*

This method is a laboratory adaptation of the field test for identification of allophanic material in soils, developed in New Zealand (Fieldes and Perrott, 1966). The test is carried out by placing a small sample of soil on a filter paper, impregnated with phenolphthalein indicator, placing 2 drops of saturated sodium fluoride solution onto the soil, and noting the intensity of the pink colour then developed. Poorly ordered inorganic material in soils is highly hydrated. The test utilizes the fact that aqueous solutions of fluorides at pH values > 7 react with hydroxyl sites, causing a corresponding increase in pH, which can be related to the relative amounts of allophanic material.

The field test has been adapted for laboratory measurement of the relative amounts of poorly ordered material in soils (Perrott *et al.*, 1976a). The method involves the reaction of 25 mg of soil (< 200 mesh) or clay, at 25°C (thermostatted), with 0.85 M sodium fluoride at pH 6.8. As the exchange between fluoride and hydroxyl ions is pH-dependent, the reaction is carried out using an automatic titrator in the pH stat mode at 6.8. The pH is kept constant by titration with a standard acid. The amount consumed, and thus the amount of hydroxyl released after 25 min, is taken as a measure of the hydroxyl release.

This technique has been shown to be very sensitive in detecting small amounts of inorganic gel material in soils. The method is unsuitable for soils with moderate to high organic matter contents (> 8%) and those which contain carbonate must be pretreated with acid or an acidified sodium acetate solution.

Perrott *et al.* (1976a) showed that selected crystalline minerals were unreactive under the above conditions, whereas gel material was very reactive (Table 8.1). They also used the method to examine 57 freely drained Scottish podzols and brown forest soils, obtaining a mean value of 80 mmol/100 g.

Perrott *et al.* (1976b) examined the effect of pH on the reaction of a series of mixed inorganic gels with sodium fluoride, and found that the reactions of the gels were pH-dependent. When the pH was raised to 7.6, the silica gel no longer

Table 8.1 Hydroxyl release values for minerals and poorly ordered materials (Perrott *et al.*, 1978a)

Material	Origin	Hydroxyl release mmol/100 g, 25 min
Quartz	Brazil	1
Goethite	Pribram, Bohemia	9
Kaolinite	Cornwall, UK	12
Hematite	Locality unknown	14
Chlorite	Wanibuchi, Japan	16
Illite	Fithian, Illinois, USA	17
Boehmite	Hungary	20
Montmorillonite	Wyoming, USA	22
Halloysite	Wagon Wheel, Colorado, USA	25
Gibbsite	Guyana	51
Allophane	Taupo, New Zealand	1075
Allophane	Kodonbaru, Japan	1900
Imogolite	Uemura, Japan	1208
Ferric oxide gel	Synthetic	547
Silica gel	Synthetic	1756
Alumina gel	Synthetic	2112

reacted, and when raised to pH 9 the iron gel showed little reaction. Therefore, by varying the pH of the reaction, which should be carried out under a nitrogen atmosphere, it is possible to separate the hydroxyl release values due to silica and iron gels from those due to alumina gel. The values obtained for the gels, however, are less than those at pH 6.8, because that reaction is pH-dependent.

8.3.5 *Trimethylsilylation*

Soil is formed through the weathering of minerals, and the type of weathering is related to the arrangement of silica units within the crystal structure. The degree of polymerization of the silica units can be quantitatively assessed by preparing organic trimethylsilyl derivatives, which can be separated and identified. Trimethylsilylation is well known in the field of organic chemistry, and was first used in silicate studies by Lentz, (1964). It was further refined by Gotz and Masson (1970), and given a quantitative basis by Gotz and Masson (1971) using an internal standard, and then by Lackowski and Glasser (1973).

Table 8.2 Monomeric silica released on trimethylsilylation from minerals, clay minerals, naturally occurring and synthetic gels. (Smith *et al.*, 1982, 1983).

Mineral	Monomeric silica mg/g	Mineral	Monomeric silica mg/g	Gels	Monomeric silica mg/g
Fayalite	247	Chlorite	2.4	Allophane, Misotsuchi, Japan	117
Olivine	207	Halloysite	2.0	Allophane, Mitaka, Japan	72
Lepidomelane	16	Montmorillonite	1.9	Allophane (A), Tirau, New Zealand	59
Hornblende	2.6	Vermiculite	1.5	Allophane (B), Taupo, New Zealand	58
Hydrobiotite	2.0	Nontronite	1.3	$SiO_2/Al_2O_3$1:7 $SiO_2/Al_2O_3$1.3:1	38 25
Augite	0.7	Serpentine	1.0	$SiO_2/Al_2O_3$1.8:1	23
Albite	0.7	Kaolinite	0.6	$SiO_2/Al_2O_3$2.8:1	8.9
Orthoclase	0.4	Illite	0.3	SiO_2/Al_2O_3	2.4
Muscovite	< 0.1				
Quartz	< 0.1				

These authors calibrated the method by a comparison with solutions of pure derivatives isolated by chromatographic separation. The procedure is based on the reaction of silicic acids with a monofunctional organosilicon compound to give stable and volatile organosilyl derivatives of the acids.

Trimethylsilylation has been used in weathering studies of rocks and their derived soils. Smith *et al.*, (1982, 1983) showed that:

(i) Of the primary minerals, the ferromagnesian ones were the most reactive

(ii) Inorganic gels, particularly those high in alumina were reactive

(iii) Crystalline clay minerals, especially kaolinite and illite were relatively stable (Table 8.2).

Gel material is certainly much more reactive than crystalline clay minerals although not as reactive as some ferromagnesian minerals. The use of the trimethylsilylation method in soil studies can, therefore, give some indication of a noncrystalline complement.

The method involves treatment of the samples with trimethylchlorosilane (TMCS) in the presence of a small amount of water. The hydrochloric acid produced by hydrolysis of the TMCS leaches the silicate minerals, replacing cations with protons. The silicic acids that are released immediately form derivatives with trimethylsilyl (TMS) groups. The TMS derivatives of the various silica polymers can then be separated, identified and quantified—using standard solutions of pure derivatives—by gas–liquid chromatography. The reaction with a monomeric orthosilicate group is as follows:

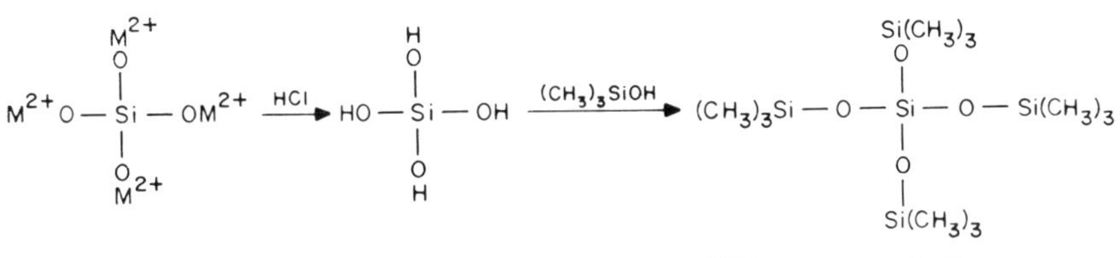

Monomer   Silicic acid   TMS monomer derivative

(a) *Trimethylsilylation method.*

Reagents:
Trimethylchlorosilane (Koch-Light)
Hexamethyldisiloxane (Koch-Light)
Isopropyl alcohol—(BDH-AR)
n-Eicosane—$CH_3(CH_2)_{18}$—CH_3 (British Drug House BDH)
Amberlyst 15 (BDH) (cation-exchange resin).

Procedure:
Weigh a lightly ground sample (25–200 mg) into a glass vial, add 9 ml hexamethyldisiloxane (HMD); 0.2 ml distilled water and 1 ml isopropyl alcohol (containing 2 mg *n*-eicosane/ml as an internal standard), stopper the vial and stir the mixture for 5 min using a magnetic stirrer. Then add 2 ml trimethylchlorosilane (TMCS) and stir continuously for 16 hours (overnight). All reagents, prior to use, are cooled in a refrigerator to minimize any side reactions. In order to hydrolyse the excess TMCS to HMD, add 2 ml distilled water

and stir the mixture for a further 1.5 h before centrifuging in 15 ml tubes at 2200 g for 5 min. Transfer the organic-solution layer to a glass vial, weigh, add ~ 1 g cation-exchange resin (Amberlyst 15) and stir the mixture for at least another 5 h to ensure complete trimethylsilylation. The organic solution is separated from the exchange resin by filtration, and the TMS derivatives of the silicate anions present are determined by gas/liquid chromatography.

Gas/liquid chromatography:
A temperature-programmed *Pye 104* chromatograph, fitted with dual flame-ionization detectors and coupled to an integrator, is used to separate and quantify the TMS derivatives. Dual stainless-steel columns (3.7 m long × 2 mm diameter) packed with silicon gum rubber (SE 30), 3% on acid-washed diatomite (80–90 mesh), are used for the separation. The carrier gas is nitrogen, the flow rate is 20 ml/min, and the following conditions are used as standard:
Injection block temperature—200°C
Detector block temperature—350°C
110°C—isothermal 5 min
110–140°C—150/min
140–220°C—80/min
220–250°C—30/min
250°C isothermal, 10 min.

The inert *n*-eicosane is added to the isopropyl alcohol as an internal standard. The greater part of the solvent is dumped by inserting a stream splitter between the column and the detector. Figure 8.1 is a typical chromatograph of the TMS derivatives. The amounts of these present in the solutions are determined from the peak areas, using pure TMS derivatives as internal standards.

The amount of monomeric silica obtained from soil-related materials is always very much larger than that of the dimer; even smaller amounts (if any) are obtained from the higher polymer forms, i.e. ring trimer, ring tetramer and chain trimer.

Since the technique involves the determination of the amount and the progress of silicate dissolution during prolonged and controlled acid attack, it can be used in soil-profile studies, particularly for temperate soils developed in a naturally acidic environment.

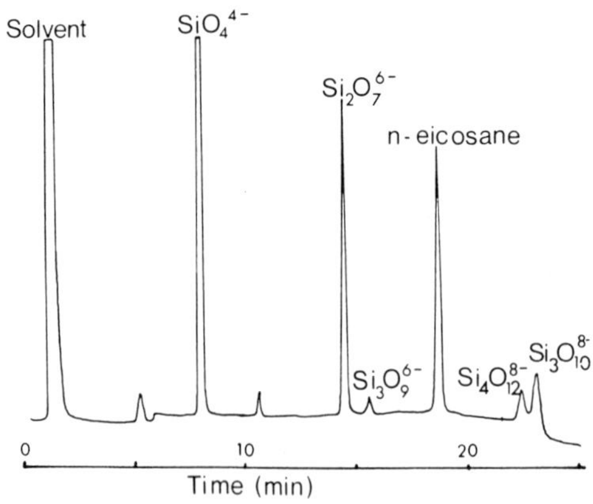

Figure 8.1 Typical gas-chromatographic separation of trimethylsilyl derivatives of lower molecular weight silicate anions

8.3.6 *Pyrophosphate extraction*

Alexsandrova (1960) showed that pyrophosphate could extract organic matter and associated aluminium and iron from soils. The method was further developed by McKeague (1967) and Bascomb (1968), the main difference between the procedures being that the former utilized sodium pyrophosphate and the latter utilized potassium pyrophosphate. It was found that with poorly ordered iron oxides and silicates and well-crystallized goethite and hematite there was virtually no reaction. On the other hand, examination of a wide range of organomineral soils showed a good correlation between the extracted iron and aluminium and the amounts of the extracted organic carbon. Pyrophosphate extraction of soils at pH 10 can, therefore, give a measure of organically complexed iron and aluminium. The method is now used as an aid to the classification of soils, in particular spodosols (McKeague *et al.*, 1971, Loveland and Bullock, 1975), i.e. those soils that are characterized by an illuvial mineral horizon resulting from the accumulation of poorly ordered material. This material is composed of aluminium and organic matter, with or without iron.

Pyrophosphate extraction:

Reagents:
Potassium pyrophosphate (0.1 M)—dissolve 38.44 g potassium pyrophosphate (AR) in 1 litre distilled water
Sodium acetate–acetic acid buffer, pH 3.8—dissolve 680 g sodium acetate (AR) (hydrated) in 3 litres of distilled water, add 1 litre glacial acetic acid (AR), adjust the pH to 3.8 ± 0.05 and dilute to 5 litres
Sodium dithionite—GPR
Superfloc flocculating agent, 0.2% aqueous.

Extraction:
Weigh 0.8 g soil (< 2 mm) into a 100 ml polypropylene centrifuge tube, add 80 ml 0.1 M potassium pyrophosphate solution and shake the mixture for 16 h (overnight). Add 5 drops flocculating agent (Schuppli *et al.*, 1983), centrifuge for 15 min at 2800 *g*, and siphon off the supernatant liquid into a 200 ml volumetric flask. Wash the residual soil with 50 ml distilled water, recentrifuge and add to the volumetric flask. Dilute to volume, reserving the solution for the determination of extractable aluminium, iron and carbon. The residue is used for the determination of the residual iron, after dithionite treatment. A reagent blank should be carried through the complete extraction procedure.

Residual iron:
Add 80 ml sodium acetate buffer at pH 3.8, and 3.2 g sodium dithionite to the soil residue in the centrifuge tube, and shake the mixture for 16 h (overnight). After centrifuging as in the last section, siphon off the supernatant into a 200 ml volumetric flask, wash the residue with 50 ml distilled water, centrifuge and add the supernatant again to the solution in the flask. Dilute to the mark with distilled water. The residual iron content is then measured using the 1, 10-phenanthroline method.

Pretreatment for Al and Fe determinations:

Treat a suitable aliquot (5–10 ml) of the pyrophosphate extract, in a conical flask, with 1 ml nitric acid (conc.) and 2 drops sulphuric acid (1:1), and evaporate to dryness on a hot-plate until white fumes of sulphuric acid appear. If the residue is still brown in colour, cool and treat again with nitric acid. Cool and add 30 ml distilled water and 2 ml 1 M hydrochloric acid to dissolve the residue, and heat on a water-bath for 2 h. Then cool the solution and reserve for the determination of iron as detailed previously, and the determination of aluminium by Xylenol orange (Pritchard, 1967).

Extractable carbon

Reagents:
Potassium dichromate (0.166 M)—dissolve 49.0 g $K_2Cr_2O_7$ (AR) in 1 litre distilled water and standardize against standard ferrous sulphate solution
Ferrous sulphate (1 M)—dissolve 278.0 g $FeSO_4 \cdot 7H_2O$ (AR) in 1 litre distilled water containing 10 ml concentrated H_2SO_4
Diphenylamine indicator—dissolve 1.60 g barium diphenylamine sulphonate by boiling in 500 ml distilled water containing 15 ml concentrated sulphuric acid. Allow mixture to cool and settle, and decant the clear supernatant liquid into a reagent bottle.

Evaporate 25 ml of the pyrophosphate extract to dryness at 80°C in a 500 ml conical flask. Then add 10 ml 0.166 M potassium dichromate to the residue, carefully add 20 ml concentrated sulphuric acid, mix thoroughly and allow to stand for 30 min. Dilute the mixture to 200 ml with distilled water, add 10 ml concentrated orthophosphoric acid and 20 drops of diphenylamine indicator solution, and titrate rapidly with 1 M ferrous sulphate until the colour changes to green. Then add a further 0.50 ml potassium dichromate, which will cause the blue colour to reappear, continue the ferrous sulphate titration slowly till the end-point, noting the amount of titrant consumed. A blank titration is carried out on 10 ml of standard potassium dichromate in a similar manner to that detailed previously, to measure the amount of standard dichromate consumed in the oxidation of the extractable carbon.

$$1 \text{ ml } 0.166 \text{ M } K_2Cr_2O_7 \equiv 3 \text{ mg C}$$

Only a 77% recovery can be obtained, so that 1 ml 0.166 $K_2Cr_2O_7$ is assumed to be equivalent to 3.9 mg C reduced.
Then

$$\%C = \frac{390 \times \text{volume (ml) } 0.166 \text{ M dichromate}}{\text{wt of sample (mg)}}$$

Aluminium—Xylenol orange (Pritchard, 1967).

Reagents:
Xylenol orange (0.15%) aqueous
Sodium acetate buffer, pH 3.8—dissolve 136 g $NaOAc \cdot 3H_2O$ in water, adjust the pH to 3.8 and dilute to 1 litre.
EDTA (ethylenediamenetetra-acetic acid) (0.05 M)—dissolve 18.6 g EDTA (disodium salt) in 1 litre distilled water
Cresol red indicator—dissolve 0.1 g cresol red in 100 ml ethanol
Al standard (stock)—dissolve 1.7584 g potassium aluminium sulphate (AR) in 1 litre distilled water: (1 ml = 0.1 mg Al)
Al standard (working)—dilute 100 ml stock Al standard to 1 litre (1 ml = 10 μg Al).

Procedure:
Pipette a suitable aliquot (0–60 μg Al) of the treated pyrophosphate extract into a 100 ml conical flask. Add 20 ml distilled water and 4 drops 1:1 sulphuric acid, and hydrolyse the residue by digesting in a boiling water-bath for 4 h. After cooling, add two drops of cresol red indicator and adjust the pH to ~2 with dropwise addition of 1:1 M NH_4OH (to a yellow end-point). Then transfer quantitatively to a 100 ml standard flask, add 25 ml buffer solution and 10 ml xylenol orange, and place in a water-bath at 40°C for 1.5 h. Cool the solutions, add 5 ml EDTA, dilute to 100 ml and measure the absorptions at 550 nm after 1 h in a 4 cm cell. The results are compared with those of a range of standard aluminium solutions prepared in a similar manner (0–60 μg Al).

8.3.7 *Other methods*

Other less commonly used techniques for the characterization of poorly ordered material in soils may be mentioned. Most can be classed as selective dissolution methods e.g. treatment with sodium EDTA at pH 10 (Borggard, 1976), alkaline tiron at pH 10.5 (Biermans and Baert, 1977), and sodium hydroxide-tetraborate (Bruckert, 1979), monitoring Al, Fe and Si in the respective extracts.

Aomine and Jackson (1959) proposed a method for the estimation of allophanic material in soils by measuring the 'cation-exchange capacity delta value'. This involved treating two separate samples of peroxidized soil, one with 2% Na_2CO_3 for 60 min and the other with boiling sodium acetate for 15 min, and measuring the difference in exchange capacity between the two samples: the CEC delta value. This was shown to be very high for allophanic clays, large for halloysitic and smectitic clays but very low for the other crystalline clay minerals.

8.4 Comparison of methods

Alkali extraction by 0.5 M Na_2CO_3 (Follett *et al.*, 1965) has been shown to remove poorly ordered aluminosilicate material from soils and to have little or no effect on well-crystallized minerals. These authors considered that the extraction method with 0.5 M NaOH (Hashimoto and Jackson, 1960) resulted in some instances in considerable attack of crystalline clay minerals. However, if the ratio of extractant to soil is high enough to prevent saturation of the extract with respect to Si and Al, and if the time of reaction is kept to a minimum, then the attack on the crystalline components can be greatly reduced. As the pH of these extracts is high (~ 11), iron is not removed from the clay or soil. The amounts of alkali-soluble material, particularly in B horizons of podzols and brown podzolic soils, are quite large, indicating a high content of poorly ordered aluminosilicates in these horizons. Improved dispersion has been noted after alkali treatment. This is probably due to the removal of poorly ordered aluminosilicates that cement the primary clay particles into aggregates or due to aluminosilica gel coatings on individual particle surfaces or perhaps to both. The amounts of aluminium and silicon extracted by alkali show a highly significant correlation with values obtained for hydroxyl release from both freely drained and poorly drained soils (Perrott *et al.*, 1976a).

Dithionite extraction removes 'free' poorly ordered and crystalline iron oxides from soils, with a minimal effect on crystalline clay minerals. Silicon and aluminium are found in the extracts, normally in smaller amounts, and are thought to derive from poorly crystalline aluminosilicate material in the soil. Dithionite extraction provides a method for determining free (non-silicate) iron in soils, dissolving finely divided hematite and goethite. However, coarse-grained hematite and magnetite, are not attacked, or at least only very slowly.

Ammonium oxalate dissolves poorly ordered inorganic- and organic-complexed iron, silicon and aluminium from soils. It has a minimal effect on crystalline silicate minerals, goethite and hematite, but does dissolve poorly ordered lepidocrocite. Oxalate extraction has been used in Canada (McKeague and Day, 1966) in the classification of soils, and in New Zealand, Saunders (1968) used it to relate the changes in amorphous iron and aluminium to the degree of weathering and leaching in soils. More recently, Farmer *et al.* (1980), and Childs *et al.* (1983) have used oxalate-soluble silicon and aluminium to support the proposal that movement of aluminium in the podzolization process can be attributed to inorganic complexes with silicon.

Hydroxyl release values can give a measure of the relative amounts of poorly ordered material in soils (Perrott *et al.*, 1976a), although the absolute amounts must remain uncertain. A major part of the reactivity can be attributed to the clay fraction, in particular to gel material. The highest reactivities for soils occur in the B_s horizons of podzols, whereas for brown earths and gley soils there is little variation between the reactivities of the genetic horizons within a profile.

An estimation of the degree of polymerization of silicon in silicates in soil which react to acid attack, can be obtained by trimethylsilylation. Crystalline clay minerals are relatively resistant, whereas poorly ordered aluminosilicate material is highly reactive, particularly gels with a high ratio of Al:Si. A good overall correlation exists (Smith *et al.*, 1983) between the amount of silica released during trimethylsilylation and hydroxyl release by fluoride exchange.

Organic-complexed iron and aluminium can be extracted from soils by pyrophosphate. The treatment has very little effect on silicate minerals, or on well-crystallized aluminium and iron oxides, but some dissolution of poorly crystalline forms has been observed. This technique has been widely used as a chemical basis identifying podzolic B horizons. More recently, Schuppli *et al.* (1983) have highlighted the fact that pyrophosphate can cause peptization of ferruginous material which, in some cases, will not centrifuge out easily unless a flocculant is used. This may have led to erroneously high values for organically bound iron and aluminium being reported in the past.

A combination of various methods can give an estimation of the different forms of iron and aluminium in soils:

(i) Free iron = dithionite iron − oxalate iron ($Fe_d - Fe_{ox}$)

(ii) Poorly ordered iron and aluminium = oxalate soluble − pyrophosphate soluble ($Fe_{ox} - Fe_{py}$ and $Al_{ox} - Al_{py}$) iron and aluminium

(iii) Organically bound iron and aluminium = pyrophosphate soluble (Fe_{py} and Al_{py}) iron and aluminium. The role of the different selective methods in soils is summarized in Table 8.3.

Selective chemical dissolution techniques play a very important role in the characterization of clay materials, but must not be considered in isolation. The various instrumental techniques and selective chemical methods are complementary. As succinctly expressed by Mackenzie (1957) 'every method, whatever its nature, will add its own quota of information and fill in some of the shadows or emphasize some of the highlights of the picture'.

Table 8.3 Characterization of poorly ordered material in soils.

poorly ordered material

METHODS	Alkali dissolution	Ammonium oxalate	Pyrophosphate	Dithionite	Trimethylsilylation	Fluoride exchange
ELEMENTS	Al, Si	Al, Fe, Mn, Si	Al, Fe_{ext}, Fe_{res}, C	Al, Fe, Si	Si polymers	relative amounts of poorly ordered material
COMPONENTS DETECTED	poorly ordered inorganic aluminosilicates	poorly ordered inorganic and organic	poorly ordered organic	poorly ordered inorganic and organic, and crystalline Fe	acid-reactive crystalline and poorly ordered silicates	poorly ordered inorganic material

References

Alexsandrova, L.N. (1960) The use of pyrophosphate for isolating free humic substances and their organic-mineral compounds from the soil. *Soviet Soil Sci.*, 190–197.

Aomine, S. and Jackson, M.L. (1959) Allophane determinations in Ando soils by cation exchange delta value. *Proc. Soil Sci. Soc. Am.*, **23**, 210–214.

Bascomb, C.L. (1968) Distribution of pyrophosphate-extractable iron and organic carbon in soils of various groups. *J. Soil Sci.*, **19**, 251–268.

Biermans, V. and Baert, L. (1977) Selective extraction of the amorphous Al, Fe and Si oxides using an alkaline tiron solution. *Clay Miner.*, **12**, 127–135.

Borggard, O.K. (1976) Selective extraction of amorphous iron oxide by EDTA from a mixture of amorphous iron oxide, goethite and hematite. *J. Soil Sci.*, **27**, 478–486.

Bruckert, S. (1979) Classification des sols bruns ocreux et cryptopodzoliques par analyse des extraits tetraborates tamponnés à pH 9.5. *Geoderma*, **22**, 205–212.

Bruckert, S. and Souchier, B. (1975) Mise au point d'un test de differentiation chimique des horizons cambiques et spodiques. *C.R. Acad. Sci. Paris*, **280**, 1361–1364.

Childs, C.W., Parfitt, R.L. and Lee, R. (1983) Movement of aluminium as an inorganic complex in some podzolised soils, New Zealand. *Geoderma*, **29**, 139–155.

Cradwick, P.D.G., Farmer, V.C., Russell, J.D., Masson, C.R., Wada, K.K. and Yoshinaga, N. (1972) Imogolite, a hydrated aluminium silicate of tubular structure. *Nature*, **240**, 187–189.

Deb, B.C. (1950) The estimation of free iron oxides in soils and clays and their removal. *J. Soil. Sci.*, **1**, 212–220.

de Endredy, A.S. (1963) Estimation of free iron oxides in soils and clays by a photolytic method. *Clay Miner. Bull.*, **5**, 209–217.

Farmer, V.C., Russell, J.D. and Berrow, M.L. (1980) Imogolite and protoimogolite allophanes in spodic horizons: evidence of a mobile aluminium silicate complex. *J. Soil Sci.*, **31**, 673–684.

Farmer, V.C., Russell, J.D. and Smith, B.F.L. (1983) Extraction of inorganic forms of translocated Al_2O_3, Fe_2O_3 and SiO_2 from a podzol Bs horizon. *J. Soil Sci.*, **34**, 571–576.

Farmer, V.C., Smith, B.F.L. and Tait, J.M. (1977) Alteration of allophane and imogolite by alkaline digestion. *Clay Miner.*, **12**, 195–198.

Fieldes, M. (1955) Clay mineralogy of New Zealand soils. Part II Allophane and related mineral colloids. *New Zealand J. Sci. & Technol.*, **B37**, 336–350.

Fieldes, M. and Perrott, K.W. (1966) Nature of allophane in soils. III Rapid field and laboratory test for allophane. *New Zealand J. Sci.*, **9**, 623–629.

Follett, E.A.C., McHardy, W.J., Mitchell, B.D. and Smith, B.F.L. (1965) Chemical dissolution techniques in the study of soil clays. *Clay Miner. Bull.*, **6**, 23–34.

Gotz, J. and Masson, C.R. (1970) Trimethylsilyl derivatives for the study of silicate structures. Part I: A direct method of trimethylsilylation. *J. Chem. Soc. (A)*, 2683–2686.

Gotz, J. and Masson, C.R. (1971) 'Trimethylsilylation of silicate anions: a method of studying the structure of crystalline silicates and glasses', in *Sci. & Techn. Comm., 9th Intern. Nat. Conf. on Glass, Versailles*, 261–275.

Guillet, B. and Souchier, B. (1982) 'Amorphous and crystalline oxyhydroxides and oxides in soils (iron, aluminium manganese, silicon)', in *Constituents and Properties of Soils* (ed. M. Bonneau and B. Souchier eds.) Academic Press, London.

Hashimoto, I. and Jackson, M.L. (1960) 'Rapid dissolution of allophane and kaolinite-halloysite after dehydration', *Proc. 7th Conf. Clays & Clay Miner. London*, Pergamon Press, Oxford.

Hendricks, S.B. and Fry, W.H. (1930) The results of X-ray and microscopical examinations of soil colloids. *Soil Sci.*, **29**, 457–479.

Hoyos, A. and Pino, C. (1958) A comparative study of soils from the Canary Islands, Spanish Guinea. *Agrochimica*, **2**, 218–235.

Jackson, M.L., Pennington, R.P. and Mackie, W.Z. (1950) Crystal chemistry of soils. I. The fundamental structure groups and families of silicate minerals. *Proc. Soil Sci. Soc. Am.*, **13**, 139–145.

Jones, R.C. and Uehura, G. (1973) Amorphous coatings on mineral surfaces. *Proc. Soil Sci. Soc. Am.*, **37**, 792–798.

Jorgensen, S.S., Birnie, A.C., Smith, B.F.L. and Mitchell, B.D. (1970) Assessment of gibbsitic material in soil clays by differential thermal analysis and alkali dissolution methods. *J. Thermal Anal.*, **2**, 277–286.

Kirkman, J.H., Mitchell, B.D. and Mackenzie, R.C. (1966) Distribution in some Scottish soils of an inorganic gel system related to 'Allophane'. *Trans. Roy. Soc. Edinburgh*, **66**, 393–418.

Lackowski, E.E. and Glasser, F.P. (1973) Application of gas chromatography to mineral chemistry: aluminium-silicon ordering in melilites. *Miner. Mag.*, **39**, 412–419.

Lentz, C.W. (1964) Silicate minerals as sources of trimethylsilyl silicates and silicate structure analysis of sodium silicate solutions. *Inorganic Chem.*, **3**, 574–579.

Loveland, P.J. and Bullock, P. (1975) Crystalline and amorphous components of the clay fractions in brown podzolic soils. *Clay Miner.*, **10**, 451–469.

Mackenzie, R.C. (1957) Modern methods for studying clays. *Agrochimica*, **1**, 305–327.

Mackenzie, R.C. (1963) '*De Natura Lutorum*', in *Proc. 11th Natl. Conf. Clays & Clay Miner., 1962,* 11–28 Pergamon Press, Oxford.

McKeague, J.A. (1967) An evaluation of 0.1 M pyrophosphate and pyrophosphate-dithionite in comparison with oxalate as extractants of the accumulation products in podzols and some other soils. *Can. J. Soil Sci.*, **47**, 95–99.

McKeague, J.A. and Day, J.H. (1966) Dithionite and oxalate extractable Fe and Al as aids in differentiating various classes of soils. *Can. J. Soil Sci.*, **46**, 13–22.

McKeague, J.A., Brydon, J.E. and Miles, N.M. (1971) Differentiation of forms of extractable iron and aluminium in soils. *Proc. Soil Sci. Soc. Am.*, **34**, 33–38.

Mehra, O.P. and Jackson, M.L. (1960) 'Iron oxide removal from soils and clays by a dithionite-citrate system buffered by sodium bicarbonate', in *Clays & Clay Miner., Proc., 7th Conf.*, Nat. Acad. Sci., Natl. Res. Council Publ. 317–327.

Mitchell, B.D. and Farmer, V.C. (1962) Amorphous clay minerals in some Scottish soil profiles. *Clay Miner. Bull.*, **5**, 128–144.

Mitchell, B.D. and Mackenzie, R.C. (1954) Removal of free iron oxide from clays. *Soil Sci.*, **77**, 173–184.

Mitchell, B.D. and Smith, B.F.L. (1974) The removal of organic matter from soil extracts by bromine oxidation. *J. Soil Sci.*, **25**, 239–241.

Mitchell, B.D., Farmer, V.C. and McHardy, W.J. (1964) Amorphous inorganic material in soils. *Adv. in Agronomy*, **16**, 327–383.

Mitchell, B.D., Smith, B.F.L. and de Endredy, A.S. (1971) The effect of buffered sodium dithionite solution and ultrasonic agitation on soil clays. *Israel J. Chem.*, **9**, 45–52.

Parfitt, R.L. and Henmi, T. (1982) Comparison of an oxalate-extraction method and an infrared spectroscopic method for determining allophane in soil clays. *Soil Sci. & Pl. Nutrition*, **28**, 183–190.

Perrott, K.W., Smith, B.F.L. and Inkson, R.E. (1976a) The reaction of fluoride with soils and soil minerals. *J. Soil Sci.*, **27**, 58–67.

Perrott, K.W., Smith, B.F.L. and Mitchell, B.D. (1976b) Effect of pH on the reaction of sodium fluoride with hydrous oxides of silicon, aluminium and iron and with poorly ordered aluminosilicates. *J. Soil Sci.*, **27**, 348–356.

Pritchard, D.T. (1967) Spectrophotometric determination of aluminium in soil extracts with xylenol orange. *Analyst*, **92**, 103–106.

Rich, C.I. and Thomas, G.W. (1960) The clay fraction of soils. *Adv. Agronomy*, **12**, 1–39.

Robertson, G. (1950) The colorimetric determination of aluminium in silicate materials. *J. Sci. Food & Agric.*, **1**, 59–63.

Ross, C.S. and Kerr, P.F., (1934) Halloysite and allophane. *U. S. Geol. Surv. Prof. Pap.*, **185-G**, 135–148.

Saunders, W.M.H. (1968) Soils of New Zealand. Part 2. *New Zealand Soil Bur., Bull.* **26**, 109–114.

Schuppli, P.A., Ross, G.J. and McKeague, J.A. (1983) The effective removal of suspended material from pyrophosphate extractions of soils from tropical and temperate regions. *Soil Sci. Soc. Am. J.*, **47**, 1026–1032.

Schwertmann, U. (1964) The differentiation of iron oxide in soils by a photochemical extraction with acid ammonium oxalate. *Zeit. Pflanz. Dueng Bodenk.*, **105**, 194–201.

Schwertmann, U. (1973) Use of oxalate for Fe extraction from soils. *Can. J. Soil Sci.*, **53**, 244–246.

Schwertmann, U. (1978) 'Non-crystalline and accessory minerals,' in *Proc. Intern. Clay Conf., 1978 Oxford.* (M.M. Mortland and V.C. Farmer eds.), (*Dev. Sedimentol.*, **27**) Elsevier, Oxford.

Smith, B.F.L. (1984) The determination of silicon in ammonium oxalate extracts of soils. *Comm. Soil Sci. & Pl. Anal.*, **15**, 199–204.

Smith, B.F.L., Mitchell, B.D. and Mackenzie, R.C. (1983) Susceptibility to weathering of some

Scottish rocks and their derived soils. *Trans. Roy. Soc. Edinburgh: Earth Sci.*, **73**, 191–203.

Smith, B.F.L., Paterson, E. and Mitchell, B.D. (1982) Trimethylsilylation of commonly occurring primary and secondary minerals in soil. *J. Soil Sci.*, **33**, 115–124.

Tait, J.M., Yoshinaga, N. and Mitchell, B.D. (1978) The occurrence of imogolite in some Scottish soils. *Soil Sci. & Pl. Nutrition*, **24**, 145–151.

Tamm, O. (1922) Eine methode zur Bestimmung der anorganischen Komponenten des Gelkomplexes. *Medd. Skogsforsoksanst Stockholm*, **19**, 385–404.

Van Bemmelen, J.M. (1910) The different modes of weathering of silicates in the Earth's crust. *Zeit. anorgan. Chem.*, **66**, 322–357.

Weaver, R.M., Syers, J.K. and Jackson, M.L. (1968) Determination of silica in citrate-bicarbonate extracts of soils. *Proc. Soil Sci. Soc. Am.*, **32**, 497–501.

Webber, A. and Wilson, A.L. (1965) The absorptiometric determination of silicon in water. Part IV·Method for determining 'Reactive' silicon in power-station waters containing phosphate. *Analyst*, **89**, 632–641.

Yoshinaga, N. (1966) Chemical composition and some thermal data of eighteen allophanes from Ando soils and weathered pumices. *Soil Sci. & Pl. Nutrition*, **12**, 47–54.

Yoshinaga, N. and Aomine, S. (1962) Imogolite in some Ando soils. *Soil Sci. & Pl. Nutrition*, **8**, 22–29.

Yoshinaga, N., Tait, J.M. and Soong, R. (1973) Occurrence of imogolite in some volcanic ash soils of New Zealand. *Clay Miner.*, **10**, 127–130.

9 Determinative methods—a brief overview

M.J. WILSON

A theme that has, it is hoped, emerged from this book, is that in examining clay materials it would be unwise to limit oneself to only one method of investigation. A detailed characterization of a clay requires information from several different techniques at least. Each of the various methods discussed has particular strengths and weaknesses and, in this final chapter, an attempt will be made to summarize briefly what these are, in the hope of encouraging a balanced approach.

X-ray diffraction (XRD) is the most widely used primary investigative technique in clay mineralogy. The reason for this is undoubtedly that the method is rapid, easy to use, generally applicable and yields a great deal of readily interpretable information on many aspects of clay minerals. The major clay-mineral groups, and frequently the particular mineral species within these groups, can be easily identified, even in complex mixtures. XRD can be used to characterize complex interstratifications, to identify particular polytypes and to provide quantitative, or at least semiquantitative, analyses of clay mixtures. Crystallinity can be assessed, and in particular instances, inferences can be drawn regarding particle size and shape. Careful study of diffraction patterns, combined with auxiliary treatments, often yields general information on the chemical composition of the major clay components present. Nevertheless, there are certain inherent weaknesses in the XRD method, of which the reader should be aware. The first is that, as pointed out in Chapter 1, the method gives only an average picture of the material being studied, one where the major components are highlighted, but the minor phases (which could be significant) disappear into the background, and go undetected. This statement may be tested by comparing the information that can be obtained by, for example, examining a fine sand fraction, first by optical microscopy and then by XRD. The microscope may reveal, in addition to quartz, feldspar and mica minerals, a variety of accessory minerals, such as hornblende, epidote, iron-ore minerals, zircon, garnet, apatite etc., that are conspicuous by their colour, high refractive index or some other optical property. These minerals may amount to 5–10% of total. If the same sample is then examined by XRD it may be found that, again, the specimen is dominated by quartz, feldspars and mica, but the minor phases are below the limit of detection. A similar disparity is often obvious when the same clay sample is examined by both XRD and transmission electron microscopy. Thus, the microscope may instantly reveal important minor components such as star-shaped goethite twins, illite laths, halloysite tubes, etc.

A second weakness of XRD is that, in common with nearly all the other investigative methods described, the technique tells us little about the distribution of the various components and the ways in which they are associated. Misleading inferences may be made if this drawback is not taken into account. For example, a recent examination by XRD of the clay mineralogy of reservoir sandstones showed that the clay ($< 2\ \mu m$) fraction was almost entirely dominated by kaolinite, with only a hint of illite on the diffractometer trace. However, an SEM study of the same sandstone showed that kaolinite tended to occur in dense, tightly packed books and vermicules, whereas the illite was finely filamentous and extended into and filling much of the available pore volume. Thus, although kaolinite was dominant in comparison with illite, the occurrence and distribution of the latter within the rock was crucial in determining the permeability, a finding that could not possibly have been achieved by XRD examination alone.

A third weakness of XRD is that it can tell us little about the occurrence of poorly ordered material. The properties of soils may be strongly influenced by even small amounts of such material, particularly where it coats soil particles, so that attempted rationalization of soil behaviour on the basis of crystalline clay components as determined by XRD would be unsuccessful.

Finally, the XRD patterns of some clay minerals may be insensitive to certain kinds of isomorphous substitution, these being more easily detected by other methods, notably infrared spectroscopy (IR). For example, montmorillonites with rather different compositions, such as those from Wyoming and Arizona, yield essentially the same XRD patterns, but can be easily distinguished by IR. Again, it is doubtful if montmorillonite and beidellite could be distinguished by XRD without the aid of auxiliary treatments.

Thermal analysis is still widely used for investigating clay materials. As shown in Chapter 3, it is best to regard thermal analysis as complementary to the main investigative techniques such as XRD. Regarded in this light rather than as a means of making absolute identifications, the technique emphasizes the differences between samples that are not apparent when using other methods. Sometimes, for example, significantly different DTA curves may be recorded for illites with similar XRD patterns and chemical compositions. A further asset is that thermal analysis can be sensitive to the presence of poorly ordered materials, although it is rarely diagnostic. Finally, for certain minerals the sensitivity of the method is superior to that of XRD; gibbsite, goethite, calcite and other carbonates can be readily detected at levels of $< 1\%$, and can be determined quantitatively. Kaolinite can also be determined quantitatively, provided that there are no interfering thermal reactions from other minerals.

The most serious limitation in applying thermal analysis to clay materials lies in the thermal variability of many of the clay minerals themselves, and this factor, combined with variations in instruments and sample preparation methods, often leads to severe difficulties in interpretation. Illitic and interstratified clays can be particularly difficult to interpret, and even in the most favourable circumstances it would be unwise to rely on only thermal methods for characterization of clay materials.

Infrared (IR) spectroscopy is a versatile and rapid method by which to investigate clays. Its greatest strength is that it can provide direct information on the structural and compositional characteristics of minerals in a relatively simple way. Chapter 4 illustrated clearly the ability of IR to identify the various polytypes of the kaolinite group minerals. The method is particularly sensitive to compositional differences in the 2:1 minerals. For example, isomorphous substitutions in the octahedral sheets of smectites and illites may be detected by IR, but overlooked by other techniques. Another advantage of IR is that it can be used to characterize noncrystalline as well as crystalline material; its use in the identification of allophane and imogolite structures in soils has recently opened up new areas of research of soil genesis. For many clay minerals the detection limits of IR are probably about the same as XRD (5–10%), although for kaolinite, IR may be the more sensitive method. Finally, IR has one outstanding advantage, which has not been discussed within the context of this book, but which nevertheless is vital in clay research. This is its use in the study of ionic and other species and how they are sorbed onto and interact with the external surfaces of clay and hydrous oxide particles. Similarly, for expansible minerals, the technique can yield a unique range of information on interactions in the interlamellar space between sorbed species, exchangeable cations, free and coordinated water molecules and silicate surfaces.

Nevertheless, IR does suffer from some of the same weaknesses as XRD, namely an inability to detect important minor phases or to emphasize the distribution and interrelationships of the various constituents in bulk material. More specifically, the method cannot be used to diagnose interstratified clay minerals, and cannot distinguish between the various kinds of order, regular or irregular, that are so clearly differentiated by XRD. Identification of vermiculite minerals is also difficult. IR is not used very extensively in clay mineral determination, although it can provide an immense range of information. Perhaps this is because an unusually high level of experience and skill is thought to be required before spectra can be interpreted with confidence. Although this idea may be true to some extent in dealing with some complex clay mixtures, it cannot be regarded as generally valid, as is illustrated in Chapter 4.

Scanning electron microscopy (SEM) enables the investigator to appreciate fully the heterogeneity of clay materials, to visualize the various mineral components in their distinct growth forms, and how they are related in terms of overall microfabric and texture, and with the help of electron microprobe analysis, to understand how different constituents may vary chemically on a micro-scale. The ability to study these heterogeneities is undoubtedly the main advantage of SEM. Suitable preparative techniques, such as critical-point drying, enable the investigator to study the most delicate clay microfabrics as they occur *in situ* and to make observations (for example, of reservoir sandstones) that may be crucially important. The increasing use of back-scattered electron imaging is opening up new perspectives for the SEM of clay materials, enabling the instant recognition in the field of view of minerals of different chemical composition.

However, it must be conceded that as a determinative method for clay minerals, SEM leaves a lot to be desired. It is true that some clay minerals may be characterized by unmistakable forms; kaolinite is an obvious example, and sometimes illite and even chlorite can be just as easily recognized. Moreover, these minerals may have an apparently distinctive chemical composition, and the investigator may be tempted to reject the other methods as unnecessary. Such a conclusion would be entirely misconceived, because SEM on its own tells us nothing about polytypes and little about interstratification, so that dickite may be interpreted as kaolinite, and what is identified as illite or chlorite may have smectite interstratified with it. For many, perhaps most, clay materials SEM will yield rather ambiguous data on form and chemical composition, so that the temptation to identify minerals without supporting evidence from other techniques will not arise. Smectites, vermiculites and interstratified minerals can be particularly difficult to recognize by SEM.

Transmission electron microscopy (TEM) is an extremely versatile and powerful technique for examining clay materials either in the dispersed or the bulk state. A major advantage is that the instrument can provide direct morphological, structural and (if a microprobe analyser is fitted) chemical information on selected clay particles. In some instances, morphology alone can be highly diagnostic and, certainly in conjunction with selected-area electron diffraction, can be used for the unambiguous identification of many clay minerals, such as halloysite, antigorite, chrysotile, palygorskite and imogolite. The physical dimensions of individual clay particles can be measured and reasonable deductions made regarding the specific surface area of the bulk material and, once the cation-exchange capacity is known, its layer-charge density. Electron-diffraction patterns can establish particular poly-types and even (particularly using the oblique texture method) a complete crystal structure analysis. (The latter is, of course, highly specialized and beyond the scope of this book.) High-resolution lattice images of layer structures can be recorded and measured on thin-sectioned or fortuitously oriented clay particles. TEM in association with the replica technique can also be used for examination of bulk specimens, a role that has been to some extent superseded by SEM because of the greater convenience of the latter. However, the high magnification attainable by TEM means that more detail, such as minute growth steps on layer surfaces, can be observed on suitably prepared replica specimens. Finally, an outstanding asset of TEM in clay investigations is its ability to identify immediately many minor phases from their distinct electron-optical characteristics. Such phases would not be detectable using any other available techniques.

Probably the most serious limitation of TEM in clay-mineral studies centres on the whether or not observations in the disperse state are representative. Only a minute amount of material from a highly diluted suspension is examined, and sometimes preferential selection of the finest particles, perhaps consisting of minor phases, could be inevitable. A further drawback is that most clay materials are composed of platy particles that may be impossible to differentiate morphologically, even when it is known that there are several distinct mineral constituents. Electron diffraction may not be very helpful here, because the patterns obtained from the uniform orientation adopted by platy clay particles may be basically similar, if not identical.

The investigation of clay materials by TEM appears to have declined somewhat in popularity in recent years, probably because, in comparison with SEM, the technique is seen as demanding a higher level of knowledge, expertise and interpretive skill and is, moreover, more time-consuming and labour-intensive. Although this is true to some extent, it is clear from Chapter 6 that TEM can yield direct information on clay materials at the individual particle level that cannot be obtained by any other method. Ultimately, therefore, the effort involved in mastering the technique is well worthwhile.

Chemical analysis of clay minerals (including determination of cation- and anion-exchange capacities and layer charge) is essential for their full characterization but, all the same, such an analysis on its own may be of poor diagnostic value. However, a complete chemical analysis, and particularly determination of cation-exchange capacity, often suggests the presence of previously unsuspected impurities. Difficulties encountered in calculation of structural formulae of supposed 'pure' minerals may lead to the same conclusion.

Selective dissolution analyses are generally effective in determination of the various forms of noncrystalline, hydrous oxides and silicates in soils. In many instances, the extracted material may be compositionally well-defined, and the presence of a discrete phase, allophane or imogolite for example, may be confirmed by other techniques. However, this is not always the case, and not infrequently the source of the material being extracted is not at all clear.

In conclusion, the techniques and applications covered in this book are outlined only at an introductory level, and the minerals described do not, by any means, form a complete list of all the possibilities that may be encountered in clay materials. It is hoped, however, that this compilation will be sufficiently comprehensive to be of real, practical use to those entering the field of clay mineralogy for the first time, and even to those who are relatively familiar with the investigation of these often puzzling and difficult, but always fascinating, materials.

Index